Dmitriy Bilyk, Josef Dick, and Friedrich Pillichshammer (Eds.)
Discrepancy Theory

Radon Series on Computational and Applied Mathematics

Volume 26

Discrepancy Theory

Edited by
Dmitriy Bilyk
Josef Dick
Friedrich Pillichshammer

DE GRUYTER

Editors
Prof. Dr. Dmitriy Bilyk
University of Minnesota
School of Mathematics
Vincent Hall, office 328
206 Church St. SE
Minneapolis, MN 55408
USA
dbilyk@math.umn.edu

Prof. Dr. Friedrich Pillichshammer
Johannes Kepler Universität Linz
Institut für Finanzmathematik und
Angewandte Zahlentheorie
Altenbergerstr. 69
4040 Linz
Austria
friedrich.pillichshammer@jku.at

Prof. Dr. Josef Dick
University of New South Wales
School of Mathematics & Statistics
Sydney, NSW 2052
Australia
josef.dick@unsw.edu.au

ISBN 978-3-11-065115-7
e-ISBN (PDF) 978-3-11-065258-1
e-ISBN (EPUB) 978-3-11-065120-1
ISSN 1865-3707

Library of Congress Control Number: 2019951574

Bibliographic information published by the Deutsche Nationalbibliothek
The Deutsche Nationalbibliothek lists this publication in the Deutsche Nationalbibliografie; detailed bibliographic data are available on the Internet at http://dnb.dnb.de.

Typesetting: VTeX UAB, Lithuania
Printing and binding: CPI books GmbH, Leck

www.degruyter.com

Preface

This book is based on several invited talks at the workshop "Discrepancy" which was part of the RICAM Special Semester on "Multivariate Algorithms and their Foundations in Number Theory." The workshop took place at the Johann Radon Institute for Computational and Applied Mathematics (RICAM) of the Austrian Academy of Sciences in Linz, Austria, on November 26–30, 2018.

Discrepancy theory deals with point distributions in compact spaces, measuring the discrepancy between the empirical distribution and the target (usually uniform) distribution. One of the most prominent directions is the study of point distributions in the multidimensional unit cube. Through the theory of quasi-Monte Carlo methods, discrepancy theory has important applications in numerical analysis, since point sets and sequences with low discrepancy are required for numerical integration as nodes in quasi-Monte Carlo rules. A classical problem in discrepancy theory is concerned with the optimal rate of convergence of the supremum of the discrepancy function when the size of point distributions increases. The so-called "inverse of the discrepancy" problem is also actively studied nowadays: it asks for explicit constructions of point sets whose discrepancy depends at most polynomially on the dimension. From the applications point of view—in particular with respect to very high-dimensional integration problems—this new approach is of utmost importance. The discrepancy of point distributions on the unit sphere (or other manifolds), as well as discrepancy with respect to nonuniform target measures are also of great interest.

The workshop "Discrepancy" focused on discrepancy theory in a broad sense and took into account aspects from number theory, geometry, combinatorics, and numerical analysis. The goal of this book is to give an overview of recent developments in discrepancy theory with its relations to other fields, like, for example, quasi-Monte Carlo integration, uniform distribution theory, and Poissonian pair correlation, presented by leading experts in these vivid fields of research.

We briefly summarize the topics in this volume. The chapter "On some recent developments in uniform distribution and discrepancy theory" surveys recent developments on the relation between Poissonian pair correlation and uniform distribution, the progress on Tusnády' s problem, Levin' s lower bounds for the discrepancy of important low-discrepancy sequences, a link between the small ball inequality and digital nets, various heuristic arguments supporting two conflicting conjectures on the growth of the star-discrepancy of d-dimensional point sets, and different versions of the Stolarsky principle for the discrepancy on the sphere. In addition, a number of open problems and conjectures are posed. The chapter "Results and problems old and new in discrepancy theory" discusses some of the main results in discrepancy theory and highlights many difficult open problems in the subject. The chap-

https://doi.org/10.1515/9783110652581-201

ter "On negatively dependent sampling schemes, variance reduction, and probabilistic upper discrepancy bounds" studies probabilistic discrepancy bounds and provides new preasymptotic bounds with explicit constants for the star discrepancy and the weighted star discrepancy of sampling schemes. The focus of this contribution is on how discrepancy depends on the dimension. The chapter "Recent advances in higher order quasi-Monte Carlo methods" studies the application of low-discrepancy point sets for numerical integration by means of quasi-Monte Carlo rules. The focus is on results on recently developed higher order quasi-Monte Carlo methods. The chapter "On the asymptotic behavior of the sine product $\prod_{r=1}^{n} |2 \sin \pi r \alpha|$" reviews recently established results on the asymptotic behavior of the sine product, which is related to the distribution of the Kronecker sequence. The chapter "Fibonacci lattices have minimal dispersion on the two-dimensional torus" studies the dispersion of a given point set in the two-dimensional torus, given by the size of the largest rectangle containing no point, which is also related to the discrepancy of the point set. The chapter "On pair correlation of sequences" surveys the concept of Poissonian pair correlation of sequences in the unit interval and discusses a quite recent extension of this concept to the multidimensional case. The chapter "Some of Jiří Matoušek' s contributions to combinatorial discrepancy theory" covers important advancements in combinatorial discrepancy theory, especially for geometric set systems, by Jiří Matoušek, who was a central figure in discrepancy theory but passed away too early in 2015. The chapter "Fourier analytic techniques for lattice point discrepancy" provides a detailed description of several discrepancy problems (integer points problems and irregularities of distribution problems) in the planar case with test sets from a particular family of convex sets.

All chapters were reviewed by renowned experts in this field. We wish to thank the anonymous referees for their precious help. We also would like to thank Annette Weihs, Melanie Traxler, and Wolfgang Forsthuber for administrative support and all the speakers of the workshop who contributed excellent talks and made the workshop a great success: Christoph Aistleitner, Bence Borda, William Chen, Ujué Etayo, Damir Ferizović, Michael Gnewuch, Takashi Goda, Peter Grabner, Sigrid Grepstad, Aicke Hinrichs, Lisa Kaltenböck, Ralph Kritzinger, Gerhard Larcher, Ryan Matzke, Mario Neumüller, Aleksandar Nikolov, Maxim Skriganov, Tetiana Stepaniuk, Kosuke Suzuki, Robert Tichy, Giancarlo Travaglini, Mario Ullrich, Alex Vlasiuk, Jan Vybíral, Christian Weiss, Marcin Wnuk, and Agamemnon Zafeiropoulos.

More details on the RICAM special semester "Multivariate Algorithms and their Foundations in Number Theory" can be found on the webpage.

https://www.ricam.oeaw.ac.at/specsem/specsem2018/

We also thank the Johann Radon Institute for Computational and Applied Mathematics (RICAM) of the Austrian Academy of Sciences for financial support. We hope that this book will be a useful resource for many people who study or apply discrepancy theory.

Dmitriy Bilyk
Josef Dick
Friedrich Pillichshammer

Linz, September 2019

Contents

Dmitriy Bilyk

1 On some recent developments in uniform distribution and discrepancy theory

Abstract: We survey some of the recent developments in uniform distribution and discrepancy theory, which include, in particular, the fact that Poissonian pair correlation implies uniform distribution, the progress on Tusnády's problem, Levin's lower bounds for the discrepancy of most low-discrepancy sequences, a link between the small ball inequality and digital nets, various heuristic arguments supporting two conflicting conjectures on the growth of the star-discrepancy of d-dimensional points set, and different versions of the Stolarsky principle for the discrepancy on the sphere. We discuss known results and pose some open problems and conjectures.

Keywords: Uniform distribution of sequences, discrepancy, low-discrepancy sets

MSC 2000: Primary 11K38

1.1 Introduction

Uniform distribution theory originated from a seminal 1916 paper of Hermann Weyl "Über die Gleichverteilung von Zahlen mod. Eins" [78]. A sequence $\omega = (\omega_n) \subset [0,1)$ is called *uniformly distributed* if for any subinterval $[a,b) \subset [0,1)$, the proportion of points $\omega_n \in [a,b)$ is asymptotically equal to the length of the interval, that is,

$$\lim_{N\to\infty} \frac{\#\{n \le N : \omega_n \in [a,b)\}}{N} = b - a. \tag{1.1}$$

A very natural example of a uniformly distributed sequence is given by the famous Kronecker sequence $\{n\alpha\}$, where α is an irrational number and $\{x\}$ denotes the fractional part of x. Uniform distribution plays an important role in sampling and numerical integration as (1.1) can be easily seen to be equivalent to the fact that

$$\lim_{N\to\infty} \frac{1}{N} \sum_{n=1}^{N} f(\omega_n) = \int_0^1 f(x)\,dx \tag{1.2}$$

Acknowledgement: The author is extremely grateful to RICAM for hospitality and for sponsoring his stay during the Special Semester on Multivariate Algorithms and their Foundations in Number Theory. This survey is partially based on the tutorial talk on discrepancy theory which the author delivered before the Discrepancy workshop. The author would also like to thank the anonymous referee for numerous useful suggestions (as well as the fastest referee report that the author had ever seen). This work is partially supported by the NSF grant DMS 1665007.

Dmitriy Bilyk, School of Mathematics, University of Minnesota, Minneapolis, MN, 55455 USA, e-mail: dbilyk@math.umn.edu

https://doi.org/10.1515/9783110652581-001

for every continuous function f on $[0,1]$.

Discrepancy theory is essentially a quantitative counterpart of the uniform distribution theory. The *discrepancy* of the sequence $\omega = (\omega_n) \subset [0,1)$ is defined as

$$D_N(\omega) = \sup_{[a,b)\subset[0,1)} \big|\#\{n \le N : \omega_n \in [a,b)\} - N(b-a)\big|, \tag{1.3}$$

in other words, $D_N(\omega)/N$ is the rate of convergence in (1.1). It is a simple exercise to show that ω is uniformly distributed if and only if $D_N(\omega)/N \to 0$ as $N \to \infty$, that is, the convergence in (1.1) is necessarily uniform with respect to $[a,b)$.

Both notions above can be extended in a straightforward way to higher-dimensional sequences with the role of intervals taken by axis-parallel rectangles. In addition, other geometries and choices of test sets lead to numerous interesting generalizations of these concepts.

Over a period of more than a hundred years since the publication of Weyl's article uniform distribution and discrepancy have grown into a well-developed area of mathematics with many connections to other fields such as analysis, combinatorics, probability, number theory, discrete geometry, approximation theory, numerical integration, etc. Several great books written on this subject [10, 29, 33, 42, 53] can provide the reader with a good introduction.

In the present survey paper, we do not make an attempt to compete with these excellent references in providing a comprehensive account of the subject. Instead, we give a gentle nontechnical exposition of some of the recent results in the field, which of course, at times requires one to recall some classical facts. We also provide ample references for the reader interested in studying these issues further. A survey of this type is doomed to be incomplete and somewhat eclectic, and naturally reflects personal tastes of the author (including some of his own results). We note that a survey by W. W. L. Chen in this volume [30] contains a remarkable array of results and open problems in discrepancy theory and has some overlap with this paper. The author has previously written several expository papers on various aspects and connections of discrepancy theory [13–15, 22], but most of the results that we present now have been obtained after the prior surveys were written.

1.2 Uniform distribution of sequences

In 1935 [75, p. 816] van der Corput implicitly conjectured that the discrepancy of any sequence cannot stay bounded. He wrote "*...Ich weisz auch nicht, ob es eine Folge ω, eine stetige Funktion $\psi(x)$ und eine Konstante K gibt, die für jedes x und jedes natürliche Zahl N der Ungleichung*

$$\big|\#\{n \le N : \omega_n < x\} - N\psi(x)\big| < K$$

genügen."[1] It is implied here that $\psi(x)$ is the distribution function of the sequence ω, that is, $\psi(x) = \lim_{N\to\infty} \{n \le N : \omega_n < x\}/N$. In the case when ω is uniformly distributed, $\psi(x) = x$ for $0 \le x < 1$ according to (1.1).

Impossibility of such a "just" distribution was first proved by van Aardenne-Ehrenfest [74], and a much stronger quantitative bound, which extends to all dimensions, was later proved by Roth [58]; see (1.11). We shall discuss this in more detail in Section 1.3, but for now we shall adhere to the one-dimensional case. A sharp discrepancy bound was given by Schmidt: he proved that *for any sequence* $\omega = (\omega_n) \subset [0,1)$ *its discrepancy satisfies*

$$D_N(\omega) \gg \log N \tag{1.4}$$

for infinitely many values of N. The fact that this estimate is sharp has been known for a long time. Most known sequences with $D_N(\omega) \approx \log N$ stem from two basic examples. The first example is of *diophantine* nature—it is the already mentioned Kronecker sequence $\{n\alpha\}$, where α is a *badly approximable* number. The second example can be called *digital*—this is the celebrated van der Corput sequence v defined by

$$v_n = \sum_{j\ge1} n_j 2^{-j}, \tag{1.5}$$

where $n = \sum_j n_j 2^{j-1}$, $n_j \in \{0,1\}$, is the binary expansion of n. See [76, p. 1062] for the original definition of this sequence and the proof that $D_N(v) \ll \log N$.

1.2.1 Pair correlation and uniform distribution

A sequence $\omega = (\omega_n) \subset [0,1)$ is said to have *Poissonian pair correlation* (PPC) if

$$\lim_{N\to\infty} \frac{1}{N} \cdot \#\left\{1 \le i,j \le N,\, i \ne j : |\omega_i - \omega_j| < \frac{s}{N}\right\} = 2s. \tag{1.6}$$

One can easily check that i. i. d. uniform random points $\omega_n \in [0,1)$ almost surely have Poissonian pair correlation. Hence, this property, just like the uniform distribution (1.1), suggests random-like behavior of the sequence ω. This notion appeared in the work of Rudnick and Sarnack [60] in 1998, but for almost 20 years a very natural question of its relation to uniform distribution has remained unanswered. Then, in a surprising twist of fate, two solutions to this question appeared independently and almost simultaneously—the papers of Aistleitner, Lachmann, and Pausinger [5]

1 We keep the author's spelling, but use our notation. The passage translates as, "*I also do not know whether there exists a sequence* ω*, a continuous function* $\psi(x)$ *and a constant* K *such that the inequality* $|\#\{n \le N : \omega_n < x\} - N\psi(x)| < K$ *is satisfied for all* x *and all natural numbers* N*.*"

and Grepstad and Larcher [37] had been posted on www.arxiv.org within a week of each other in December 2016. Moreover, in only a month, in January 2017, Steinberger posted yet another solution [67]. They have proved the following.

Theorem 1.1. *If a sequence* $\omega = (\omega_n) \subset [0,1)$ *has Poissonian pair correlation, then it is uniformly distributed.*

It is easy to see that the converse of this implication is not true: indeed, the Kronecker sequence $\{n\alpha\}$ is uniformly distributed for irrational α, but does not have Poissonian pair correlation for any value of α.

The three known proofs of Theorem 1.1 are very different, yet all of them are very accessible and easy to read. The proof in [37] is very *ad hoc* and uses elementary methods, it also provides a quantitative discrepancy estimate in terms of the rate of convergence in (1.6). The argument in [5] proceeds by bounding the pair correlation from below by a quadratic form associated to a circulant matrix, which upon averaging has nonnegative eigenvalues, thus leading to a contradiction if one assumes that ω is not uniformly distributed. Finally, in [67] the author obtains more general quantitative criteria for uniform distribution, which can be viewed as quadratic analogues of the celebrated Weyl criterion with the Jacobi θ-functions (or Gaussians) in place of the complex exponentials (which makes these criteria more localized)—Theorem 1.1 is then obtained as their corollary.

A lot of other work related to Poissonian pair correlation has been done in the recent years. For example, the definition of PPC sequences (1.6) naturally extends to sequences in $[0,1)^d$, but it depends on the metric that one uses. Theorem 1.1 has been generalized and extended to the higher-dimensional setting: for the ℓ^∞-norm (i. e., with respect to boxes) [40], and subsequently for the Euclidean distance [68]. It had also been extended to smooth Riemannian manifolds with nonnegative Ricci curvature; see [52] and also [36].

Among other recent publications related to the pair correlation, we would like to specially point out the article of Aistleitner, Larcher, and Lewko [4] (with an Appendix written by late Jean Bourgain, whose tragic passing in December 2018 is a tremendous loss for the mathematical community). In this paper, the authors show that the metric property that *the sequence* $\{n_k\alpha\}$ *has Poissonian pair correlation for almost every* α *(the sequence* (n_k) *is metric Poissonian)* is closely related to the arithmetic structure of the sequence (n_k), namely, the *additive energy* of the truncations of this sequence. The additive energy of a finite N-element subset A of an additive group is defined as

$$E(A) = \#\{(a,b,c,d) \in A^4 : a+b=c+d\}, \tag{1.7}$$

and it is obvious that $N^2 \le E(A) \le N^3$. The authors prove, in particular, that if the additive energy of the truncations $A_N = (n_k)_{k\le N}$ is at least a little smaller than maximal, that is, $E(A_N) \ll N^{3-\varepsilon}$, then the sequence $\{n_k\alpha\}$ has Poissonian pair correlation for almost every α. Other recent work related to the interplay of the metric Poissonian property and additive energy includes [1, 6, 25, 43, 44, 77]: some of these papers deal, in particular, with the delicate situation when the additive energy is of the order very close to N^3, for example, [77] proves that primes, whose additive energy is roughly $N^3(\log N)^{-1}$, are *not* metric Poissonian, while at the same time, there exists a constant $C > 0$ such that whenever $E(A_N) \ll N^3(\log N)^{-C}$, the sequence (n_k) *is* metric Poissonian, and it is conjectured that this holds for any $C > 1$ [26].

In the end, we would like to raise the following interesting question concerning Poissonian pair correlation and uniform distribution: *What are the optimal discrepancy bounds for a sequence with Poissonian pair correlation?* Some upper bounds have been obtained in [37, 69], but this question is really about the lower bounds. It is known that both Kronecker and van der Corput sequences fail to have Poissonian pair correlation, the former due to the famous three gap theorem, and for the latter, as well as the more general LS-sequences and digital $(t, 1)$-sequences; see [45]. Therefore, it is reasonable to assume that, while PPC implies equidistribution, the optimal bound of Schmidt (1.4) cannot be achieved by sequences with PPC, and a lower bound of order strictly greater than $\log N$ should exist in this case.

Another important question concerns explicit deterministic constructions of sequences with PPC. At present, apart from the sequence $\{\sqrt{n}\}$ (see [34]), very few such examples are known.

In conclusion, we would like to mention that the survey by Larcher and Stockinger [46] in this volume is dedicated entirely to the Poissonian pair correlation and contains much more information than our short two-page description.

1.3 Star-discrepancy and related topics

We now turn to the circle of questions about the discrepancy of multidimensional point distributions in the unit cube. Let $\mathcal{P}_N = \{p_1, \dots, p_N\} \subset [0,1)^d$ be a finite set of N points. We define the discrepancy function associated to this set as

$$D_N(x) = \#\{\mathcal{P}_N \cap [0,x)\} - N x_1 \cdot \dots \cdot x_N, \tag{1.8}$$

where $x = (x_1, \dots, x_d) \in [0,1)^d$ and $[0,x) = [0,x_1) \times \dots \times [0,x_d)$ is an axis-parallel box with corners at 0 and x. This functions measures the local discrepancy of $\mathcal{P}_N$ with respect to the box $R = [0,x)$, much like in (1.3). Various norms of D_N provide information about the extent of equidistribution of the point set $\mathcal{P}_N$. Perhaps, the most natural is its L^∞ norm, also know as the *star-discrepancy*:

$$D^*(\mathcal{P}_N) = \|D_N\|_\infty = \sup_{x\in[0,1)^d} \left|\#\{\mathcal{P}_N \cap [0,x)\} - N x_1 \cdot \dots \cdot x_N\right|, \tag{1.9}$$

(the reason for the name simply comes from the fact that historically this quantity was denoted by D^*). The aforementioned result of Schmidt (1.4) for the discrepancy of sequences easily translates to the following statement for two-dimensional point sets (see, e. g., [10, 58]): for any $\mathcal{P}_N \subset [0,1)^2$, we have

$$\|D_N\|_\infty \gg \log N. \tag{1.10}$$

More generally, there is a certain "transference" between estimates akin to (1.4) for infinite sequences in $[0,1)^d$ and uniform bounds for N-point sets in $[0,1)^d$, and we adopt the latter setting in this section.

1.3.1 The main conjectures: $(\log N)^{d-1}$ vs. $(\log N)^{d/2}$

Schmidt's bound (1.10) has long been known to be sharp, hence the question of the optimal asymptotic behavior of the star discrepancy is essentially closed in dimension $d = 2$. However, in higher dimensions, despite years of research, it is still completely open, even at the level of conjectures. In his foundational 1954 paper [58] Klaus Roth proved that in all dimensions $d \geq 2$,

$$\|D_N\|_\infty \geq \|D_N\|_2 \gg (\log N)^{\frac{d-1}{2}}, \tag{1.11}$$

where the implicit constant depends only on the dimension. The L^2 bound above is sharp; see [31, 59]. But what is the correct optimal order of magnitude for the L^∞ norm?! Beck and Chen [10] dubbed this question the "great open problem" and called it "excruciatingly difficult."

It is conjectured, heuristically, that the true L^∞ (extremal) bound should grow faster than the L^2 (average) bound—in other words, the discrepancy function of well-distributed sets cannot be "too flat." Schmidt's estimate (1.10) confirms this heuristic in dimension $d = 2$: indeed, the L^2 bound (1.11) is $\sqrt{\log N}$ in this case, while the optimal bound (1.10) is $\log N$.

The precise conjectural rate of growth of the star-discrepancy in higher dimensions is a subject of (sometimes heated) debate between the experts. Two main conjectures have crystallized after years of research. The first conjecture has been around for a long time and may be regarded as classical.

Conjecture 1.2. *For all dimensions $d \geq 2$, there exists a constant $c_d > 0$ such that*

$$\|D_N\|_\infty \geq c_d (\log N)^{d-1}. \tag{1.12}$$

The second conjecture is much younger; it has been first put forth by Pollington in an unpublished preprint and then stated again and popularized by the author and collaborators [21, 24, 22, 13–15].

Conjecture 1.3. *For all dimensions $d \geq 2$, there exists a constant $c_d > 0$ such that*

$$\|D_N\|_\infty \geq c_d (\log N)^{d/2}. \tag{1.13}$$

The two conjectures differ in the power of the logarithm, and they coincide with Schmidt's result (1.10) in dimension $d = 2$.

A shift in attitudes toward these conjectures may be traced in the literature. In the classical 1987 reference [10, p. 6], Beck and Chen essentially state Conjecture 1.2, while much later, in his 2010 book [53, p. 176] Matoušek poses the conjecture in a much more ambiguous way: *Schmidt's result improves Roth's lower bound in the plane by a factor of $\sqrt{\log N}$. A similar improvement in higher dimensions turns out to be much more challenging (although it is widely believed that it should be possible).*

Both conjectures have their own pros and cons. Below we attempt to present and impartial list of reasons supporting each of the conjectures: some of these reasons are heuristic or numerological, some anecdotal, and some stem from related proven facts.

Pros for Conjecture 1.2: $(\log N)^{d-1}$

- This conjecture is older and much more established. The popular belief in this conjecture is evidenced by the fact that the literature often uses the term *low-discrepancy sets* to denote point sets with $\|D_N\|_\infty \ll (\log N)^{d-1}$, thus implicitly suggesting that sets with lower discrepancy cannot exist.
- The best among all the known constructions of well-distributed sets (digital nets, lattices, Kronecker sequences, Halton–Hammersley sets) indeed have discrepancy of this order $\|D_N\|_\infty \ll (\log N)^{d-1}$. Moreover, in a recent series of papers [47]–[51] Levin had shown that, for each of these sets, the upper bound cannot be improved, that is, they satisfy $\|D_N\|_\infty \gg (\log N)^{d-1}$. In other words, if Conjecture 1.2 were not sharp, one should be able to find absolutely different constructions with much smaller discrepancy—it would be strange that they have eluded us so far. This is further discussed in Section 1.3.3.
- Certain other "smoother" notions of discrepancy do satisfy a bound similar to (1.12) with $(\log N)^{d-1}$ and a negative power of N depending on the smoothness; see, for example, [73]. There is however a barrier which does not allow one to extend the arguments all the way down to smoothness zero.
- Recent progress on Tusnády's problem [54, 55] about the order of magnitude of the combinatorial discrepancy generated by axis-parallel rectangles provides a lower bound of the order $(\log N)^{d-1}$, which is conjectured to be sharp. Combinatorial discrepancy is an upper bound for the geometric discrepancy, which in many situations is sharp. See Section 1.3.4 for a more detailed discussion.

Pros for Conjecture 1.3: $(\log N)^{d/2}$

- This conjecture is newer, but has gained acclaim in the recent years.
- The *small ball conjecture* (see (1.14) in Section 1.3.2 below) about the L^∞ norm of some linear combinations of the multivariate Haar functions, which can be viewed as a linear model for the discrepancy function D_N, yields power $\frac{d}{2}$. This conjecture, if true, would be sharp. However, no general "transference" technique is available in order to translate such inequalities into discrepancy inequalities.

- The aforementioned small ball conjecture, if proven, would also provide lower bounds in conjectures in probability (small deviation for the Brownian sheet) and approximation theory (metric entropy for spaces with mixed smoothness), whereas matching upper bounds are already known in these cases. A detailed exposition of the small ball inequality, its connections to discrepancy, and related topics can be read in the author's papers [13, 14, 22, 15].
- A difference of $\sqrt{\log N}$ between the L^∞ and L^2 bounds occurs often in the subject of discrepancy in other geometric settings: for example, discrepancy with respect to rotated boxes, balls, spherical caps: see, for example, (1.25) and (1.26). It usually stems from the sub-Gaussian large deviation estimates; see, for example, [8]. However, these geometric settings are rather different, and besides this $\sqrt{\log N}$ difference is still conjectural (it is present only in the upper, not lower bounds—see open problems in the end of Section 1.4.1).

Other conjectures have been proposed as to the correct power of the logarithm: $\frac{d-1}{2} + \frac{d-1}{d}$ [63] and $\frac{3}{4}d - \frac{1}{2}$ [39]: both match (1.10) for $d = 2$, and the second one is the average between Conjectures 1.2 and 1.3, but the author is not aware of more serious evidence in their support. We conclude by mentioning that the best currently known bounds in dimensions $d \geq 3$ slightly improve on Roth's L^2 estimate (1.11): $\|D_N\| \gg (\log N)^{\frac{d-1}{2}+\eta_d}$, where η_d is a small constant; see [21, 24].

1.3.2 The small ball inequality, discrepancy, and digital nets

The proof of Roth's L^2 lower bound (1.11) involves a convenient choice of an orthogonal basis on $L^2([0,1]^d)$—the so-called Haar functions (although Roth in [58] never uses this term). In dimension $d = 1$, these functions were introduced by Haar in 1911 [38] and historically they are the first example of wavelets. Let $\mathcal{D}$ denote the system of dyadic subintervals of $[0,1)$, that is, intervals of the form $[\frac{k}{2^n}, \frac{k+1}{2^n})$, where $n \in \mathbb{Z}_+$ and $k = 0, 1, \ldots, 2^n - 1$. The Haar function on $I \in \mathcal{D}$ is defined as $h_I(x) = \mathbf{1}_{I_L}(x) - \mathbf{1}_{I_R}(x)$, where I_L and I_R are the left and right halves of I. Together with the constant function, this system forms an orthogonal basis of $L^2([0,1])$. For a dyadic box $R = R_1 \times \cdots \times R_d \in \mathcal{D}^d$, the multidimensional Haar function h_R is defined as a tensor product of one-dimensional counterparts, $h_R(x) = h_{R_1}(x_1) \cdot \cdots \cdot h_{R_d}(x_d)$.

The *small ball conjecture*, in a simplified form, states that

$$\Big\| \sum_{|R|=2^{-n}} \varepsilon_R h_R \Big\|_\infty \gg n^{d/2}, \tag{1.14}$$

where the coefficients $\varepsilon_R = \pm 1$ and the sum is extended over all $R \in \mathcal{D}^d$ with volume $|R| = 2^{-n}$. Such sums with $n \approx \log N$ serve as models for the discrepancy function D_N and the similarity between (1.14) and (1.13) of Conjecture 1.3 is apparent. Moreover, for the L^2 norm of this sum one easily obtains the lower bound of the order $n^{\frac{d-1}{2}}$, which is consistent with Roth's bound (1.11). This analog also explains that the exponent $d-1$ is combinatorially natural, there are $\approx n^{d-1}$ different shapes of dyadic rectangles R with $|R| = 2^{-n}$ for a fixed $n \geq 0$.

The small ball conjecture is known to be true in dimension $d = 2$: it was initially proved by Talagrand [71] and several other proofs have been given subsequently [72, 20, 41]. In higher dimensions, this conjecture is still wide open; see [21, 24] for known estimates. Much has been written about the connections of this conjecture to discrepancy, as well as to open problems in probability and approximation theory [14, 22].

While the latter connections are formal, the relation to discrepancy is only heuristic. The first rigorous link was established by the author and Feldheim, who showed that the extremal sets in the two-dimensional small ball inequality coincide with a class of low-discrepancy sets. A dyadic (t, m, d)-net is a set of points 2^m points in $[0,1)^d$ such that any dyadic rectangle $R \in \mathcal{D}^d$ of volume $|R| = 2^{-m+t}$ contains a fair share of points of the net, that is, 2^t points (m is referred to as the "order" of the net, and t is called "deficiency"). The idea of such sets stems from the van der Corpus sequence, and it is well known that, for fixed d and t, they satisfy the bound $\|D_N\|_\infty \ll_{t,d} (\log N)^{d-1}$. A lot of information about these constructions can be found in the book by Dick and Pillichshammer [32]. In [20], the following was proved: *in dimension $d = 2$ the set of points $x \in [0,1)^2$, where the sum on the left-hand side of* (1.14) *achieves its maximal value, consists of 2^{n+1} squares, whose lower left corners form a $(0, n+1, 2)$-net. Moreover, every such net can be obtained by changing the coefficients ε_R.*

1.3.3 Known low-discrepancy sets do not break the $(\log N)^{d-1}$ barrier: recent results of M. Levin

A whole array of examples of N-point sets in $[0,1)^d$ with discrepancy of the order $(\log N)^{d-1}$ (or almost equivalently, d-dimensional sequences with discrepancy $\mathcal{O}(\log^d N)$) is available in the literature. We do not describe them in detail here, but just mention that most of them arise from two classical examples of one-dimensional sequences mentioned in the beginning of Section 1.2: Kronecker and van der Corput sequences, although higher-dimensional extensions may be quite elaborate and nontrivial. Generalizations of the van der Corput sequence to higher dimensions include the Halton sequence, where the digit-reversing procedure is done in different coprime bases, as well as the so-called (t, d)-sequences.

Infinite sequences $\omega = (\omega_n) \subset [0,1)^{d-1}$ can be converted to the finite N-point sets in $[0,1)^d$ by considering collections $\{(\frac{n}{N}, \omega_n)\}_{n=0}^{N-1}$. Under this identification, the Halton sequence becomes the Hammersley set, and the (t, d)-sequences correspond to the aforementioned (t, m, d)-nets. More complicated extensions are also known.

For a while only the upper discrepancy bounds were known for all these sets and sequences—these bounds were consistent with Conjecture 1.2, but there was still hope that their discrepancy could be even smaller, which perhaps could lead to Conjecture 1.3. However, in the last few years, this hope was shattered by a series of papers of Levin.

In a nutshell, Levin had proved that almost all known low-discrepancy examples also satisfy matching lower bounds consistent with Conjecture 1.3. Namely, he proved the estimate

$$\limsup_{N\to\infty} D_N(\omega)\,(\log N)^{-d} > 0, \tag{1.15}$$

whenever ω is the Halton sequence [48], generalized Halton sequence [49], shifted Niederreiter sequence [50], Niederreiter–Xing sequence, or other explicit construction of (t,d)-sequences [51]. This translates to the estimate

$$\|D_N\|_\infty \gg (\log N)^{d-1} \tag{1.16}$$

for the Hammersley set and various (t,m,d)-nets, showing that Conjecture 1.3 cannot be improved for these sets. He also proved that the same lower bound holds for certain lattices, obtained from modules in a totally real algebraic number field [47]; the matching upper bound in this setting was earlier obtained by Skriganov [62].

These results give a lot of weight to Conjecture 1.2 by effectively ruling out most of the potential candidates that might have broken the $(\log N)^{d-1}$ barrier.

1.3.4 Combinatorial discrepancy: Tusnády's problem and the interplay with geometric discrepancy

Let $\mathcal{A}$ be a collection of geometric sets in the unit cube $[0,1)^d$. For a N-point set $\mathcal{P}_N \subset [0,1)^d$, its combinatorial discrepancy with respect to $\mathcal{A}$ is defined as

$$\operatorname{disc}(\mathcal{P}_N, \mathcal{A}) = \inf_{\chi:\mathcal{P}_N\to\{\pm1\}} \sup_{A\in\mathcal{A}} \Big|\sum_{p\in\mathcal{P}_N\cap A} \chi(p)\Big|. \tag{1.17}$$

The map χ is usually thought of as a red-blue coloring of the points of $\mathcal{P}_N$, that is, we are looking for a largest disbalance of colors within sets in $\mathcal{A}$, and then optimize over all colorings. The combinatorial discrepancy of $\mathcal{A}$ is

$$\operatorname{disc}(N, \mathcal{A}) = \sup_{\substack{\mathcal{P}\subset[0,1)^d \\ \#\mathcal{P}_N = N}} \operatorname{disc}(\mathcal{P}_N, \mathcal{A}), \tag{1.18}$$

which, in some sense, is a measure of complexity of the collection $\mathcal{A}$.

It is well known that, under some mild conditions, the combinatorial discrepancy $\mathrm{disc}(N, \mathcal{A})$ serves as an upper bound for the optimal geometric discrepancy. Namely, if one defines

$$D_N(\mathcal{A}) = \inf_{\substack{\mathcal{P}_N \subset [0,1)^d \\ \#\mathcal{P}_N = N}} \sup_{A \in \mathcal{A}} \left| \#\{\mathcal{P}_N \cap A\} - N \cdot \mathrm{vol}(A) \right|, \tag{1.19}$$

then the following *"transference principle"* holds between the two discrepancies

$$D_N(\mathcal{A}) \ll \mathrm{disc}(\mathcal{P}_N, \mathcal{A}). \tag{1.20}$$

This fact in various forms can be found in [7, 53, 2].

In the case when $\mathcal{A} = \mathcal{R}_d$ is the system of axis-parallel boxes, anchored at the origin, the question of finding the correct order of magnitude of $\mathrm{disc}(N, \mathcal{R}_d)$ is known as *Tusnády's problem*. It is, in fact, conjectured that $\mathrm{disc}(N, \mathcal{R}_d) \approx (\log N)^{d-1}$. Observe that

$$D_N(\mathcal{R}_d) = \inf_{\substack{\mathcal{P}_N \subset [0,1)^d \\ \#\mathcal{P}_N = N}} \|D_N\|_\infty .$$

For a long time, the only lower bounds available for this problems were obtained through (1.20) and the known lower bounds for the star-discrepancy $\|D_N\|_\infty$.

However, in the past few years this problem came very close to being solved. The following estimate is currently known:

$$(\log N)^{d-1} \ll \mathrm{disc}(N, \mathcal{R}_d) \ll (\log N)^{d-\frac{1}{2}}. \tag{1.21}$$

The lower bound was proved by Nikolov and Matoušek [54] (unfortunately, this was one of the last papers of Jiří Matoušek, who died in 2015 at the age of 51); see also [55]. The upper bound was just recently proved by Nikolov [56]. Much more information about combinatorial discrepancy and about Matoušek's contributions is contained in Nikolov's survey in this volume [57].

We conclude this subsection with a question on combinatorial discrepancy: for most classes of sets $\mathcal{A}$, the transference inequality (1.20) is sharp, that is, the optimal combinatorial and geometric discrepancies are of the same order. Does this statement hold in general for reasonable geometric collections? The only known exception to this pattern is a rather rich class consisting of all convex sets, in which case the optimal geometric discrepancy $D_N(\mathcal{A})$ is of the order $N^{1-\frac{2}{d+1}}$ (up to logarithms), while the combinatorial discrepancy $\mathrm{disc}(\mathcal{P}_N, \mathcal{A}) \approx N$.

1.4 Geometric discrepancy: spheres and more

So far, we have mostly discussed discrepancy with respect to intervals or axis-parallel rectangles for point sets contained in the unit cube, but it is also natural to consider discrepancies with respect to other collections of sets and in other geometrical settings. One of the most popular such settings that has received considerable attention recently is discrepancy for points on the sphere.

1.4.1 Discrepancy on the sphere

Let $\mathbb{S}^d \subset \mathbb{R}^{d+1}$ denote the d-dimensional unit sphere, and let σ represent the (Hausdorff) surface measure on $\mathbb{S}^d$, normalized so that $\sigma(\mathbb{S}^d) = 1$. Consider a set of N points on the sphere $Z = \{z_1, \ldots, z_N\} \subset \mathbb{S}^d$. For any reasonable collection of geometric subsets of $\mathbb{S}^d$, we can define the discrepancy of Z, much like it was done in (1.19). The most standard choice of the geometric test sets are the *spherical caps*. For $x \in \mathbb{S}^d$, $t \in [-1, 1]$, the spherical cap centered at x with height t is defined as

$$C(x, t) = \{z \in \mathbb{S}^d : x \cdot y > t\}. \tag{1.22}$$

The spherical cap discrepancy of $Z = \{z_1, \ldots, z_N\} \subset \mathbb{S}^d$ is the quantity

$$D_{\text{cap}}(Z) = \sup_{x \in \mathbb{S}^d, t \in [-1,1]} \left|\#(Z \cap C(x,t)) - N\sigma(C(x,t))\right|, \tag{1.23}$$

and if one replaces the supremum with the quadratic average over all caps, one gets the L^2 spherical cap discrepancy:

$$D^2_{L^2,\text{cap}}(Z) = \int_{-1}^{1} \int_{\mathbb{S}^d} \left|\#(Z \cap C(x,t)) - N\sigma(C(x,t))\right|^2 d\sigma(x)\, dt. \tag{1.24}$$

Unlike the discrepancy with respect to axis-parallel boxes, which grows logarithmically, the optimal spherical cap discrepancy grows as a power of N. The following estimates have been proved by Beck [8, 9].

Theorem 1.4. *The optimal spherical cap discrepancy of N-point sets $Z = \{z_1, \ldots, z_N\} \subset \mathbb{S}^d$ satisfies*

$$c_d N^{\frac{1}{2} - \frac{1}{2d}} \le \inf_{\#Z=N} D_{\text{cap}}(Z) \le C_d N^{\frac{1}{2} - \frac{1}{2d}} \sqrt{\log N}, \tag{1.25}$$

and the L^2 spherical discrepancy satisfies

$$c_d N^{\frac{1}{2} - \frac{1}{2d}} \le \inf_{\#Z=N} D_{L^2,\text{cap}}(Z) \le C_d N^{\frac{1}{2} - \frac{1}{2d}}. \tag{1.26}$$

This effect also propagates to other settings in which curvature or rotational invariance is involved, for example, balls or rotated rectangles—in these cases, the estimates are exactly the same as above. The lower bounds in (1.25) and (1.26) are proved using L^2 methods (Fourier or spherical harmonics), and for the L^2 discrepancy they are actually sharp as (1.26) shows.

The lower bound in (1.26) has been recently refined in [18], it was shown that

$$D_{L^2,\mathrm{cap}}(Z) \gg N^{\frac{1}{2}-\frac{1}{2d}} \left(\frac{1}{N} \sum_{i,j=1}^{N} \frac{\log(2 + N^{1/d}\|z_i - z_j\|)}{(1 + N^{1/d}\|z_i - z_j\|)^{d+1}} \right)^{\frac{1}{2}}. \tag{1.27}$$

By only keeping the diagonal terms in the sum above, one recovers Beck's lower bound (1.26), but for specific sets the discrete energy on the right may give more precise information about the order of the discrepancy. An even deeper connection between discrepancy and discrete energy is discussed in Section 1.4.2.

The upper bound in both (1.25) and (1.26) is obtained using a semirandom construction known as *"jittered (or stratified) sampling."* This construction starts with an equal area partition of the sphere, that is, a partition $\mathbb{S}^d = \bigcup_{i=1}^N R_i$, such that the components R_i:

(a) are essentially disjoint, that is, $\sigma(R_i \cap R_j) = 0$;
(b) have the same size $\sigma(R_i) = \frac{1}{N}$, $i = 1, \ldots, N$;
(c) have small diameters, $\operatorname{diam}(R_i) \ll N^{-1/d}$, $i = 1, \ldots, N$.

Then one chooses independent random points in each of the components $z_i \in R_i$. This construction provides upper bounds in both (1.25) and (1.26). The additional $\sqrt{\log N}$ in for the extremal discrepancy emerges as an inverse of the Gaussian function $e^{-\lambda^2}$ through the use of sub-Gaussian large deviation bounds, for example, Hoeffding's inequality. This construction is also useful for other problems of optimal point distributions, in particular, for optimal estimates of the Riesz energy of N points on the sphere. Another random construction that almost satisfies the upper bound in (1.25), albeit with a larger power of logarithm, is the so-called determinantal point process [11].

Several questions arise naturally in this respect and are still wide open:

- What is the correct order of growth of the optimal spherical cap discrepancy $D_{\mathrm{cap}}(Z)$? In other words, is the $\sqrt{\log N}$ necessary? It is quite possible that this logarithmic factor is, in fact, needed, that is, $N^{\frac{1}{2}-\frac{1}{2d}}\sqrt{\log N}$ may be the correct estimate for the discrepancy. If it is so, this would be a manifestation of the $\sqrt{\log N}$ difference between the extremal and average (quadratic) discrepancy estimates, which was alluded to in Section 1.3.1, and this would support Conjecture 1.3 for the star-discrepancy. However, this would also require a novel approach to lower bounds.
- Another important question is about explicit constructions of optimal or almost optimal point sets on the sphere. All the constructions that come close to satisfying the upper bounds in (1.25) and (1.26) are partially random: for example, the aforementioned jittered sampling, or the so-called determinantal point process. *Is there an explicit construction of a set of points with the (almost) optimal order of*

the spherical cap discrepancy? Even in the case of $\mathbb{S}^2$ such constructions are not known. Several attempts have been made, for example, [3], however, the problem still remains unsolved.

1.4.2 Stolarsky principle

It turns out that the L^2 spherical cap discrepancy is closely related to another object, which arises frequently in discrete and metric geometry—the sum of Euclidean distances. In 1973, Stolarsky [70] proved the following remarkable identity:

$$c_d D^2_{L^2,\mathrm{cap}}(Z) = N^2 \int_{\mathbb{S}^d}\int_{\mathbb{S}^d} \|x-y\|\, d\sigma(x)\, d\sigma(y) - \sum_{i,j=1}^{N} \|z_i - z_j\|, \tag{1.28}$$

where c_d is an explicit dimensional constant. This relation (which came to be known as the *Stolarsky invariance principle*) implies that minimizing the L^2 spherical cap discrepancy is equivalent to the problem of maximizing the sum of pairwise Euclidean distances between the points of $Z = \{z_1, \dots, z_N\} \subset \mathbb{S}^d$. New simplified proofs of this identity have been given in [28, 17]. The idea became quite popular in the recent years, and a number of papers have appeared exploring extensions and generalizations of (1.28).

On a heuristic level, given a notion of discrepancy, it is fairly straightforward to obtain an analog of (1.28): one can expand the square in the definition (1.24) of $D^2_{L^2,\mathrm{cap}}(Z)$, and the "cross terms" depending on z_i and z_j, after (sometimes technically complicated) integration, yield an object similar to the distance sum, or discrete energy. In this vein, the following versions of the Stolarsky principle have been obtained (rather than stating the exact identities, we shall just describe the notions of discrepancy and the discrete energies that arise from them):

- For the L^2 discrepancy with respect to *hemispheres*, that is, caps with height $t = 0$, one obtains a complete analog of (1.28) with the Euclidean distance replaced by the geodesic distance $d(x, y)$ on the sphere. This identity yielded a complete solution of the conjecture by Fejes Tóth [35] about configurations that maximize the sum of geodesic distances. See [17, 65] for details.
- Consider *spherical wedges*, that is, sets of points which lie "between" the hyperplanes $x^\perp$ and $y^\perp$ for some $x, y \in \mathbb{S}^d$:

  $$W_{xy} = \{z \in \mathbb{S}^d : \operatorname{sgn} x \cdot z \neq \operatorname{sgn} y \cdot z\}.$$

  These sets arise naturally in the problem of uniform tessellation of the sphere by hyperplanes. The L^2 discrepancy with respect to the spherical wedges yields the discrete energy of the form

  $$\sum_{i,j=1}^{N} \left(\frac{\pi}{2} - d(z_i, z_j)\right)^2,$$

where d again represents the geodesic distance; see [23]. This energy is very similar to the so-called *frame potential* $\sum_{i,j} |z_i \cdot z_j|^2$, which plays an important role in frame theory and signal processing [12].

- *Spherical slices* are "halves" of spherical wedges: for $x, y \in \mathbb{S}^d$ they are defined as $S_{xy} = \{z \in \mathbb{S}^d : \operatorname{sgn} x \cdot z > 0, \operatorname{sgn} y \cdot z < 0\}$. Discrepancy with respect to these sets was studied by Blümlinger [27]. They give rise to the Stolarsky identity with the discrete energy of the form (see [17]),

$$\sum_{i,j=1}^{N} d(z_i, z_j)(\pi - d(z_i, z_j))$$

- On the circle $\mathbb{S}^1 \simeq \mathbb{T}$, the L^2 discrepancy with respect to unions of opposite quadrants, that is, rotations of the set $[0, \pi/2) \cup [3\pi/2, 2\pi)$, leads to the Stolarsky principle with the discrete energy

$$\sum_{i,j} \min\left\{d(z_i, z_j), \frac{\pi}{2} - d(z_i, z_j)\right\},$$

in other words, this is the sum of (nonobtuse) angles between the lines generated by the vectors z_i. This sum is the subject of yet another conjecture by Fejes Tóth [35], and this version of the Stolarsky principle leads to a new proof of this conjecture in dimension $d = 1$. Unfortunately, this idea does not easily extend to higher dimensions, and the conjecture is still wide open in this case; see [19].

In addition, Skriganov [64] had proved an extension of the Stolarsky principal (1.28) to general distance invariant spaces $\mathcal{M}$ with a measure μ, for a "symmetric difference" metric

$$\rho^*(x, y) = \int_0^{\operatorname{diam}(\mathcal{M})} \mu(B_r(x) \Delta B_r(y))\, d\xi(r),$$

where ξ is some positive Borel measure on $[0, \operatorname{diam}(\mathcal{M})]$. He also proved probabilistic extensions of the Stolarsky principle to general metric spaces, as well as a version of the Stolarsky identity on projective spaces [66].

Finally, we would like to mention that, although it is generally harder to start with a discrete energy and find a notion of discrepancy, which would give rise to a Stolarsky-type identity, a fairly general result in this direction was obtained by the author, Dai, and Matzke in [16, 17]. Let $F : [-1, 1] \to \mathbb{R}$ be a continuous positive definite function on the sphere $\mathbb{S}^d$ in the sense that the matrix $[F(z_i \cdot z_j)]_{i,j=1}^N$ is positive semidefinite for each collection $\{z_1, \ldots, z_N\} \subset \mathbb{S}^d$. This is known to be equivalent to the fact that all the coefficients in the Gegenbauer polynomial expansion of $F(t) = \sum_{n=0}^{\infty} \widehat{F}_n \frac{2n+d-1}{d-1} C_n^{\frac{d-1}{2}}(t)$

are nonnegative, that is, $\widehat{F}_n \geq 0$, $n \in \mathbb{N}$; see [61]. Then let us define the function f through its Gegenbauer coefficients by the identity

$$(\widehat{f}_n)^2 = \widehat{F}_n, \quad \text{for all } n \in \mathbb{N}, \tag{1.29}$$

or, equivalently,

$$F(x \cdot y) = \int_{\mathbb{S}^d} f(x \cdot z) f(z \cdot y)\, d\sigma(z) \quad \text{for all } x, y \in \mathbb{S}^d, \tag{1.30}$$

that is, F is a "spherical convolution" of f with itself. Notice that f is not unique, as we are free to choose the signs of $\widehat{f}_n$. We can now define the L^2 discrepancy of $Z = \{z_1, \ldots, z_N\} \subset \mathbb{S}^d$ with respect to f as follows:

$$D^2_{L^2,f}(Z) = \int_{\mathbb{S}^d} \left| \sum_{i=1}^N f(x \cdot z_i) - N \int_{\mathbb{S}^d} f(x \cdot z)\, d\sigma(z) \right|^2 d\sigma(x). \tag{1.31}$$

Observe that if $f(\tau) = \mathbf{1}_{[t,1]}(\tau)$, one obtains precisely the discrepancy with respect to spherical caps of height t, that is, the inner integral in (1.24).

If F and f are related by identities (1.29)–(1.30), one obtains a very general analog of the Stolarsky principle on the sphere:

$$D^2_{L^2,f}(Z) = \sum_{i,j}^N F(z_i, z_j) - N^2 \int_{\mathbb{S}^d} \int_{\mathbb{S}^d} F(x \cdot y)\, d\sigma(x)\, d\sigma(y). \tag{1.32}$$

While the relation between the functions F and f is somewhat implicit, it still allows one to translate between discrepancy estimates and bounds for the discrete energy, in particular one has the bound

$$\min_{1 \leq k \ll N^{1/d}} N\widehat{F}_k \ll \inf_{\#Z=N} D^2_{L^2,f}(Z) \ll \max_{|t| \ll N^{-1/d}} (F(1) - F(\cos t)); \tag{1.33}$$

see [16] for the details.

Some natural examples of functions to which the generalized Stolarsky principle may be applied (i. e., positive definite functions on $\mathbb{S}^d$, up to constant terms) include $F(x \cdot y) = -\|x - y\|^\alpha$, where $\alpha \in (0, 2]$, or $F(x \cdot y) = -d^\alpha(x, y)$ for $\alpha \in (0, 1]$, that is, this covers both the case of the classical Stolarsky principle (1.28) and the aforementioned version for the geodesic distances. However, unfortunately, the explicit relation between the functions F and f is only known in the latter case. It would be interesting to understand how to rewrite the original Stolarsky principle in the form (1.32): observe that the left-hand side of (1.28) is not of the same form as in (1.32), as there is an additional integration in t in the definition (1.24) of $D^2_{L^2,\mathrm{cap}}(Z)$ compared to that of $D^2_{L^2,f}(Z)$ in (1.31). It is possible to write out the Gegenbauer coefficients of f, but the closed form of f is still elusive. A similar question can be formulated for many other interesting energies.

Bibliography

[1] I. Aichinger, C. Aistleitner, G. Larcher. On quasi-energy-spectra, pair correlations of sequences and additive combinatorics. In: Contemporary Computational Mathematics – A Celebration of the 80th Birthday of Ian Sloan. Vol. 1, 2, pp. 1–16. Springer, Cham (2018).

[2] C. Aistleitner, D. Bilyk, A. Nikolov. Tusnády's problem, the transference principle, and non-uniform QMC sampling. In: Monte Carlo and Quasi-Monte Carlo Methods, pp. 169–180. Springer Proc. Math. Stat., vol. 241. Springer (2018).

[3] C. Aistleitner, J. S. Brauchart, J. Dick. Point sets on the sphere $\mathbb{S}^2$ with small spherical cap discrepancy, Discrete Comput. Geom. **48**(4), 990–1024 (2012).

[4] C. Aistleitner, G. Larcher, M. Lewko. Additive energy and the Hausdorff dimension of the exceptional set in metric pair correlation problems. With an appendix by Jean Bourgain. Isr. J. Math. **222**(1), 463–485 (2017).

[5] C. Aistleitner, T. Lachmann, F. Pausinger. Pair correlations and equidistribution, J. Number Theory **182**, 206–220 (2018).

[6] C. Aistleitner, T. Lachmann, N. Technau. There is no Khintchine threshold for metric pair correlations, preprint (2018), available at https://www.arxiv.org/pdf/1802.02659.pdf.

[7] J. Beck. Balanced two-colorings of finite sets in the square. I. Combinatorica **1**(4), 327–335 (1981).

[8] J. Beck. Some upper bounds in the theory of irregularities of distribution, Acta Arith. **43**(2), 115–130 (1984).

[9] J. Beck. Sums of distances between points on a sphere—an application of the theory of irregularities of distribution to discrete geometry, Mathematika **31**(1), 33–41 (1984).

[10] J. Beck, W. W. L. Chen. Irregularities of Distribution. Cambridge University Press, Cambridge (1987).

[11] C. Beltrán, J. Marzo, J. Ortega-Cerdà. Energy and discrepancy of rotationally invariant determinantal point processes in high dimensional spheres, J. Complex. **37**, 76–109 (2016).

[12] J. J. Benedetto, M. Fickus. Finite normalized tight frames. Adv. Comput. Math. **18**(2–4), 357–385 (2003).

[13] D. Bilyk. On Roth's orthogonal function method in discrepancy theory. Unif. Distrib. Theory **6**(1), 143–184 (2011).

[14] D. Bilyk. Roth's orthogonal function method in discrepancy theory and some new connections. In: A Panorama of Discrepancy Theory, pp. 71–158. Lecture Notes in Math., vol. 2107. Springer (2014).

[15] D. Bilyk. Discrepancy theory and harmonic analysis. In: Uniform Distribution and Quasi-Monte Carlo Methods, pp. 45–61. Radon. Ser. Comput. Appl. Math., vol. 15. De Gruyter, Berlin (2014).

[16] D. Bilyk, F. Dai. Geodesic distance Riesz energy on the sphere, Trans. Am. Math. Soc. **372**(5), 3141–3166 (2019).

[17] D. Bilyk, F. Dai, R. Matzke. The Stolarsky principle and energy optimization on the sphere. Constr. Approx. **48**(1), 31–60 (2018).

[18] D.Bilyk, F.Dai, S.Steinerberger. General and refined Montgomery Lemmata, Math. Ann. **373**(3–4), 1283–1297 (2018).

[19] D. Bilyk, R. Matzke. On the Fejes Tóth problem about the sum of angles between lines. Proc. Am. Math. Soc. **147**(1), 51–59 (2019).

[20] D. Bilyk, N. Feldheim. The two-dimensional small ball inequality and binary nets. J. Fourier Anal. Appl. **23**(4), 817–833 (2017).

[21] D. Bilyk, M. T. Lacey. On the small ball inequality in three dimensions. Duke Math. J. **143**(1), 81–115 (2008).

[22] D. Bilyk, M. T. Lacey. The supremum norm of the discrepancy function: recent results and connections. In: Monte Carlo and Quasi-Monte Carlo Methods 2012, pp. 23–38. Springer Proc. Math. Stat., vol. 65. Springer, Heidelberg (2013).
[23] D. Bilyk, M. T. Lacey. One bit sensing, discrepancy, and the Stolarsky principle. Sb. Math. **208**(5–6), 744–763 (2017).
[24] D. Bilyk, M. T. Lacey, A. Vagharshakyan. On the small ball inequality in all dimensions. J. Funct. Anal. **254**(9), 2470–2502 (2008).
[25] T. Bloom, S. Chow, A. Gafni, A. Walker. Additive energy and the metric Poissonian property, Mathematika **64**, 679–700 (2018).
[26] T. Bloom, A. Walker. GCD sums and sum-product estimates, preprint (2018), available at https://www.arxiv.org/pdf/1806.07849.pdf.
[27] M. Blümlinger. Slice discrepancy and irregularities of distribution on spheres, Mathematika **38**(1), 105–116 (1991).
[28] J. S. Brauchart, J. Dick. A simple proof of Stolarsky's invariance principle. Proc. Am. Math. Soc. **141**, 2085–2096, (2013).
[29] B. Chazelle. The Discrepancy Method. Cambridge University Press, Cambridge (2000).
[30] W. W. Chen. Results and problems old and new in discrepancy theory. In: Discrepancy Theory. De Gruyter (2019), in this volume.
[31] H. Davenport. Note on irregularities of distribution. Mathematika **3**, 131–135 (1956).
[32] J. Dick, F. Pillichshammer. Digital Nets and Sequences. Discrepancy Theory and Quasi-Monte Carlo Integration. Cambridge University Press, Cambridge (2010).
[33] M. Drmota, R. Tichy. Sequences, Discrepancies and Applications. Lecture Notes in Mathematics, vol. 1651. Springer-Verlag, Berlin (1997).
[34] D. El-Baz, J. Marklof, I. Vinogradov. The two-point correlation function of the fractional parts of $\sqrt{n}$ is Poisson, Proc. Am. Math. Soc. **143** 2815–2828 (2015).
[35] L. Fejes Tóth. On the sum of distances determined by a pointset. Acta Math. Acad. Sci. Hung. **7**, 397–401 (1956).
[36] P. Grabner, T. Stepanyuk. Poissonian pair correlation on manifolds via the heat kernel, preprint (2019), available at https://arxiv.org/pdf/1904.08286.pdf.
[37] S. Grepstad, G. Larcher. On pair correlation and discrepancy, Arch. Math. (Basel) **109**(2), 143–149 (2017).
[38] A. Haar. Zur Theorie der orthogonalen Funktionensysteme. Math. Ann. **69**, 331–371 (1910).
[39] A. Hinrichs. private communication and a talk at MCQMC 2012, Sydney (2012).
[40] A. Hinrichs, L. Kaltenböck, G. Larcher, W. Stockinger, M. Ullrich. On a multi-dimensional Poissonian pair correlation concept and uniform distribution, Monatshefte Math. 1–20 (2019).
[41] N. Kravitz. Refining the Two-Dimensional Signed Small Ball Inequality, preprint (2017), available at https://www.arxiv.org/pdf/1712.01206.pdf.
[42] L. Kuipers, H. Niederreiter. Uniform Distribution of Sequences. John Wiley & Sons, New York–London–Sydney (1974).
[43] T. Lachmann, N. Technau. On exceptional sets in the metric Poissonian pair correlations problem, Monatshefte Math. **189**(1), 137–156 (2019).
[44] G. Larcher, W. Stockinger. Pair correlation of sequences $(a_n\alpha)$ with maximal order of additive energy, preprint (2018), available at https://www.arxiv.org/pdf/1802.02901.pdf.
[45] G. Larcher, W. Stockinger. On some negative results related to the Poissonian pair correlation problems, preprint (2018), available at https://arxiv.org/pdf/1803.05236.pdf.
[46] G. Larcher, W. Stockinger. On pair correlation of sequences. In: Discrepancy Theory. De Gruyter (2019), in this volume.
[47] M. B. Levin. On the lower bound in the lattice point remainder problem for a parallelepiped. Discrete Comput. Geom. **54**(4), 826–870 (2015).

[48] M. B. Levin. On the lower bound of the discrepancy of Halton's sequence I. C. R. Math. Acad. Sci. Paris **354**(5), 445–448 (2016).
[49] M. B. Levin. On the lower bound of the discrepancy of Halton's sequence II. Eur. J. Math. **2**(3), 874–885 (2016).
[50] M. B. Levin. On the lower bound of the discrepancy of (t,s)-sequences: I. C. R. Math. Acad. Sci. Paris **354**(6), 562–565 (2016).
[51] M. B. Levin. On the lower bound of the discrepancy of (t,s)-sequences: II. Online J. Anal. Comb. **12**, 74 pp. (2017).
[52] J. Marklof. Pair correlation and equidistribution on manifolds, preprint (2019), available at https://www.arxiv.org/pdf/1903.00670.pdf.
[53] J. Matoušek. Geometric Discrepancy. Algorithms and Combinatorics, vol. 18. Springer-Verlag, Berlin (2010).
[54] J. Matoušek, A. Nikolov. Combinatorial discrepancy for boxes via the γ_2 norm. In: 31st International Symposium on Computational Geometry (SoCG 2015) (LIPIcs, Leibniz International Proceedings in Informatics 34). pp. 1–15. Schloss Dagstuhl–Leibniz Center for Informatics (2015).
[55] J. Matoušek, A. Nikolov, K.Talwar. Factorization norms and hereditary discrepancy. CoRR Preprint (2015), available at https://arxiv.org/pdf/1408.1376v2.pdf.
[56] A. Nikolov. Tighter bounds for the discrepancy of boxes and polytopes, Mathematika **63**(3), 1091–1113 (2017).
[57] A. Nikolov. Some of Jiří Matoušek's contributions to combinatorial discrepancy theory. In: Discrepancy Theory. De Gruyter (2019), in this volume.
[58] K. F. Roth. On irregularities of distribution. Mathematika **1**, 73–79 (1954).
[59] K. F. Roth. On irregularities of distribution. II. Commun. Pure Appl. Math. **29**(6), 739–744 (1976).
[60] Z. Rudnick, P. Sarnak. The pair correlation function of fractional parts of polynomials, Commun. Math. Phys. **194**, 61–70 (1998).
[61] I. Schoenberg. Some extremal problems for positive definite sequences and related extremal convex conformal maps of the circle. Nederl. Akad. Wetensch. Proc. Ser A **20**, 28–37 (1958).
[62] M. M. Skriganov. Construction of uniform distributions in terms of geometry of numbers. Algebra Anal. **6**(3), 200–230 (1994). Reprinted in St. Petersburg Math. J. **6** (3), 635–664 (1995).
[63] M. M. Skriganov. private communication, Palo Alto (2008).
[64] M. M. Skriganov. Point distributions in compact metric spaces. Mathematika **63**(3), 1152–1171 (2017).
[65] M. M. Skriganov. Point distributions in two-point homogeneous spaces. Mathematika **65**(3), 557–587 (2019).
[66] M. M. Skriganov. Stolarsky's invariance principle for projective spaces, preprint (2018), available at https://arxiv.org/pdf/1805.03541.pdf.
[67] S. Steinerberger. Localized quantitative criteria for equidistribution, Acta Arith. **180**(2), 183–199, (2017).
[68] S. Steinerberger. Poissonian Pair Correlation in Higher Dimensions, preprint (2018), available at https://arxiv.org/pdf/1812.10458.pdf.
[69] S. Steinerberger. Poissonian pair correlation and discrepancy, Indag. Math. **29**, 1167–1178 (2018).
[70] K. B. Stolarsky. Sums of distances between points on a sphere. II, Proc. Am. Math. Soc. **41**, 575–582 (1973).
[71] M. Talagrand. The small ball problem for the Brownian sheet, Ann. Probab. **22**, 1331–1354 (1994).

[72] V. N. Temlyakov. Some inequalities for multivariate haar polynomials, East J. Approx. **1**, 61–72 (1995).
[73] V. N. Temlyakov. Cubature formulas, discrepancy, and nonlinear approximation. J. Complex. **19**(3), 352–391 (2003).
[74] T. van Aardenne-Ehrenfest. Proof of the impossibility of a just distribution of an infinite sequence of points over an interval. Proc. Kon. Ned. Akad. v. Wetensch. **48**, 266–271 (1945).
[75] J. G. van der Corput. Verteilungsfunktionen I. Akad. Wetensch. Amsterdam, Proc. **38**, 813–821 (1935), in German.
[76] J. G. van der Corput. Verteilungsfunktionen II. Akad. Wetensch. Amdterdam, Proc. **38**, 1058–1068 (1935), in German.
[77] A. Walker. The primes are not metric Poissonian, Mathematika **64**, 230–236 (2018).
[78] H. Weyl. Über die Gleichverteilung von Zahlen mod. Eins. Math. Ann. **77**(3), 441–479 (1916), in German.

W. W. L. Chen

2 Results and problems old and new in discrepancy theory

Dedicated to the memory of my grandson Alexander

Abstract: In this brief survey, we discuss some of the main results in discrepancy theory and highlight many of the very difficult open problems that remain in the subject.

Keywords: Discrepancy theory, irregularities of distribution

MSC 2000: Primary 11K38

2.1 Introduction

The subject of discrepancy theory, or irregularities of point distribution, began with the conjecture of van der Corput [29, 30] in 1935 and the pioneering results of van Aardenne-Ehrenfest [1, 2] in 1945 and 1949, and took on a geometric flavor through the groundbreaking early work of Roth [43] in 1954. Today, many of the problems are formulated in the following way.

Let U be a bounded region in the k-dimensional Euclidean space $\mathbb{R}^k$, where $k \geqslant 2$, endowed with a measure μ, usually the Lebesgue measure, and let $\mathscr{P}$ be a set of N points in U. The irregularity of the distribution of the point set $\mathscr{P}$ is usually described in terms of an infinite collection $\mathcal{B}$ of measurable sets in U. For any such measurable set B in $\mathcal{B}$, we consider the discrepancy function

$$D[\mathscr{P};B] = |\mathscr{P} \cap B| - N\mu(B).$$

Often the collection $\mathcal{B}$ is endowed with an integral geometric measure $\mathrm{d}B$. Then for any real number q satisfying $0 < q < \infty$, we can consider the L_q-discrepancy

$$\|D_{\mathcal{B}}(\mathscr{P})\|_q = \left(\int_{\mathcal{B}} |D[\mathscr{P};B]|^q \,\mathrm{d}B \right)^{1/q}.$$

Acknowledgement: The author is indebted to Dmitriy Bilyk for pointing out the elementary proof of Lev's theorem by Kolountzakis, and to Mihalis Kolountzakis for permission to include the details in this survey. He also owes Giancarlo Travaglini a word of thanks for pointing out several extra references.

W. W. L. Chen, Department of Mathematics and Statistics, Macquarie University, Sydney NSW 2109, Australia, e-mail: william.chen@mq.edu.au

https://doi.org/10.1515/9783110652581-002

Here, the values $q = 1$ and $q = 2$ are often of particular interest. We also consider the L_∞-discrepancy, or extreme discrepancy,

$$\|D_{\mathcal{B}}(\mathcal{P})\|_\infty = \sup_{B\in\mathcal{B}} |D[\mathcal{P}; B]|.$$

Our goal is then to find lower and upper bounds for the quantities

$$\inf_{|\mathcal{P}|=N} \|D_{\mathcal{B}}(\mathcal{P})\|_q \quad \text{and} \quad \inf_{|\mathcal{P}|=N} \|D_{\mathcal{B}}(\mathcal{P})\|_\infty,$$

where each infimum is taken over all points sets $\mathcal{P}$ of N points in U.

Notation. For any function f and positive function g, we write $f \ll g$ to denote that there exists a positive constant C such that $|f| \leqslant Cg$. In particular, if f is a positive function, then we also write $f \gg g$ to indicate that $g \ll f$, and write $f \asymp g$ to indicate that both $f \ll g$ and $f \gg g$ hold. The signs $\ll$, $\gg$ and $\asymp$ may contain subscripts, denoting that any implicit constants that arise may depend on these parameters. For any finite set S, we write $|S|$ to denote the cardinality of S.

2.2 The classical problem

The classical problem in discrepancy theory was formulated by Roth [43] in 1954. Here, $U = [0,1]^k$, the unit cube in $\mathbb{R}^k$, where $k \geqslant 2$, and the irregularity of a point set $\mathcal{P}$ in U is described in terms of the infinite collection

$$\mathcal{B} = \{B(\mathbf{x}) = [0, x_1) \times \cdots \times [0, x_k) : \mathbf{x} \in [0,1]^k\}$$

of aligned rectangular boxes in the unit cube anchored at the origin.

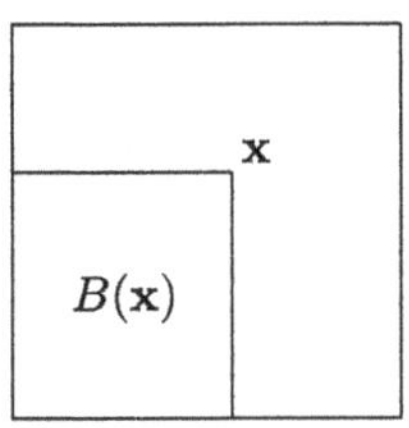

The integral geometric measure in $\mathcal{B}$ is given by the usual Lebesgue volume measure $\mathrm{d}B = \mathrm{d}\mathbf{x}$.

The L_q-discrepancy in this problem is well understood for every real number q satisfying $1 < q < \infty$, and we have the estimates

$$(\log N)^{(k-1)/2} \ll_{k,q} \inf_{|\mathcal{P}|=N} \|D_{\mathcal{B}}(\mathcal{P})\|_q \ll_{k,q} (\log N)^{(k-1)/2}. \tag{2.1}$$

Here, the lower bound is due to Schmidt [51] in 1977, following the earlier work of Roth [43] in 1954 on the special case $q = 2$ using an orthogonal function technique. The upper bound is due to Chen [19, 20], following the earlier work of Davenport [31] in 1956 on the special case $k = q = 2$ and the big breakthrough of Roth [45] in 1980 on the special case $q = 2$.

We make here a few comments concerning the special case $q = 2$.

The proof of the lower bound is given in Roth [43] for the case $k = 2$ only, although generalization to arbitrary dimensions $k \geqslant 2$ presents no extra difficulties. In fact, the ideas are much more clearly presented in Schmidt [51]. A complete proof of these results of Roth and Schmidt in arbitrary dimensions can be found in the monograph of Beck and Chen [9, Section 2.1]. However, a simple description of the ideas along these lines for the case $k = 2$ can be found in the survey of Chen and Travaglini [26, Section 1]. The idea is that sets where the expectation is a small fraction between 0 and 1 can be found in abundance, and they give rise to what we call *trivial discrepancies*. We need to combine these and not allow them to cancel each other. The tool is given by Roth's auxiliary function, of the form

$$F(\mathbf{x}) = \sum_{\mathbf{r}} f_{\mathbf{r}}(\mathbf{x}),$$

a sum of orthogonal functions over a suitable collection of vectors $\mathbf{r}$, where each $f_{\mathbf{r}}(\mathbf{x})$ is either a Rademacher-type function with values ± 1 or zero. Writing $D(\mathbf{x})$ for $D[\mathcal{P}; B(\mathbf{x})]$, the Cauchy–Schwarz inequality then gives

$$\left| \int_U D(\mathbf{x}) F(\mathbf{x})\, d\mathbf{x} \right| \leqslant \|D\|_2 \|F\|_2. \tag{2.2}$$

A lower bound for $\|D\|_2$ will result from a lower bound for the left-hand side of (2.2) and an upper bound for $\|F\|_2$.

The proof of the upper bound in Roth [45] is probabilistic, with no explicitly given point sets, as are subsequent proofs by Chen [20] in 1983 and Skriganov [53] in 1994. The first proof of the upper bound with explicitly given point sets can be found in Chen and Skriganov [24] in 2002. A different proof is given by Dick and Pillichshammer [32] in 2014. For some comments on the differences between these two explicit proofs, see also the paper of Dick and Pillichshammer [33].

On the other hand, for the case $q = 1$, we have the estimates

$$(\log N)^{1/2} \ll_k \inf_{|\mathcal{P}|=N} \|D_{\mathcal{B}}(\mathcal{P})\|_1 \ll_k (\log N)^{(k-1)/2}.$$

Here, the upper bound is a simple consequence of the upper bound in (2.1), while the lower bound is due to Halász [34] in 1981, using a variant of Roth's lower bound technique that only works when $k = 2$. Indeed, Halász uses the auxiliary function

$$H(\mathbf{x}) = \prod_{\mathbf{r}} (1 + \mathrm{i} n^{-1/2} f_{\mathbf{r}}(\mathbf{x})) - 1,$$

where $\log N \ll n \ll \log N$. Then $H(\mathbf{x}) \ll 1$, and so

$$\left| \int_U D(\mathbf{x})H(\mathbf{x})\,\mathrm{d}\mathbf{x} \right| \ll \|D\|_1. \tag{2.3}$$

A lower bound for $\|D\|_1$ will result from a lower bound for the left-hand side of (2.3).

Thus the problem of the L_q-discrepancy in this classical setting is completely solved for all finite $q > 1$ and for the case $(k, q) = (2, 1)$.

Clearly, the upper bound in (2.1) remains valid for every natural number $k \geqslant 2$ and every finite positive real number q.

Open Problem 1. *In the classical discrepancy problem, is it true that for every natural number $k \geqslant 2$ and every finite positive real number q, the estimate*

$$\inf_{|\mathcal{P}|=N} \|D_{\mathcal{B}}(\mathcal{P})\|_q \gg_{k,q} (\log N)^{(k-1)/2} \tag{2.4}$$

holds?

It is interesting to observe that for every real number q satisfying $0 < q < 1$, the currently known best lower bound is precisely zero.

Much less is known for the L_∞-discrepancy. We have the estimates

$$(\log N)^{(k-1)/2} \ll_k \inf_{|\mathcal{P}|=N} \|D_{\mathcal{B}}(\mathcal{P})\|_\infty \ll_k (\log N)^{k-1}. \tag{2.5}$$

Here, the lower bound is a simple consequence of the lower bound in (2.1), while the upper bound is due to Halton [35] in 1960. The lower bound has been improved in the intervening years, and we have the estimate

$$\inf_{|\mathcal{P}|=N} \|D_{\mathcal{B}}(\mathcal{P})\|_\infty \gg \log N \tag{2.6}$$

in the special case $k = 2$, due first to Schmidt [49] in 1972, using a combinatorial argument, with an alternative proof given by Halász [34] in 1981, using Roth's technique with yet another auxiliary function, as well as the estimate

$$\inf_{|\mathcal{P}|=N} \|D_{\mathcal{B}}(\mathcal{P})\|_\infty \gg_k (\log N)^{(k-1)/2+\delta(k)} \tag{2.7}$$

for some $\delta(k) \in (0, 1/2)$, due to Bilyk, Lacey, and Vagharshakyan [15] in 2008.

There remains a rather big gap between the lower and upper bounds when $k > 2$.

Open Problem 2 (Great open problem). *In the classical discrepancy problem, for every natural number $k \geqslant 3$, find the correct order of magnitude of*

$$\inf_{|\mathcal{P}|=N} \|D_{\mathcal{B}}(\mathcal{P})\|_\infty.$$

At the very least, try to prove or disprove the following conjectures:

(i) (Old conjecture) *For every natural number $k \geqslant 3$, the estimate*

$$\inf_{|\mathscr{P}|=N} \|D_{\mathcal{B}}(\mathscr{P})\|_{\infty} \gg_k (\log N)^{k-1}$$

holds, so that Halton's upper bound in (2.5) *is sharp.*

(ii) (New conjecture) *For every natural number $k \geqslant 3$, the estimate*

$$\inf_{|\mathscr{P}|=N} \|D_{\mathcal{B}}(\mathscr{P})\|_{\infty} \gg_k (\log N)^{k/2} \tag{2.8}$$

holds, corresponding to the estimate in (2.7) *with $\delta(k) = 1/2$.*

We comment that both estimates (2.4) and (2.8) hold *on average* over digit shifts, as shown by Skriganov [54] in 2016. Digit shifts, since its introduction to discrepancy theory by Chen [20] in 1983, have always been used to study upper bound questions. This recent work of Skriganov is the first instance that they have been used in lower bound considerations.

Before we make our concluding remarks, we mention a very interesting piece of work of Lev [38] in 1996 which shows that our estimates are rather delicate.

Suppose that the real number q is fixed, where $1 \leqslant q < \infty$. In view of the upper estimate in (2.1), clearly there exists sets $\mathscr{P}$ of N points such that the L_q-discrepancy satisfies the upper bound

$$\|D_{\mathcal{B}}(\mathscr{P})\|_q \ll_{k,q} (\log N)^{(k-1)/2}. \tag{2.9}$$

Let us treat the unit cube $U = [0,1]^k$ as a torus. For every $\mathbf{t} = (t_1, \dots, t_k) \in [0,1]^k$, we now consider the translate

$$\mathscr{P} - \mathbf{t} = \{\mathbf{p} - \mathbf{t} : \mathbf{p} \in \mathscr{P}\}$$

of the point set $\mathscr{P}$. Then

$$\sup_{\mathbf{t}\in[0,1]^k} \|D_{\mathcal{B}}[\mathscr{P} - \mathbf{t}]\|_q \asymp_k \|D_{\mathcal{B}}[\mathscr{P}]\|_{\infty}, \tag{2.10}$$

where the implicit constants may depend on the dimension k, but not on q.

In view of the great open problem, the inequality (2.10) tells us that the sharp upper bound (2.9) can be destroyed by a simple translation on the point set $\mathscr{P}$.

The original proof of Lev of the inequality (2.10) is an ingenious *tour de force* involving a number of different ideas. The main thrust is an induction argument up the dimensions. However, to make this work, one has to first consider the case $k = 1$, usually dismissed by most experts as trivial. Also, to make the induction work, it is necessary to introduce weights in order to hide some extra quantities that arise.

An elementary proof the inequalities (2.10), given later by Kolountzakis [36], is no less ingenious. We shall only discuss the case $k = 2$, as the proof generalizes naturally to higher dimensions. Suppose that $\mathscr{P}$ is a distribution of N points in the unit square $[0,1]^2$, treated as a torus. Note, first of all, that instead of shifting the set $\mathscr{P}$, we may equivalently shift the origin and the coordinate system and leave the set $\mathscr{P}$ in place. Suppose that

$$\|D_{\mathcal{B}}[\mathscr{P}]\|_\infty = M.$$

Then there exists a point $\mathbf{a} = (a_1, a_2) \in [0,1]^2$ such that $|D[\mathscr{P}; B(\mathbf{a})]| > M/2$. We assume that the set $\mathscr{A}$ of points $\mathbf{x} = (x_1, x_2) \in [0,1]^2$ such that $x_1 \geqslant a_1$ and $x_2 \geqslant a_2$ has measure at least 1/10; the proof can be easily adjusted in other cases. For each such point $\mathbf{x}$, consider the following picture:

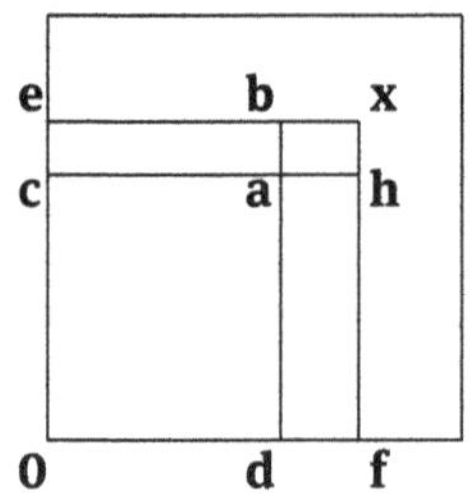

Let

$$\begin{aligned} R_1 &= \text{rectangle with vertices } \mathbf{0}, \mathbf{e}, \mathbf{x}, \mathbf{f}, \\ R_2 &= \text{rectangle with vertices } \mathbf{c}, \mathbf{e}, \mathbf{x}, \mathbf{h}, \\ R_3 &= \text{rectangle with vertices } \mathbf{d}, \mathbf{b}, \mathbf{x}, \mathbf{f}, \\ R_4 &= \text{rectangle with vertices } \mathbf{a}, \mathbf{b}, \mathbf{x}, \mathbf{h}. \end{aligned}$$

Then

$$D[\mathscr{P}; B(\mathbf{a})] = D[\mathscr{P}; R_1] - D[\mathscr{P}; R_2] - D[\mathscr{P}; R_3] + D[\mathscr{P}; R_4],$$

and clearly

$$\max\{|D[\mathscr{P}; R_1]|, |D[\mathscr{P}; R_2]|, |D[\mathscr{P}; R_3]|, |D[\mathscr{P}; R_4]|\} > \frac{M}{8}.$$

Let $f(\mathbf{x}) = i$, where $i \in \{1,2,3,4\}$ and

$$|D[\mathscr{P}; R_i]| = \max\{|D[\mathscr{P}; R_1]|, |D[\mathscr{P}; R_2]|, |D[\mathscr{P}; R_3]|, |D[\mathscr{P}; R_4]|\},$$

with the convention that if there is more than one such value of i, then we choose the smallest such value. Clearly, there exists one value $i^* \in \{1,2,3,4\}$ for which the set

$$\{\mathbf{x} \in \mathscr{A} : f(\mathbf{x}) = i^*\}$$

has measure at least 1/40. Accordingly, we shift the origin to the point

$$\begin{cases} \mathbf{0}, & \text{if } i^* = 1, \\ \mathbf{c}, & \text{if } i^* = 2, \\ \mathbf{d}, & \text{if } i^* = 3, \\ \mathbf{a}, & \text{if } i^* = 4. \end{cases}$$

This implies that there exists $\mathbf{t} \in [0,1]^2$ such that $\|D[\mathscr{P} + \mathbf{t}]\|_1 \geqslant M/320$, and completes the proof.

Note that all the estimates in this classical setting are logarithmic in size in terms of the cardinality N of the point sets $\mathscr{P}$ in question. We sometimes refer to this as a *small discrepancy phenomenon.*

2.3 Some work of Schmidt

There are many interesting discrepancy problems when we move away from the classical problem concerning aligned rectangular boxes anchored at the origin. The pioneering work in this direction is due to Schmidt [46–48] in 1969, using an integral equation technique and involving tilted rectangular boxes and balls as well as other geometric objects. The paper [48] is of particular interest. Let $U = [0,1]^k$, treated as a torus and with $k \geqslant 2$.

In the case when $\mathcal{B}$ is the collection of all rectangular boxes, we have the estimates

$$\inf_{|\mathscr{P}|=N} \|D_{\mathcal{B}}(\mathscr{P})\|_\infty \gg_\epsilon \begin{cases} N^{1/4-\epsilon}, & \text{if } k = 2, \\ N^{1/3-\epsilon}, & \text{if } k \geqslant 3. \end{cases} \tag{2.11}$$

In the case when $\mathcal{B}$ is the collection of all balls, we have the estimate

$$\inf_{|\mathscr{P}|=N} \|D_{\mathcal{B}}(\mathscr{P})\|_\infty \gg_\epsilon N^{1/2-1/2k-\epsilon}. \tag{2.12}$$

Note that the estimates in these new settings are now powers of the cardinality N of the point sets $\mathscr{P}$ in question. We sometimes refer to these as a *large discrepancy phenomena.*

Indeed, apart from the term ϵ in the exponent, both estimates are essentially sharp, with the exception of (2.11) if $k > 3$. We shall demonstrate this observation by Beck [4] in 1981 only in the special case $k = 2$, as the argument generalizes to higher dimensions without any extra difficulties.

For simplicity, let us suppose that $N = M^2$ is a perfect square. Then we partition $U = [0,1]^2$ into N little squares in the usual way.

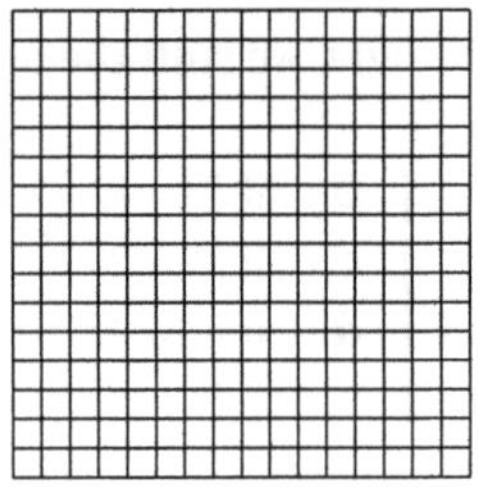

Let $\mathcal{S}$ denote the collection of all little squares, and we place one point anywhere in each such little square $S \in \mathcal{S}$, and denote by $\mathscr{P}$ the collection of all these points. This is a deterministic point set of precisely N points in U.

Now take any convex set $B \in \mathcal{B}$. This can be a tilted rectangle or a circular disc; the latter case is shown below on the left.

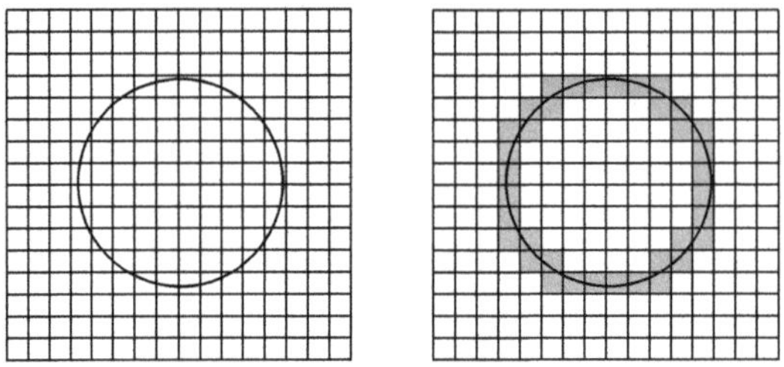

Clearly, $D[\mathscr{P}; S \cap B] = 0$ whenever $S \cap B = \emptyset$ or $S \subseteq B$, and so

$$D[\mathscr{P}; B] = \sum_{S \in \mathcal{S}} D[\mathscr{P}; S \cap B] = \sum_{\substack{S \in \mathcal{S} \\ S \cap \partial B \neq \emptyset}} D[\mathscr{P}; S \cap B];$$

see the picture above on the right. The triangle inequality now leads to the estimate

$$|D[\mathscr{P}; B]| \leqslant \sum_{\substack{S \in \mathcal{S} \\ S \cap \partial B \neq \emptyset}} |D[\mathscr{P}; S \cap B]| \ll M = N^{1/2}.$$

This is rather crude, and clearly not good enough.

To get a better upper bound, we randomize the point set $\mathscr{P}$ by making the point in any little square $S \in \mathcal{S}$ a random point, uniformly distributed within that little square S and independently of the random points in the other little squares in $\mathcal{S}$. Applying large deviation techniques due to Bernstein–Chernoff or Hoeffding, this crude upper bound $N^{1/2}$ can then be converted to an upper bound of the form $N^{1/4}(\log N)^{1/2}$. The logarithmic factor represents the cost of using probability theory.

For slightly more details and related problems, see the survey article by Chen [23, Section 2].

2.4 Beck's Fourier transform technique

Let $\mathbb{T}^k = [0,1]^k$, treated as a torus and with $k \geqslant 2$. For any convex and compact set $B \subseteq [0,1]^k$, it is easy to see that $\mathbf{y} \in B + \mathbf{x}$ if and only if $\chi_{-B}(\mathbf{x}-\mathbf{y}) = 1$, where $-B = \{-\mathbf{y} : \mathbf{y} \in B\}$ and χ_{-B} denotes its characteristic function, and so

$$\begin{aligned} D[\mathscr{P}; B+\mathbf{x}] &= \sum_{\mathbf{p}\in\mathscr{P}} \chi_{-B}(\mathbf{x}-\mathbf{p}) - N \int_{\mathbb{T}^k} \chi_{-B}(\mathbf{x}-\mathbf{y})\,\mathrm{d}\mu(\mathbf{y}) \\ &= \int_{\mathbb{T}^k} \chi_{-B}(\mathbf{x}-\mathbf{y})\,(\mathrm{d}Z - N\mathrm{d}\mu)(\mathbf{y}). \end{aligned}$$

This can be summarized in the form

$$D = \chi_{-B} * (\mathrm{d}Z - N\mathrm{d}\mu). \tag{2.13}$$

In other words, under translation, discrepancy is a convolution of geometry and measure.

As lower bound estimates apply to arbitrary point sets $\mathscr{P}$, there is very limited information on the measure part of this convolution, and so we wish to concentrate on the geometry part. To separate the geometry part from the measure part, we apply Fourier transform. Then the convolution (2.13) becomes an ordinary product

$$\widehat{D} = \widehat{\chi_{-B}} \cdot (\widehat{\mathrm{d}Z - N\mathrm{d}\mu})$$

of the Fourier transforms of the constituent parts.

This is the basis of Beck's Fourier transform technique, motivated by the work of Roth [44] in 1964 on irregularities of distribution of integer sequences.

Indeed, the similarity of the bounds (2.11) and (2.12) is no coincidence.

Let $U = [0,1]^k$, treated as a torus and with $k \geqslant 2$, and let A be a fixed convex and compact set in U satisfying some mild technical condition.

Suppose that the irregularity of a point set $\mathscr{P}$ in U is described in terms of the infinite collection

$$\mathcal{B} = \{A(\lambda, \tau, \mathbf{x}) : \lambda \in [0,1], \tau \in \mathcal{T}, \mathbf{x} \in [0,1]^k\}, \tag{2.14}$$

where $A(\lambda,\tau,\mathbf{x}) = \{\tau(\lambda\mathbf{y}) + \mathbf{x} : \mathbf{y} \in A\}$ denotes a similar copy obtained from the set A under a contraction λ, an orthogonal transformation τ and a translation $\mathbf{x}$, and where $\mathcal{B}$ is endowed with the integral geometric measure $\mathrm{d}B = \mathrm{d}\lambda\,\mathrm{d}\tau\,\mathrm{d}\mathbf{x}$.

Here, we have the lower bounds

$$\inf_{|\mathscr{P}|=N} \|D_{\mathcal{B}}(\mathscr{P})\|_\infty \geqslant \inf_{|\mathscr{P}|=N} \|D_{\mathcal{B}}(\mathscr{P})\|_2 \gg_A N^{1/2-1/2k}, \tag{2.15}$$

due to Beck [6] in 1987.

A discussion of the special case $k = 2$ with a square A can be found in the survey by Chen [22, Section 2]. The special case $k = 2$ is discussed in generality in the lecture notes of Montgomery [41, Chapter 6]. There it is shown how estimates concerning the decay of the Fourier transform of the characteristic function of A lead to the lower bounds (2.15). Here, Montgomery also discusses the special case $k = 2$ with a circular disc A of radius 1/2. Note that rotation is irrelevant here. Note also that the Fourier transform of the characteristic function of A involves a Bessel function of the first kind, and so it does appear that contraction is essential. However, Montgomery can show that the contraction parameter λ can be restricted to a very small set. Indeed, he can show that the inequality

$$\int_{[0,1]^2} |D[\mathscr{P}; A(1,\mathbf{x})]|^2 \, d\mathbf{x} + \int_{[0,1]^2} |D[\mathscr{P}; A(1/2,\mathbf{x})]|^2 \, d\mathbf{x} \gg N^{1/2}$$

holds for every set $\mathscr{P}$ of N points in $[0,1]^2$. Here, we have omitted reference to the unnecessary orthogonal transformation τ in our notation. Indeed, some average over contractions is necessary in view of the work of Parnovski and Sobolev [42]. See also the paper of Travaglini and Tupputi [55].

For an introduction to the relationship between the average decay of the Fourier transform and discrepancy theory, the interested reader is referred to the survey of Brandolini, Gigante, and Travaglini [18].

Returning to our original problem, we also have the upper bounds

$$\inf_{|\mathscr{P}|=N} \|D_{\mathcal{B}}(\mathscr{P})\|_\infty \ll_A N^{1/2-1/2k} (\log N)^{1/2}, \tag{2.16}$$

obtained by Beck [4] in 1981, and

$$\inf_{|\mathscr{P}|=N} \|D_{\mathcal{B}}(\mathscr{P})\|_2 \ll_A N^{1/2-1/2k},$$

due to Beck and Chen [11] in 1990 and then improved to

$$\inf_{|\mathscr{P}|=N} \|D_{\mathcal{B}}(\mathscr{P})\|_q \ll_{A,q} N^{1/2-1/2k} \tag{2.17}$$

for any fixed positive real number q by Chen [21] in 2000.

Combining (2.15) and (2.17), it is clear that the L_q-discrepancy in this problem concerning all similar copies of a given convex and compact set in U is completely solved for every finite real number $q \geqslant 2$. However, comparing (2.15) and (2.16), we see that there is a gap in our knowledge for the L_∞-discrepancy in this problem.

Open Problem 3. *Let $U = [0,1]^k$, treated as a torus and with $k \geqslant 2$, and let A be a fixed convex and compact set in U satisfying some mild technical condition. Suppose that the set $\mathcal{B}$ is given by (2.14). Does an estimate of the form*

$$\inf_{|\mathscr{P}|=N} \|D_{\mathcal{B}}(\mathscr{P})\|_\infty \ll_A N^{1/2-1/2k}$$

hold?

Suppose next that we no longer permit orthogonal transformation, and that the irregularity of a point set $\mathscr{P}$ in U is described in terms of the infinite collection

$$\mathcal{B} = \{A(\lambda, \mathbf{x}) : \lambda \in [0,1], \mathbf{x} \in [0,1]^k\}, \tag{2.18}$$

where $A(\lambda, \mathbf{x}) = \{\lambda\mathbf{y} + \mathbf{x} : \mathbf{y} \in A\}$ denotes a homothetic copy obtained from the set A under a contraction λ and a translation $\mathbf{x}$, and where $\mathcal{B}$ is endowed with the integral geometric measure $\mathrm{d}B = \mathrm{d}\lambda\,\mathrm{d}\mathbf{x}$. Then much less is known, and our limited knowledge is essentially restricted to the case $k = 2$, where we have a lower bound of the form

$$\inf_{|\mathscr{P}|=N} \|D_{\mathcal{B}}(\mathscr{P})\|_\infty \gg_A \max\{(\log N)^{1/2}, \xi(A, N)\}, \tag{2.19}$$

obtained by Beck [7] in 1988 from the corresponding estimate for the L_2-discrepancy. Here, $\xi(A, N)$ is a function which depends on the boundary curve of the fixed set A. In particular, $\xi(A, N)$ is finite if A is a convex polygon, and $\xi(A, N) = N^{1/4}$ if A is a circular disc.

Open Problem 4 (Greater open problem). *Let $U = [0,1]^k$, treated as a torus and with $k \geqslant 2$, and let A be a fixed convex and compact set in U satisfying some mild technical condition. Suppose that the set $\mathcal{B}$ is given by* (2.18)*:*
(i) (Generalization of the bound (2.6)) *In the case $k = 2$, can the term* $(\log N)^{1/2}$ *in the estimate* (2.19) *be improved to* $\log N$*?*
(ii) *What can we say when $k \geqslant 3$?*

We complete this section by making a digression and discussing a result obtained in part by Fourier transform considerations.

Let $U = [0,1]^k$, treated as a torus and with $k \geqslant 1$. Suppose that the irregularity of a point set $\mathscr{P}$ in U is described in terms of the infinite collection $\mathcal{B}$ of all balls of diameter 1/2. Let $\mathscr{P}$ denote the perfect square grid of $N = M^k$ points in U.

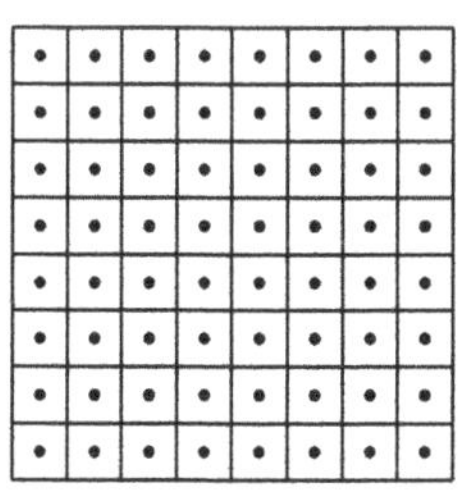

We now consider the quantity

$$\mathrm{DET}_k(M^k) = \int_{\mathcal{B}} |D[\mathscr{P}; B]|^2 \,\mathrm{d}B, \tag{2.20}$$

where the integral geometric measure $\mathrm{d}B$ is given by Lebesgue translation measure.

Next, let $\widetilde{\mathscr{P}}$ denote the random point set obtained by replacing each fixed point of $\mathscr{P}$ by a random point which is uniformly distributed in its own little cube and independently of any other random point in any other little cube. We consider the corresponding quantity

$$\mathrm{PROB}_k(M^k) = \mathbb{E}\Big(\int_{\mathcal{B}} |D[\widetilde{\mathscr{P}}; B]|^2 \,\mathrm{d}B \Big). \tag{2.21}$$

Which of the quantities (2.20) and (2.21) is smaller? We have the following surprising result, obtained by Chen and Travaglini [27] in 2009.

Suppose that $k \not\equiv 1 \bmod 4$. Then:

- $\mathrm{DET}_2(M^2) \leqslant \mathrm{PROB}_2(M^2)$ for all large M; and
- for all large k, $\mathrm{PROB}_k(M^k) \leqslant \mathrm{DET}_k(M^k)$ for all large M.

Suppose that $k \equiv 1 \bmod 4$. Then

- for all large k, $\mathrm{PROB}_k(M^k) \leqslant \mathrm{DET}_k(M^k)$ for infinitely many M;
- for all k, $\mathrm{DET}_k(M^k) \leqslant \mathrm{PROB}_k(M^k)$ for infinitely many M; and
- $\mathrm{DET}_1(M) \leqslant \mathrm{PROB}_1(M)$ for every M.

This is consistent with the work of Konyagin, Skriganov, and Sobolev [37] in 2003 on lattice points in balls.

2.5 Roth's disc segment problem

Let U be the circular disc in $\mathbb{R}^2$ of area 1 and centered at the origin. Suppose that the irregularity of a point set $\mathscr{P}$ in U is described in terms of the infinite collection

$$\mathcal{B} = \{S(r,\theta) : 0 \leqslant \theta \leqslant 2\pi,\ 0 \leqslant r \leqslant \pi^{-1/2}\} \tag{2.22}$$

of disc segments in U.

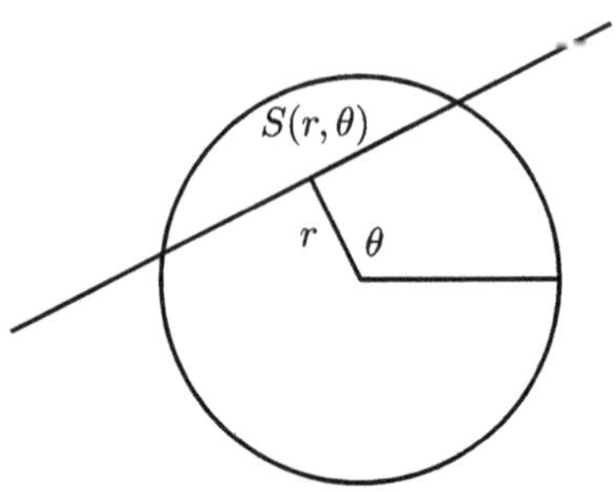

A question of Roth concerns whether the quantity

$$\inf_{|\mathcal{P}|=N} \|D_{\mathcal{B}}(\mathcal{P})\|_{\infty}$$

is unbounded as a function of N.

Although this question never appeared in any of Roth's writings, it was recorded in Schmidt [48, Section I, last paragraph] as well as in Schmidt [52, Chapter II, Section 16].

To describe the results, it is useful to introduce the integral geometric measure $\mathrm{d}B = (2\pi^{1/2})^{-1}\,\mathrm{d}\theta\,\mathrm{d}r$, appropriately normalized so that the total measure equals unity.

Roth's question was answered in the affirmative by Beck [5] in 1983. Using his Fourier transform approach, suitably adapted, one can establish the lower bound

$$\inf_{|\mathcal{P}|=N} \|D_{\mathcal{B}}(\mathcal{P})\|_{\infty} \geqslant \inf_{|\mathcal{P}|=N} \|D_{\mathcal{B}}(\mathcal{P})\|_{2} \gg N^{1/4}(\log N)^{-7/2}.$$

A stronger lower bound

$$\inf_{|\mathcal{P}|=N} \|D_{\mathcal{B}}(\mathcal{P})\|_{\infty} \geqslant \inf_{|\mathcal{P}|=N} \|D_{\mathcal{B}}(\mathcal{P})\|_{2} \gg N^{1/4}, \tag{2.23}$$

via a new approach involving integral geometry, is due to Alexander [3] in 1990. On the other hand, it can be shown that

$$\inf_{|\mathcal{P}|=N} \|D_{\mathcal{B}}(\mathcal{P})\|_{2} \leqslant \inf_{|\mathcal{P}|=N} \|D_{\mathcal{B}}(\mathcal{P})\|_{\infty} \ll N^{1/4}(\log N)^{1/2},$$

using the idea of Beck [4] in 1981. Here, the factor $(\log N)^{1/2}$ arises for precisely the same reason as the corresponding factor in the estimate (2.16). However, there is no analogue of open problem 3 in this setting. Courtesy of an extraordinary piece of work by Matoušek [39] in 1995, it is now known that

$$\inf_{|\mathcal{P}|=N} \|D_{\mathcal{B}}(\mathcal{P})\|_{2} \leqslant \inf_{|\mathcal{P}|=N} \|D_{\mathcal{B}}(\mathcal{P})\|_{\infty} \ll N^{1/4}. \tag{2.24}$$

Combining the bounds (2.23) and (2.24), we see that the problem of the L_{∞}-discrepancy in this setting is completely solved, as is the problem of the L_q-discrepancy for any finite real number $q \geqslant 2$.

The situation is rather different if one studies the problem of the L_q-discrepancy in this setting when $1 \leqslant q < 2$.

Here, in particular when $q = 1$, the problem takes on some flavor of the classical discrepancy problem. Indeed, one can establish an upper bound of the form

$$\inf_{|\mathcal{P}|=N} \|D_{\mathcal{B}}(\mathcal{P})\|_{1} \ll (\log N)^{2}, \tag{2.25}$$

as demonstrated by Beck and Chen [12] in 1993. The majority of the points of $\mathcal{P}$ come from a square grid.

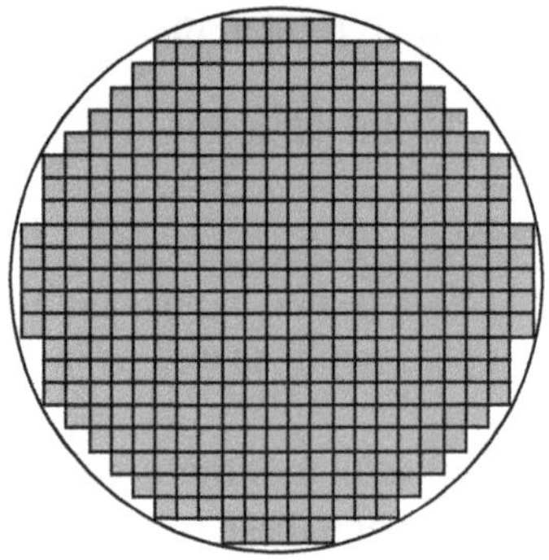

The remaining points give rise to a *one-dimensional* discrepancy function along the boundary of the disc, and contribute only to the error terms. Thus for a fixed disc segment, the size of the discrepancy depends on the diophantine approximation properties of the slope of the boundary of the disc segment. What the estimate (2.25) tells us, therefore, is that in L_1-average, these properties of the slope are reasonably close to those of a badly approximable number, whereas the estimate (2.23) tells us that this is far from the case when we look at the corresponding L_2-average.

The argument can also be modified to show that

$$\inf_{|\mathcal{P}|=N} \|D_{\mathcal{B}}(\mathcal{P})\|_q \ll N^{(q-1)/2q}$$

for every real number q satisfying $1 < q \leqslant 2$. Note that the exponent drops from 1/4 to 0 as q drops from 2 to 1.

Open Problem 5A. *Let U be the circular disc in $\mathbb{R}^2$ of area 1 and centered at the origin. Suppose that the irregularity of a point set $\mathcal{P}$ in U is described in terms of the infinite collection $\mathcal{B}$ given by* (2.22) *and endowed with the integral geometric measure* $\mathrm{d}B = (2\pi^{1/2})^{-1}\,\mathrm{d}\theta\,\mathrm{d}r$.

(i) *Improve the upper bound* (2.25) *if possible.*

(ii) *Find a lower bound for the problem of the L_1-discrepancy.*

An analogue of the Roth disc segment problem is the half-plane problem in the unit cube. It is almost identical to the disc segment problem, except that we take $U = [-1/2, 1/2]^2$.

This problem can be extended to higher dimensions. Let $U = [-1/2, 1/2]^k$, with $k \geqslant 2$. Suppose that each half-space $H(r, \mathbf{v})$ is characterized by its perpendicular distance r to the origin and its unit normal $\mathbf{v}$. Then we write

$$\mathcal{B} = \{S(r, \mathbf{v}) = H(r, \mathbf{v}) \cap U : \mathbf{v} \in S^{k-1}, r \geqslant 0\}, \tag{2.26}$$

where S^{k-1} denotes the surface of the sphere of radius 1 in $\mathbb{R}^k$, endowed with the integral geometric measure $\mathrm{d}B = \mathrm{d}\mathbf{v}\,\mathrm{d}r$, suitably normalized. In the special case $k = 2$, the bounds (2.23)–(2.25) remain valid. For arbitrary $k \geqslant 2$, an analogue of the bound (2.25) is given by

$$\inf_{|\mathcal{P}|=N} \|D_{\mathcal{B}}(\mathcal{P})\|_1 \ll_k (\log N)^k, \tag{2.27}$$

obtained by Chen and Travaglini [28] in 2011. A key ingredient in the argument is the divergence theorem which allows us to *climb* the dimensions.

Open Problem 5B. *Let* $U = [-1/2, 1/2]^k$, *with* $k \geqslant 2$. *Suppose that the irregularity of a point set* $\mathcal{P}$ *in* U *is described in terms of the infinite collection* $\mathcal{B}$ *given by* (2.26) *and endowed with the integral geometric measure* $\mathrm{d}B = \mathrm{d}\mathbf{v}\,\mathrm{d}r$, *suitably normalized.*
(i) *Improve the upper bound* (2.27) *if possible.*
(ii) *Find a lower bound for the problem of the* L_1*-discrepancy.*

Unfortunately, the technique of Chen and Travaglini has not so far been shown to work if we study the direct analogue of the Roth disc segment problem in higher dimensions.

Open Problem 6. *For every natural number* $k \geqslant 3$, *study the analogue of the Roth disc segment problem when* U *is the ball in* $\mathbb{R}^k$ *of volume* 1 *and centered at the origin.*

2.6 Problem of convex polygons

We all know that a convex polygon can be viewed as the intersection of finitely many half-planes. This suggests that the idea surrounding the Roth disc segment problem can perhaps be transported over to this setting.

Let $U = [0, 1]^2$, treated as a torus, and let A be a fixed convex polygon in U. The irregularity of a point set $\mathcal{P}$ in U is described in terms of the infinite collection

$$\mathcal{B} = \{A(\lambda, \tau, \mathbf{x}) : \lambda \in [0, 1], \tau \in \mathcal{T}, \mathbf{x} \in [0, 1]^2\}, \tag{2.28}$$

where $A(\lambda, \tau, \mathbf{x}) = \{\tau(\lambda\mathbf{y}) + \mathbf{x} : \mathbf{y} \in A\}$ denotes a similar copy obtained from the polygon A under a contraction λ, a rotation τ, and a translation $\mathbf{x}$, and where $\mathcal{B}$ is endowed with the integral geometric measure $\mathrm{d}B = \mathrm{d}\lambda\,\mathrm{d}\tau\,\mathrm{d}\mathbf{x}$.

We see that this is a special case of a problem studied in Section 2.4. Corresponding to the bounds (2.15) and (2.17), we have

$$N^{1/4} \ll_{A,q} \inf_{|\mathcal{P}|=N} \|D_{\mathcal{B}}(\mathcal{P})\|_q \ll_{A,q} N^{1/4}$$

for every finite real number $q \geqslant 2$, so that the L_q-discrepancy in this problem is completely solved for these values of q.

As for the Roth disc segment problem, the situation is again rather different if one studies the problem of the L_q-discrepancy in this setting when $1 \leqslant q < 2$. Indeed, one can establish an analogous upper bound of the form

$$\inf_{|\mathcal{P}|=N} \|D_{\mathcal{B}}(\mathcal{P})\|_1 \ll_A (\log N)^2, \tag{2.29}$$

as demonstrated by Beck and Chen [13] in 1993. The argument there can, as before, be modified to show that

$$\inf_{|\mathcal{P}|=N} \|D_{\mathcal{B}}(\mathcal{P})\|_{A,q} \ll N^{(q-1)/2q}$$

for every real number q satisfying $1 < q \leqslant 2$.

Corresponding to open problem 5A, we have the following.

Open Problem 5C. *Let $U = [0,1]^2$, treated as a torus, and let A be a fixed convex polygon in U. Suppose that the irregularity of a point set $\mathcal{P}$ in U is described in terms of the infinite collection $\mathcal{B}$ given by* (2.28) *and endowed with the integral geometric measure* $\mathrm{d}B = \mathrm{d}\lambda\,\mathrm{d}\tau\,\mathrm{d}\mathbf{x}$.
(i) *Improve the upper bound* (2.29) *if possible.*
(ii) *Find a lower bound for the problem of the L_1-discrepancy.*

One can also study exceedingly difficult higher dimensional analogues.

Open Problem 5D. *Let $U = [0,1]^k$, treated as a torus and with $k \geqslant 2$, and let A be a fixed convex polytope in U. Suppose that the irregularity of a point set $\mathcal{P}$ in U is described in terms of the infinite collection*

$$\mathcal{B} = \{A(\lambda, \tau, \mathbf{x}) : \lambda \in [0,1], \tau \in \mathcal{T}, \mathbf{x} \in [0,1]^k\},$$

where $A(\lambda, \tau, \mathbf{x}) = \{\tau(\lambda \mathbf{y}) + \mathbf{x} : \mathbf{y} \in A\}$ denotes a similar copy obtained from the polytope A under a contraction λ, a rotation τ, and a translation $\mathbf{x}$, and endowed with the integral geometric measure $\mathrm{d}B = \mathrm{d}\lambda\,\mathrm{d}\tau\,\mathrm{d}\mathbf{x}$. *What can one say about the L_1-discrepancy?*

Again, let $U = [0,1]^2$, treated as a torus, and let A be a fixed convex polygon in U. Suppose that we no longer permit rotation, and that the irregularity of a point set $\mathcal{P}$ in U is described in terms of the infinite collection

$$\mathcal{B} = \{A(\lambda, \mathbf{x}) : \lambda \in [0,1], \mathbf{x} \in [0,1]^2\}$$

where $A(\lambda, \mathbf{x}) = \{\lambda \mathbf{y} + \mathbf{x} : \mathbf{y} \in A\}$ denotes a homothetic copy obtained from the polygon A under a contraction λ and a translation $\mathbf{x}$, and where $\mathcal{B}$ is endowed with the integral geometric measure $\mathrm{d}B = \mathrm{d}\lambda\,\mathrm{d}\mathbf{x}$.

Note that if A is a rectangle, then this is somewhat analogous to the classical problem. Indeed, corresponding to the classical problem, we have the bound

$$\inf_{|\mathcal{P}|=N} \|D_{\mathcal{B}}(\mathcal{P})\|_2 \ll_A (\log N)^{1/2},$$

due to Beck and Chen [14] in 1997. Funny enough, this paper contains no new ideas, as all the major ingredients are known to Davenport and Roth, but not all of them to both.

We do not know whether the lower bound

$$\inf_{|\mathcal{P}|=N} \|D_{\mathcal{B}}(\mathcal{P})\|_2 \gg_A (\log N)^{1/2},$$

analogous to Roth's classical result, the lower bound in (2.1) with $k = q = 2$, holds.

Open Problem 7. *Let $U = [0,1]^k$, treated as a torus and with $k \geqslant 2$, and let A be a fixed convex polytope in U. Suppose that the irregularity of a point set $\mathcal{P}$ in U is described in terms of the infinite collection*

$$\mathcal{B} = \{A(\lambda, \mathbf{x}) : \lambda \in [0,1], \mathbf{x} \in [0,1]^k\},$$

where $A(\lambda, \mathbf{x}) = \{\lambda \mathbf{y} + \mathbf{x} : \mathbf{y} \in A\}$ denotes a homothetic copy obtained from the polytope A under a contraction λ and a translation $\mathbf{x}$, and endowed with the integral geometric measure $\mathrm{d}B = \mathrm{d}\lambda\,\mathrm{d}\mathbf{x}$. Is it true that

$$(\log N)^{(k-1)/2} \ll_A \inf_{|\mathcal{P}|=N} \|D_{\mathcal{B}}(\mathcal{P})\|_2 \ll_A (\log N)^{(k-1)/2}?$$

2.7 Rotations of rectangles

Throughout this section, let $U = [0,1]^2$. We first consider the problem concerning discrepancy of finite point sets in U with respect to various collections of convex polygons in U.

Suppose that $\mathcal{B}$ is the collection of all convex polygons in U with sides in Θ, where Θ is a fixed finite set of directions. Then

$$\log N \ll_\Theta \inf_{|\mathcal{P}|=N} \|D_{\mathcal{B}}(\mathcal{P})\|_\infty \ll_\Theta \log N. \tag{2.30}$$

The lower bound is due to Beck and Chen [10] in 1989, whereas the upper bound is due to Chen and Travaglini [25] in 2007.

We can expand the collection $\mathcal{B}$ in a number of ways. For instance, if $\mathcal{B}$ is the collection of all convex polygons in U of at most k sides, then

$$N^{1/4} \ll_k \inf_{|\mathcal{P}|=N} \|D_{\mathcal{B}}(\mathcal{P})\|_\infty \ll_k N^{1/4}(\log N)^{1/2}.$$

Here, the lower bound is a simple consequence of the lower bound (2.15) when $k = 2$, and the upper bound is again due to Chen and Travaglini [25] in 2007. On the other hand, if $\mathcal{B}$ is the collection of all convex polygons in U, then

$$N^{1/3} \ll \inf_{|\mathcal{P}|=N} \|D_{\mathcal{B}}(\mathcal{P})\|_\infty \ll N^{1/3}(\log N)^4.$$

Here, the upper bound is due to Beck [8] in 1988, while the lower bound arises from an adaptation by Chen and Travaglini [25] in 2007 of an ingenious argument of Schmidt [50] in 1975.

Our original problem concerning convex polygons in U with sides in a finite set Θ has an analogous problem concerning finite rotations of rectangles. More precisely, suppose that $\mathcal{B}$ is the collection of all rectangles in U tilted by angles in a finite set Θ. Then the inequalities in (2.30) remain valid with this choice of $\mathcal{B}$.

A natural question is what happens if Θ is no longer finite. Some answers can be found in the work of Bilyk, Ma, Pipher, and Spencer [16, 17] in 2011 and 2016. Using powerful results in diophantine approximation, it can be shown that

$$\inf_{|\mathscr{P}|=N} \|D_{\mathcal{B}}(\mathscr{P})\|_\infty \begin{cases} \ll_\Theta \log N, & \text{if } \Theta \text{ is a finite set,} \\ \ll_\Theta (\log N)(\log\log N)^2, & \text{if } \Theta \text{ is a superlacunary set,} \\ \ll_\Theta (\log N)^3, & \text{if } \Theta \text{ is a lacunary sequence,} \\ \ll_\Theta (\log N)^{M+2}, & \text{if } \Theta \text{ is a lacunary set of order } M. \end{cases}$$

Furthermore, if Θ has upper Minkowski dimension $d \in [0,1)$, then

$$\inf_{|\mathscr{P}|=N} \|D_{\mathcal{B}}(\mathscr{P})\|_\infty \ll_{\Theta,\epsilon} N^{d/(d+1)+\epsilon}.$$

The following problem seems to be rather hard.

Open Problem 8. *Study the problem of discrepancy of points sets in $U = [0,1]^3$ with respect to polytopes in some suitable formulation.*

2.8 Cartesian products

We complete this survey by discussing a problem motivated by some interesting work of Matoušek prompted by a question posed to him by the author in the first ever workshop on discrepancy theory in Kiel in 1998.

Consider first an example involving the well-known classical discrepancy problem. Let $U_1 = [0,1]^k$, and let

$$\mathcal{B}_1 = \{B_1(\mathbf{x}) = [0,x_1) \times \cdots \times [0,x_k) : \mathbf{x} \in [0,1]^k\},$$

endowed with integral geometric measure $\mathrm{d}B_1 = \mathrm{d}\mathbf{x}$. We know that

$$(\log N)^{(k-1)/2} \ll_k \inf_{|\mathscr{P}|=N} \|D_{\mathcal{B}_1}(\mathscr{P})\|_2 \ll_k (\log N)^{(k-1)/2}. \tag{2.31}$$

Next, let $U_2 = [0,1]^\ell$, and let

$$\mathcal{B}_2 = \{B_2(\mathbf{y}) = [0,y_1) \times \cdots \times [0,y_\ell) : \mathbf{y} \in [0,1]^\ell\},$$

endowed with integral geometric measure $\mathrm{d}B_2 = \mathrm{d}\mathbf{y}$. We know that

$$(\log N)^{(\ell-1)/2} \ll_\ell \inf_{|\mathscr{P}|=N} \|D_{\mathcal{B}_2}(\mathscr{P})\|_2 \ll_\ell (\log N)^{(\ell-1)/2}. \tag{2.32}$$

Now let $U = U_1 \times U_2 = [0,1]^{k+\ell}$, and let

$$\mathcal{B} = \mathcal{B}_1 \times \mathcal{B}_2 = \{B_1(\mathbf{x}) \times B_2(\mathbf{y}) : \mathbf{x} \in [0,1]^k, \mathbf{y} \in [0,1]^\ell\},$$

endowed with integral geometric measure $\mathrm{d}B = \mathrm{d}\mathbf{x}\,\mathrm{d}\mathbf{y}$. We know that

$$(\log N)^{(k+\ell-1)/2} \ll_{k,\ell} \inf_{|\mathscr{P}|=N} \|D_{\mathcal{B}}(\mathscr{P})\|_2 \ll_{k,\ell} (\log N)^{(k+\ell-1)/2}. \tag{2.33}$$

Note that the order of magnitude in the estimates in (2.33) is roughly the product of the order of magnitude of the estimates in (2.31) and the order of magnitude of the estimates in (2.32). Clearly, both $\mathcal{B}_1$ and $\mathcal{B}_2$ play important roles.

Consider a second example. Let $U_1 = [0,1]^k$, treated as a torus, and let

$$\mathcal{B}_1 = \{A(\lambda, \tau, \mathbf{x}) : \lambda \in [0,1], \tau \in \mathcal{T}, \mathbf{x} \in [0,1]^k\},$$

endowed with integral geometric measure $\mathrm{d}B_1 = \mathrm{d}\lambda\,\mathrm{d}\tau\,\mathrm{d}\mathbf{x}$. We know from (2.15) and (2.17) that

$$N^{1/2-1/2k} \ll_A \inf_{|\mathscr{P}|=N} \|D_{\mathcal{B}_1}(\mathscr{P})\|_2 \ll_A N^{1/2-1/2k}. \tag{2.34}$$

Next, let U_2, $\mathcal{B}_2$ and the integral geometric measure be the same as in the previous example, so that the estimates (2.32) remain valid. Now let $U = U_1 \times U_2 = [0,1]^{k+\ell}$, and let

$$\mathcal{B} = \mathcal{B}_1 \times \mathcal{B}_2 = \{A(\lambda, \tau, \mathbf{x}) \times B_2(\mathbf{y}) : \lambda \in [0,1], \tau \in \mathcal{T}, \mathbf{x} \in [0,1]^k, \mathbf{y} \in [0,1]^\ell\},$$

endowed with integral geometric measure $\mathrm{d}B_1 = \mathrm{d}\lambda\,\mathrm{d}\tau\,\mathrm{d}\mathbf{x}\,\mathrm{d}\mathbf{y}$. We can show that

$$N^{1/2-1/2k} \ll_{A,\ell} \inf_{|\mathscr{P}|=N} \|D_{\mathcal{B}}(\mathscr{P})\|_2 \ll_{A,\ell} N^{1/2-1/2k}. \tag{2.35}$$

The lower bound follows from work of Beck [6] in 1987, while the upper bound is due to Beck and Chen [11] in 1990. We observe that the order of magnitude of the estimates in (2.34) and the order of magnitude of the estimates in (2.35) are identical, and the only contribution that the classical problem part of this Cartesian product problem makes to the estimates in (2.35) is in the implicit constants. In other words, $\mathcal{B}_1$ dominates and $\mathcal{B}_2$ hardly matters.

To understand the situation a little better, we next consider a third example. Let $U_1 = [0,1]^2$, treated as a torus, and let $\mathcal{B}_1$ be the collection of all circular discs in U_1.

Then it follows from (2.15) and (2.16) with $k = 2$ and noting that circular discs are invariant under rotation that

$$N^{1/4} \ll \inf_{|\mathscr{P}|=N} \|D_{\mathcal{B}_1}(\mathscr{P})\|_\infty \ll N^{1/4}(\log N)^{1/2}. \tag{2.36}$$

Next, let $U_2 = [0,1]^4$, treated as a torus, and let $\mathcal{B}_2$ be the collection of all circular balls in U_2. Then it follows from (2.15) and (2.16) with $k = 4$ and noting that circular balls are invariant under orthogonal transformation that

$$N^{3/8} \ll \inf_{|\mathscr{P}|=N} \|D_{\mathcal{B}_2}(\mathscr{P})\|_\infty \ll N^{3/8}(\log N)^{1/2}. \tag{2.37}$$

Finally, let $U = [0,1]^4 = U_1 \times U_1$, treated as a torus, and let $\mathcal{B} = \mathcal{B}_1 \times \mathcal{B}_1$ be the collection in U of Cartesian products of two circular discs in U_1. Then

$$\inf_{|\mathscr{P}|=N} \|D_{\mathcal{B}}(\mathscr{P})\|_\infty \ll_\epsilon N^{1/4+\epsilon}, \tag{2.38}$$

as shown by Matoušek [40] in 2000. Comparing the order of magnitude of the terms in (2.36) and (2.38), we see that the Cartesian product of two copies of the 2-dimensional problem in U_1 does not produce any estimate that is substantially greater than the estimates produced by a single copy, and certainly nothing as large the estimates in (2.37) produced by the corresponding 4-dimensional problem in U_2. Indeed, in the paper of Matoušek, it is shown that, under certain conditions, the discrepancy estimates for a cartesian product problem is governed by the largest bound among the constituent parts.

Open Problem 9 (Matoušek's problem). *Try to obtain a better understanding concerning the discrepancy of Cartesian products.*

Bibliography

[1] T. van Aardenne-Ehrenfest. Proof of the impossibility of a just distribution of an infinite sequence of points over an interval. Proc. Kon. Ned. Akad. v. Wetensch. **48**, 266–271 (1945).

[2] T. van Aardenne-Ehrenfest. On the impossibility of a just distribution. Proc. Kon. Ned. Akad. v. Wetensch. **52**, 734–739 (1949).

[3] J. R. Alexander. Geometric methods in the study of irregularities of distribution. Combinatorica **10**, 115–136 (1990).

[4] J. Beck. Balanced two-colourings of finite sets in the square I. Combinatorica **1**, 327–335 (1981).

[5] J. Beck. On a problem of K. F. Roth concerning irregularities of point distribution. Invent. Math. **74**, 477–487 (1983).

[6] J. Beck. Irregularities of point distribution I. Acta Math. **159**, 1–49 (1987).

[7] J. Beck. Irregularities of point distribution II. Proc. Lond. Math. Soc. **56**, 1–50 (1988).

[8] J. Beck. On the discrepancy of convex plane sets. Monatshefte Math. **105**, 91–106 (1988).

[9] J. Beck, W. W. L. Chen. Irregularities of Distribution. Cambridge Tracts in Mathematics, vol. 89. Cambridge University Press (1987).

[10] J. Beck, W. W. L. Chen. Irregularities of point distribution relative to convex polygons. In: G. Halász, V. T. Sós (eds.) Irregularities of Partitions. Algorithms and Combinatorics, vol. 8, pp. 1–22. Springer (1989).

[11] J. Beck, W. W. L. Chen. Note on irregularities of distribution II. Proc. Lond. Math. Soc. **61**, 251–272 (1990).

[12] J. Beck, W. W. L. Chen. Irregularities of point distribution relative to half-planes I. Mathematika **40**, 102–126 (1993).

[13] J. Beck, W. W. L. Chen. Irregularities of point distribution relative to convex polygons II. Mathematika **40**, 127–136 (1993).

[14] J. Beck, W. W. L. Chen. Irregularities of point distribution relative to convex polygons III. J. Lond. Math. Soc. **56**, 222–230 (1997).

[15] D. Bilyk, M. T. Lacey, A. Vagharshakyan. On the small ball inequality in all dimensions. J. Funct. Anal. **254**, 2470–2502 (2008).

[16] D. Bilyk, X. Ma, J. Pipher, C. Spencer. Directional discrepancy in two dimensions. Bull. Lond. Math. Soc. **43**, 1151–1166 (2011).

[17] D. Bilyk, X. Ma, J. Pipher, C. Spencer. Diophantine approximations and directional discrepancy of rotated lattices. Trans. Am. Math. Soc. **368**, 3871–3897 (2016).

[18] L. Brandolini, G. Gigante, G. Travaglini. Irregularities of distribution and average decay of Fourier transform. In: W. W. L. Chen, A. Srivastav, G. Travaglini (eds.) A Panorama of Discrepancy Theory. Lecture Notes in Mathematics, vol. 2107, pp. 159–220. Springer (2014).

[19] W. W. L. Chen. On irregularities of distribution. Mathematika **27**, 153–170 (1980).

[20] W. W. L. Chen. On irregularities of distribution II. Quart. J. Math. Oxford **34**, 257–279 (1983).

[21] W. W. L. Chen. On irregularities of distribution IV. J. Number Theory **80**, 44–59 (2000).

[22] W. W. L. Chen. Fourier techniques in the theory of irregularities of point distribution. In: L. Brandolini, L. Colzani, A. Iosevich, G. Travaglini (eds.) Fourier Analysis and Convexity, pp. 59–82. Applied and Numerical Harmonic Analysis, Birkhäuser (2004).

[23] W. W. L. Chen. Upper bounds in discrepancy theory. In: L. Plaskota, H. Woźniakowski (eds.) Monte Carlo and Quasi-Monte Carlo Methods 2010. Springer Proceedings in Mathematics and Statistics, vol. 23, pp. 23–41. Springer (2012).

[24] W. W. L. Chen, M. M. Skriganov. Explicit constructions in the classical mean squares problem in irregularities of point distribution. J. Reine Angew. Math. **545**, 67–95 (2002).

[25] W. W. L. Chen, G. Travaglini. Discrepancy with respect to convex polygons. J. Complex. **23**, 662–672 (2007).

[26] W. W. L. Chen, G. Travaglini. Some of Roth's ideas in discrepancy theory. In: W. W. L. Chen, W. T. Gowers, H. Halberstam, W. M. Schmidt, R. C. Vaughan (eds.) Analytic Number Theory: Essays in Honour of Klaus Roth, pp. 150–163. Cambridge University Press (2009).

[27] W. W. L. Chen, G. Travaglini. Deterministic and probabilistic discrepancies. Ark. Mat. **47**, 273–293 (2009).

[28] W. W. L. Chen, G. Travaglini. An L^1 estimate for half-space discrepancy. Acta Arith. **146**, 203–214 (2011).

[29] J. G. van der Corput. Verteilungsfunktionen I. Proc. Kon. Ned. Akad. v. Wetensch. **38**, 813–821 (1935).

[30] J. G. van der Corput. Verteilungsfunktionen II. Proc. Kon. Ned. Akad. v. Wetensch. **38**, 1058–1066 (1935).

[31] H. Davenport. Note on irregularities of distribution. Mathematika **3**, 131–135 (1956).

[32] J. Dick, F. Pillichshammer. Optimal $\mathcal{L}_2$ discrepancy bounds for higher order digital sequences over the finite field $\mathbb{F}_2$. Acta Arith. **162**, 65–99 (2014).

[33] J. Dick, F. Pillichshammer. Explicit constructions of point sets and sequences with low discrepancy. In: P. Kritzer, H. Niederreiter, F. Pillichshammer, A. Winterhof (eds.) Uniform Distribution and Quasi-Monte Carlo Methods: Discrepancy, Integration and Applications. Radon Series on Computational and Applied Mathematics, vol. 15, pp. 63–86. de Gruyter (2014).
[34] G. Halász. On Roth's method in the theory of irregularities of point distributions. In: H. Halberstam, C. Hooley (eds.) Recent Progress in Analytic Number Theory, vol. 2, pp. 79–94. Academic Press (1981).
[35] J. H. Halton. On the efficiency of certain quasirandom sequences of points in evaluating multidimensional integrals. Numer. Math. **2**, 84–90 (1960).
[36] Kolountzakis, M.: Private communication to D. Bilyk.
[37] S. V. Konyagin, M. M. Skriganov, A. V. Sobolev. On a lattice point problem arising in the spectral analysis of periodic operators. Mathematika **50**, 87–98 (2003).
[38] V. F. Lev. Translations of nets and relationship between supreme and L^{κ}-discrepancies. Acta Math. Hung. **70**, 1–12 (1996).
[39] J. Matoušek. Tight upper bounds for the discrepancy of half-spaces. Discrete Comput. Geom. **13**, 593–601 (1995).
[40] J. Matoušek. On the discrepancy of cartesian products. J. Lond. Math. Soc. **61**, 737–747 (2000).
[41] H. L. Montgomery. Ten Lectures on the Interface between Analytic Number Theory and Harmonic Analysis. CBMS Regional Conference Series in Mathematics, vol. 84. American Mathematical Society (1994).
[42] L. Parnovski, A. Sobolev. On the Bethe–Sommerfeld conjecture for the polyharmonic operator. Duke Math. J. **107**, 209–238 (2001).
[43] K. F. Roth. On irregularities of distribution. Mathematika **1**, 73–79 (1954).
[44] K. F. Roth. Remark concerning integer sequences. Acta Arith. **9**, 257–260 (1964).
[45] K. F. Roth. On irregularities of distribution IV. Acta Arith. **37**, 67–75 (1980).
[46] W. M. Schmidt. Irregularities of distribution II. Trans. Am. Math. Soc. **136**, 347–360 (1969).
[47] W. M. Schmidt. Irregularities of distribution III. Pac. J. Math. **29**, 225–234 (1969).
[48] W. M. Schmidt. Irregularities of distribution IV. Invent. Math. **7**, 55–82 (1969).
[49] W. M. Schmidt. Irregularities of distribution VII. Acta Arith. **21**, 45–50 (1972).
[50] W. M. Schmidt. Irregularities of distribution IX. Acta Arith. **27**, 385–396 (1975).
[51] W. M. Schmidt. Irregularities of distribution X. In: H. Zassenhaus (ed.) Number Theory and Algebra, pp. 311–329. Academic Press (1977).
[52] W. M. Schmidt. Lectures on Irregularities of Distribution. Tata Institute of Fundamental Research (1977).
[53] M. M. Skriganov. Constructions of uniform distributions in terms of geometry of numbers. St. Petersburg Math. J. **6**, 200–230 (1994).
[54] M. M. Skriganov. Dyadic shift randomisation in classical discrepancy theory. Mathematika **62**, 183–209 (2016).
[55] G. Travaglini, M. R. Tupputi. A Characterization theorem for the L^2-discrepancy of integer points in dilated polygons. J. Fourier Anal. Appl. **22**, 675–693 (2016).

Marcin Wnuk, Michael Gnewuch, and Nils Hebbinghaus

3 On negatively dependent sampling schemes, variance reduction, and probabilistic upper discrepancy bounds

Abstract: We study some notions of negative dependence of a sampling scheme that can be used to derive variance bounds for the corresponding estimator or discrepancy bounds for the underlying random point set that are at least as good as the corresponding bounds for plain Monte Carlo sampling. We provide new preasymptotic bounds with explicit constants for the star discrepancy and the weighted star discrepancy of sampling schemes that satisfy suitable negative dependence properties. Furthermore, we compare the different notions of negative dependence and give several examples of negatively dependent sampling schemes, including mixed sequences.

Keywords: Star discrepancy, randomized quasi-Monte Carlo methods, mixed sequences, variance reduction

MSC 2010: 65C05

3.1 Introduction

Plain Monte Carlo (MC) sampling is a method frequently used in stochastic simulation and multivariate numerical integration. Let $p_1, \dots, p_N$ be independent random points, each uniformly distributed in the d-dimensional unit cube $[0,1)^d$. For an arbitrary integrable random variable (or function) $f : [0,1)^d \to \mathbb{R}$, we consider the MC estimator (or quadrature)

$$\mu^{\mathrm{MC}}(f) = \frac{1}{N} \sum_{i=1}^{N} f(p_i) \tag{3.1}$$

for the expected value (or integral)

$$I(f) = \int_{[0,1)^d} f(u)\, du.$$

An advantage of the MC estimator is that already under the very mild assumption on f to be square integrable, it converges to $I(f)$ for $N \to \infty$ with convergence rate

Marcin Wnuk, Michael Gnewuch, Mathematisches Seminar, Osnabrück University, Germany, e-mails: marcin.wnuk@math.uni-osnabrueck.de, michael.gnewuch@math.uni-osnabrueck.de
Nils Hebbinghaus, CAU Kiel, Mathematisches Seminar, Ludwig–Meyn-Strasse 4, 24118 Kiel, Schleswig-Holstein, Germany, e-mail: hebbing@gmx.de

https://doi.org/10.1515/9783110652581-003

1/2. Even though the convergence rate is not very impressive, it has the invaluable advantage that it does not depend on the number of variables d.

However, there are many dependent sampling schemes (i. e., random sample points p_i, $i = 1,\dots,N$, that are still uniformly distributed in $[0,1)^d$, but not necessarily independent any more) known that are superior to plain MC sampling with respect to certain objectives. An example is suitably randomized quasi-Monte Carlo (RQMC) point sets. They ensure, for instance, higher convergence rates for numerical integration of sufficiently smooth functions, they lead to much smaller asymptotic discrepancy measures, their sample points do not tend to cluster, and have more evenly distributed lower dimensional projections (see, e. g., [4, 5, 18, 21]). It would be desirable to be able to construct dependent sampling schemes that have some of these or other favorable properties, and that are, with respect to other objectives, at least as good as MC sampling schemes.

Recently, in this direction some research has been done. In [19], Christiane Lemieux showed that a negative dependence property of RQMC points ensures that the variance of the corresponding RQMC estimator for functions f that are monotone with respect to each variable is never larger than the variance of the corresponding MC estimator $\mu^{\mathrm{MC}}f$. She also proved that a different negative dependence property yields that the variance of the RQMC estimator for an arbitrary bounded quasi-monotone f is never larger than the variance of $\mu^{\mathrm{MC}}f$. Those negative dependence properties rely solely on the marginals and the bivariate copulas of the RQMC points (i. e., on the distribution of single points and on the common distribution of pairs of points). Related results can be found in [33].

In a different line of research, the second and the third author of this book chapter showed in [11, 13] that a specific negative dependence property of RQMC points guarantees that they satisfy the same preasymptotic probabilistic discrepancy bounds (with explicitly revealed dependence on the number of points N as well as on the dimension d) as MC points. Here, the negative dependence property relies on the common distribution of all sample points. Related results can be found in [6].

For more extensive motivations of both lines of research, we refer to the elaborate introductions of [19] and [6, 11], respectively. The aim of this book chapter is to survey and compare the approaches mentioned above and to provide several new results.

This chapter is organized as follows: In Section 3.2, we introduce some notions of negatively dependent sampling schemes and discuss how one can benefit from them. In Section 3.3, we provide new probabilistic upper discrepancy bounds for sampling schemes. The discrepancy measures we consider are the star discrepancy and the weighted star discrepancy. These bounds are "plug-in results" in the following sense: One just has to check whether a sampling scheme satisfies the sufficient negative dependence condition and—if this is the case—obtains immediately a probabilistic discrepancy bound with explicitly given constants. In the Section 3.4, we give several examples of sampling schemes that satisfy the one or the other notion of negative dependence, including a generalized notion of stratified sampling schemes and mixed randomized sequences. Finally, in the Section 3.5 we elaborate on relations between different notions of negative dependence.

We finish the introduction by stating some notation. Let $d, N \in \mathbb{N}$. If not stated otherwise, we are always considering a randomized point set $(p_j)_{j=1}^N := \mathcal{P} \subset [0,1)^d$ consisting of N points. For $a, b \in \mathbb{R}^d$, $a = (a_1, \dots, a_d)$, $b = (b_1, \dots, b_d)$, we write $a \le b$ if $a_i \le b_i$, $i = 1, \dots, d$. All other inequalities are also to be understood componentwise. Moreover, $[a,b) := [a_1, b_1) \times \cdots \times [a_d, b_d)$. Via $\mathcal{C}_0^d$, we denote the set of boxes ("corners") anchored at 0,

$$\mathcal{C}_0^d := \{[0,a) \mid a \in [0,1)^d\},$$

and by $\mathcal{C}_1^d$ the set of boxes anchored at 1,

$$\mathcal{C}_1^d := \{[a,1) \mid a \in [0,1)^d\}.$$

We write $\mathcal{D}_0^d$ for the set of differences of boxes anchored at 0,

$$\mathcal{D}_0^d := \{Q \setminus R \mid Q, R \in \mathcal{C}_0^d\}.$$

For $m \in \mathbb{N}$, we denote the set $\{1, 2, \dots, m\}$ by $[m]$, λ^d stands for the d-dimensional Lebesgue measure on $\mathbb{R}^d$, in case $d = 1$ we just write λ. If not specified, all random variables are defined on a probability space $(\Omega, \Sigma, \mathbf{P})$.

3.2 Review of notions of negative dependance of sampling schemes

3.2.1 γ-negative dependence of binary random variables and sampling schemes

The concept of negative dependence was introduced by Lehmann [17] for pairs of random variables. In the literature, one finds several contributions on rather demanding notions of negative dependence as, for example, negative association introduced in [15]; a survey can be found in [30]. Sufficient for our purpose is the following notion for *Bernoulli* or *binary random variables*, that is, random variables that only take values in $\{0,1\}$.

Definition 3.1. Let $\gamma \ge 1$. We call binary random variables $T_1, T_2, \dots, T_N$ *upper γ-negatively dependent* if

$$\mathbf{P}\Big(\bigcap_{j \in u} \{T_j = 1\}\Big) \le \gamma \prod_{j \in u} \mathbf{P}(T_j = 1) \quad \text{for all } u \subseteq [N], \tag{3.2}$$

and *lower γ-negatively dependent* if

$$\mathbf{P}\Big(\bigcap_{j\in u}\{T_j = 0\}\Big) \le \gamma \prod_{j\in u} \mathbf{P}(T_j = 0) \quad \text{for all } u \subseteq [N]. \tag{3.3}$$

We call $T_1, T_2, \ldots, T_N$ *γ-negatively dependent* if both conditions (3.2) and (3.3) are satisfied. If $\gamma = 1$, we usually suppress the explicit reference to γ.

1-Negative dependence is usually called negative orthant dependence; cf. [3].

Notice that, in particular, independent binary random variables are negatively dependent. Furthermore, it is easily seen that for $N = 2$ and $\gamma = 1$ the notions of upper and lower γ-negative dependence are equivalent; cf. [17].

We are interested in binary random variables T_i, $i = 1, \ldots, N$, of the form $T_i = \mathbf{1}_A(p_i)$, where A is a Lebesgue-measurable subset of $[0,1)^d$ (whose characteristic function is denoted by $\mathbf{1}_A$), and $p_1, \ldots, p_N$ are randomly chosen points in $[0,1)^d$.

We will use the following bound of Hoeffding-type; for a proof see, for example, [13].

Theorem 3.2. *Let $\gamma \ge 1$, and let $T_1, \ldots, T_N$ be γ-negatively dependent binary random variables. Put $S := \sum_{i=1}^N (T_i - \mathbb{E}[T_i])$. We have*

$$\mathbf{P}(|S| \ge t) \le 2\gamma \exp\Big(-\frac{2t^2}{N}\Big) \quad \textit{for all } t > 0. \tag{3.4}$$

Definition 3.3. A randomized point set $\mathcal{P} = (p_j)_{j=1}^N$ is called a *sampling scheme* if every single $p \in \mathcal{P}$ is distributed uniformly in $[0,1)^d$ and the vector $(p_1, \ldots, p_N)$ is exchangeable, meaning that for any permutation π of $[N]$ it holds that the law of $(p_1, \ldots, p_N)$ is the same as the law of $(p_{\pi(1)}, \ldots, p_{\pi(N)})$.

The assumption of exchangeability is only of technical nature and, if we consider $\mathcal{P}$ as a randomized point set, it may be always obtained in the process of symmetrization. Indeed, let $\tilde{\mathcal{P}}$ be a point set such that every $\tilde{p} \in \tilde{\mathcal{P}}$ is uniformly distributed in $[0,1)^d$ and let π be a random uniformly chosen permutation of $[N]$. Then $(\tilde{p}_{\pi(1)}, \ldots, \tilde{p}_{\pi(N)})$ is already a sampling scheme.

3.2.2 Pairwise negative dependence and variance reduction

Definition 3.4. We say that a sampling scheme $\mathcal{P}$ is *pairwise negatively dependent* if for every $Q, R \in \mathcal{C}_1^d$ it holds that the random variables

$$\mathbf{1}_Q(p_1), \mathbf{1}_R(p_2)$$

are negatively dependent. In other words, a sampling scheme $\mathcal{P}$ is pairwise negatively dependent if for every $Q, R \in \mathcal{C}_1^d$ we have

$$\mathbf{P}(p_1 \in Q, p_2 \in R) \le \mathbf{P}(p_1 \in Q)\,\mathbf{P}(p_2 \in R), \tag{3.5}$$

$$\mathbf{P}(p_1 \notin Q, p_2 \notin R) \leq \mathbf{P}(p_1 \notin Q)\,\mathbf{P}(p_2 \notin R). \tag{3.6}$$

Note that (3.5) implies (3.6) and vice versa, therefore, one of the conditions is in fact redundant. In [19], this is known as negatively upper orthant dependent—or NUOD—sampling schemes.

Our interest lies in numerical integration of functions from some class $\mathcal{F} \subset L^2([0,1)^d)$ using RQMC. A QMC quadrature is just a quadrature consisting of N nodes, such that the evaluation in every node is given the same weight $\frac{1}{N}$. By randomizing the set of nodes, we obtain an RQMC quadrature. Let $\mu_{\mathcal{P}} f$ be an RQMC estimator of $I(f) := \int_{[0,1)^d} f(u)\,du$ based on the sampling scheme $\mathcal{P} = (p_i)_{i=1}^N$, that is,

$$\mu_{\mathcal{P}} f = \frac{1}{N}\sum_{i=1}^{N} f(p_i).$$

Moreover, let $\mu^{\mathrm{MC}} f$ be an estimator of $I(f)$ based on a Monte Carlo sample consisting of N points (i. e., the integration nodes are chosen independently and uniformly from $[0,1)^d$, see (3.1)).

It turns out that randomized QMC quadratures based on pairwise negatively dependent sampling schemes may lead to variance reduction in comparison to the simple MC quadratures. Here, we describe shortly one of such cases, namely when integrands are bounded quasimonotone functions. The following exposition is based on [19].

To define what a quasimonotone function is we need first to introduce the notion of quasivolume. For $a, b \in [0,1)^d$, $J \subset [d]$, and a function $f : [0,1)^d \to \mathbb{R}$, we write $f(a_J, b_{-J})$ to represent the evaluation of f at the point $(x_1, \dots, x_d)$, where $x_j = a_j$ for $j \in J$ and $x_j = b_j$ otherwise. The quasivolume of f over an interval $A = [a, b) \subset [0,1)^d$ is given by

$$\Delta^d(f, A) := \sum_{J \subset [d]} (-1)^{|J|} f(a_J, b_{-J}).$$

We say that the function f is *quasimonotone* if

$$\Delta^d(f, A) \geq 0$$

for every interval A. Note that if we define a content $\nu_f([0,a)) := f(a)$, $a \in [0,1)^d$ then quasimonotonicity of f means exactly that for any axis-parallel rectangle $R \subset [0,1)^d$ it holds $\nu_f(R) \geq 0$.

Apart from pairwise negative dependence there are a few similar notions which are also of interest. Let $p_j = (p_j^{(1)}, \dots, p_j^{(d)})$, $j = 1, \dots, N$. If for every $i = 1, \dots, d$, and every measurable $A, B \subset [0,1)^{i-1}$, $\alpha, \beta \in [0,1)$,

$$\begin{aligned}
&\mathbf{P}(p_1^{(i)} \geq \alpha, p_2^{(i)} \geq \beta \,|\, p_1^{(1:i-1)} \in A, p_2^{(1:i-1)} \in B) \\
&\leq \mathbf{P}(p_1^{(i)} \geq \alpha \,|\, p_1^{(1:i-1)} \in A, p_2^{(1:i-1)} \in B)\,\mathbf{P}(p_2^{(i)} \geq \beta \,|\, p_1^{(1:i-1)} \in A, p_2^{(1:i-1)} \in B),
\end{aligned}$$

we say that the sampling scheme $(p_j)_{j=1}^N$ is *conditionally negatively quadrant dependent* (conditionally NQD). Here, $p^{(1:i-1)}$ denotes the orthogonal projection of p onto its first $i-1$ coordinates. Note that the conditional NQD property holds in particular if $(p_1^{(i)}, p_2^{(i)})_{i=1}^d$ are independent, and for every $i = 1, \dots, d$, and every $q, r \in [0,1)$, we have

$$\mathbf{P}(p_1^{(i)} \in [q,1), p_2^{(i)} \in [r,1)) \le \mathbf{P}(p_1^{(i)} \in [q,1))\, \mathbf{P}(p_2^{(i)} \in [r,1)),$$

in which case we talk of a *coordinatewise independent NQD sampling scheme*. Christiane Lemieux showed in [19, Corollary 2] that conditionally NQD sampling schemes provide RQMC estimators of integrals with variance no bigger then the variance of the MC estimator if the integrand is monotone in each coordinate.

The following is basically a combination of Proposition 3, Remark 8, and Corollary 2 from [19].

Theorem 3.5. *Let $f : [0,1)^d \to \mathbb{R}$ and $\mathcal{P}$ be a sampling scheme. Then if either:*

1. *The function f is bounded and f or $-f$ is quasimonotone and $\mathcal{P}$ is pairwise negatively dependent,*
2. *The function f is monotone in each coordinate and $\mathcal{P}$ is conditionally negatively quadrant dependent,*

it holds

$$\mathrm{Var}(\mu_{\mathcal{P}} f) \le \mathrm{Var}(\mu^{\mathrm{MC}} f).$$

In Section 3.5, we discuss relations between the introduced notions of negative dependence.

Let us note that the aforementioned paper provides actually more general results. Interested readers will find details in Sections 3 and 4 of [19].

For examples of pairwise negatively dependent and conditionally NQD sampling schemes, see Sections 3.4.2 and 3.4.3.

3.2.3 Negatively dependent sampling schemes and discrepancy

Definition 3.6. We say that a sampling scheme $(p_j)_{j=1}^N = \mathcal{P}$ is $\mathcal{S} - \gamma$*–negatively dependent* if for every $Q \in \mathcal{S}$ the random variables

$$(\mathbf{1}_Q(p_j))_{j=1}^N$$

are γ-negatively dependent. In other words, for every $t \le N$ we require

$$\mathbf{P}\left(\bigcap_{j=1}^t \{p_j \in Q\}\right) \le \gamma \prod_{j=1}^t \mathbf{P}(p_j \in Q), \tag{3.7}$$

$$\mathbf{P}\left(\bigcap_{j=1}^t \{p_j \notin Q\}\right) \le \gamma \prod_{j=1}^t \mathbf{P}(p_j \notin Q). \tag{3.8}$$

Note that differently from the case of pairwise negative dependence, for $N > 2$ one indeed needs to check both inequalities as they do not, in general, imply one another. If $\gamma = 1$ and $\mathcal{S} = \mathcal{C}_0^d$, we usually talk just of negatively dependent sampling schemes. Moreover, if (3.7) is satisfied we speak of upper γ-negatively dependent sampling schemes, and if (3.8) is satisfied, we speak of lower γ-negatively dependent sampling schemes.

To motivate the interest in negatively dependent sampling schemes, we introduce the notion of discrepancy. Discrepancy is meant to quantify how far is a finite point set $P \subset [0,1)^d$ consisting of N points from being equidistributed in $[0,1)^d$. It plays an important role in fields like numerical integration, computer graphics, empirical process theory, and many more. Let $x \in [0,1]^d$ and $Q_x := [0,x) \in \mathcal{C}_0^d$. We define the *discrepancy function* $D_N(P,\cdot)$ for the point set P at the point x via

$$D_N(P,x) := D_N(P,Q_x) := \left|\frac{1}{N}|P \cap Q_x| - \lambda^d(Q_x)\right|$$

and the *star discrepancy* $D_N^*(P)$ by

$$D_N^*(P) := \sup_{x\in[0,1]^d} D_N(P,x).$$

Making a connection to numerical integration, we note one of the versions of the Koksma–Hlawka inequality, which states that for every point set P consisting of N points it holds

$$\left|\int_{[0,1)^d} f\, d\lambda^d(x) - \frac{1}{N}\sum_{p\in P} f(p)\right| \le D_N^*(P)\,\mathrm{Var}_{\mathrm{HK}}(f),$$

where $\mathrm{Var}_{\mathrm{HK}}(f)$ is the Hardy–Krause variation of f. The inequality is actually sharp; cf. [23].

It has been shown in [11] that $\mathcal{D}_0^d$–γ-negatively dependent sampling schemes have with large probability star discrepancy of the order $\sqrt{\frac{d}{N}}$. More precisely, the following theorem holds.

Theorem 3.7. *Let $d,N \in \mathbb{N}$, and $\rho \in [0,\infty)$. Let $\mathcal{P} = (p_j)_{j=1}^N$ be a negatively $\mathcal{D}_0^d$-$e^{\rho d}$-dependent sampling scheme.*

Then for every $c > 0$,

$$D_N^*(\mathcal{P}) \le c\sqrt{\frac{d}{N}} \tag{3.9}$$

holds with probability at least $1 - e^{-(1.6741\cdot c^2 - 10.7042-\rho)\cdot d}$. Moreover, for every $\theta \in (0,1)$,

$$\mathbf{P}\left(D_N^*(\mathcal{P}) \le 0.7729\sqrt{10.7042 + \rho + \frac{\ln((1-\theta)^{-1})}{d}}\sqrt{\frac{d}{N}}\right) \ge \theta. \tag{3.10}$$

Notice that these bounds depend only mildly on ρ or $\gamma = e^{\rho d}$. In particular, $\mathcal{D}_0^d$-1-negatively dependent sampling schemes satisfy the same preasymptotic discrepancy bound as Monte Carlo point sets do. For more details, see [11].

In Remark 3.23, we present a bound similar to (3.10) under a bit different assumptions that can be applied to so-called mixed randomized sequences.

3.3 New probabilistic discrepancy bounds

3.3.1 Bound on the star discrepancy for negatively dependent sampling schemes

Proving that a given sampling scheme is $\mathcal{D}_0^d$-γ-negatively dependent may turn out to be a difficult task. One of the problems lies in the fact that elements of $\mathcal{D}_0^d$ may in general not be represented as Cartesian products of one-dimensional intervals; cf. also Remark 3.23. With this in mind, we would like to weaken the assumptions on the sampling scheme $\mathcal{P}$. In the following result, we show that by requiring the sampling scheme $\mathcal{P}$ only to be $\mathcal{C}_0^d$-γ-negatively dependent one already gets with high probability a discrepancy of the order $\sqrt{\frac{d}{N}\log(e+\frac{N}{d})}$.

Theorem 3.8. *Let $d, N \in \mathbb{N}$, and $\rho \in [0,\infty)$. Let $\mathcal{P} = (p_j)_{j=1}^N$ be a $\mathcal{C}_0^d$-$e^{\rho d}$-negatively dependent sampling scheme in $[0,1)^d$. Then for every $c > 0$,*

$$D_N^*(\mathcal{P}) \le c\sqrt{\frac{d}{N}}\max\left\{1, \log\left(\frac{N}{d}\right)\right\} \tag{3.11}$$

holds with probability at least $1 - 2e^{(-\frac{1}{2}(c^2-1)\xi + \rho + \log(2e(\frac{2}{c}+1)))d}$, where $\xi = \max\left\{1, \log\left(\frac{N}{d}\right)\right\}$. Moreover, for every $\theta \in (0,1)$,

$$\mathbf{P}\left(D_N^*(\mathcal{P}) \le \sqrt{\frac{2}{N}}\sqrt{d\log(\eta) + \rho d + \log\left(\frac{2}{1-\theta}\right)}\right) \ge \theta, \tag{3.12}$$

where $\eta := \eta(N,d) = 6e(\max(1, \frac{N}{2d\log(6e)}))^{\frac{1}{2}}$.

The proof of Theorem 3.8 requires some preparation. To "discretize" the star discrepancy, we define δ–covers as in [7]: for any $\delta \in (0,1]$, a finite set Γ of points in $[0,1)^d$ is called a *δ–cover* of $[0,1)^d$, if for every $y \in [0,1)^d$ there exist $x, z \in \Gamma \cup \{0\}$ such that $x \le y \le z$ and $\lambda^d([0,z]) - \lambda^d([0,x]) \le \delta$. The number $\mathcal{N}(d,\delta)$ denotes the smallest cardinality of a δ–cover of $[0,1)^d$.

The following theorem was stated and proved in [9].

Theorem 3.9. *For any $d \ge 1$ and $\delta \in (0,1]$, we have*

$$\mathcal{N}(d,\delta) \le 2^d \frac{d^d}{d!}(\delta^{-1}+1)^d.$$

Notice that due to Stirling's formula we have $d^d/d! \leq e^d/\sqrt{2\pi d}$ and so the cardinality of the δ-cover may be bounded from above by $(2e)^d(1+\delta^{-1})^d$. Furthermore, it is easy to verify that in the case $d = 1$ the identity

$$\mathcal{N}(1,\delta) = \lceil \delta^{-1} \rceil \tag{3.13}$$

is established with the help of the δ-cover $\Gamma := \{1/\lceil\delta^{-1}\rceil, 2/\lceil\delta^{-1}\rceil, \ldots, 1\}$.

With the help of δ-covers, the star discrepancy can be approximated in the following sense.

Lemma 3.10. *Let $P \subset [0,1)^d$ be an N-point set, $\delta > 0$, and Γ be a δ-cover of $[0,1)^d$. Then*

$$D_N^*(P) \leq \max_{x\in\Gamma} D_N(P,[0,x)) + \delta.$$

The proof of Lemma 3.10 is straightforward; cf., for example, [7, Lemma 3.1].

Now we are ready to prove Theorem 3.8.

Proof. For $\delta \in (0,1)$ to be chosen later, let Γ be a δ-cover consisting of at most $(2e)^d(1+\delta^{-1})^d$ elements. Such a Γ exists due to Theorem 3.9 and discussion thereafter.

Define

$$D_{N,\Gamma}^*(\mathcal{P}) = \max_{\beta\in\Gamma}\left|\lambda^d([0,\beta)) - \frac{1}{N}\sum_{j=1}^{N}\mathbf{1}_{[0,\beta)}(p_j)\right|.$$

Now Lemma 3.10 gives us

$$D_N^*(\mathcal{P}) \leq D_{N,\Gamma}^*(\mathcal{P}) + \delta.$$

For every $\beta \in \Gamma$ and $j \in [N]$ put

$$\xi_\beta^{(j)} = \lambda^d([0,\beta)) - \mathbf{1}_{[0,\beta)}(p_j).$$

Let $\epsilon = 2\delta$. Due to Hoeffding's inequality applied to random variables $(\xi_\beta^{(j)})_{j=1}^N$ (applicable since $(p_j)_{j=1}^N$ is $e^{\rho d}$-negatively dependent), we obtain for every $\beta \in \Gamma$,

$$\mathbf{P}\left(\left|\frac{\sum_{j=1}^N \xi_\beta^{(j)}}{N}\right| \geq \delta\right) \leq 2e^{\rho d}e^{-2N\delta^2}.$$

With the help of a simple union bound, we get

$$\begin{aligned}
\mathbf{P}(D_N^*(\mathcal{P}) < \epsilon) &= 1 - \mathbf{P}(D_N^*(\mathcal{P}) \geq \epsilon) \\
&\geq 1 - \mathbf{P}(D_{N,\Gamma}^*(\mathcal{P}) \geq \epsilon - \delta) = 1 - \mathbf{P}\left(\max_{\beta\in\Gamma}\left|\frac{\sum_{j=1}^N \xi_\beta^{(j)}}{N}\right| \geq \delta\right) \\
&\geq 1 - 2e^{\rho d}|\Gamma|e^{-\frac{N}{2}\epsilon^2}.
\end{aligned} \tag{3.14}$$

We first prove (3.12). Using (3.14) we would like to find a bound on the discrepancy of the sampling scheme $\mathcal{P}$ which holds with probability at least $\theta \in (0,1)$. We are looking for ϵ_θ such that

$$\mathbf{P}(D_N^*(\mathcal{P}) < \epsilon_\theta) \geq \theta. \tag{3.15}$$

Put $\epsilon_\theta = C_\theta(\frac{d}{N}\log(1+\frac{N}{d}))^{\frac{1}{2}} = 2\delta_\theta$. Inequality (3.15) holds true if

$$\delta_\theta \geq \left(\frac{1}{2N}\right)^{\frac{1}{2}}\left[\log(|\Gamma|) + \rho d + \log\left(\frac{2}{1-\theta}\right)\right]^{\frac{1}{2}}.$$

Our problem boils now down to finding possibly small $\delta_\theta \in (0,1)$ for which

$$\delta_\theta \geq \left(\frac{1}{2N}\right)^{\frac{1}{2}}\left[d\log(2e[1+\delta_\theta^{-1}]) + \rho d + \log\left(\frac{2}{1-\theta}\right)\right]^{\frac{1}{2}}. \tag{3.16}$$

Specifying δ_θ to be of the form

$$\delta_\theta = \left(\frac{1}{2N}\right)^{\frac{1}{2}}\left[d\log(\eta) + \rho d + \log\left(\frac{2}{1-\theta}\right)\right]^{\frac{1}{2}}$$

we get that (3.16) is satisfied if

$$\eta \geq 2e(1+\delta_\theta^{-1}).$$

Expanding δ_θ in dependence of η, it suffices to find η for which

$$\left(\frac{\eta}{2e} - 1\right)\log(\eta)^{\frac{1}{2}} \geq \left(\frac{2N}{d}\right)^{\frac{1}{2}}$$

and one easily sees that this is satisfied for η given in the statement of the theorem.

To prove (3.11) one only needs to plug in $\epsilon := c\sqrt{\frac{d}{N}}\xi$ into (3.14) and then consider the two cases $\xi = 1$ and $\xi = \log\left(\frac{N}{d}\right)$ separately. □

3.3.2 Bound on the weighted star discrepancy for $\mathcal{D}_\text{U}^d - \gamma$-negatively dependent sampling schemes

One of the reasons why the QMC integration may be successfully applied in many high-dimensional problems is the fact that quite often only a small number of coordinates is really important. This observation led to the introduction of weighted function spaces and weighted discrepancies by Sloan and Woźniakowski in [31]. The above concepts are closely related to the theory of weighted spaces of Sobolev type, in particular the integration error in those spaces obeys a Koksma–Hlawka-type upper bound, which may be phrased using the norm of the function and the weighted star discrepancy.

By weights, we understand a set of nonnegative numbers $\gamma = (\gamma_u)_{u \in [d]\setminus\emptyset}$, where γ_u is interpreted as the weight of the coordinates from u. Let $|u|$ denote the cardinality of u. For $x \in [0,1]^d$, we write $(x(u),1)$ to denote the point in $[0,1]^d$ agreeing with x on the coordinates from u and having all the other coordinates set to 1.

The weighted star discrepancy of a point set $X = (x_1, \ldots, x_N)$ and weights γ is defined by

$$D^*_{N,\gamma}(X) := \sup_{z \in [0,1]^d} \max_{u \in [d]\setminus\emptyset} \gamma_u \big| D_N(X, (z(u),1)) \big|.$$

The following theorem is similar in flavor to Theorem 1 from [1].

Theorem 3.11. *Let $N, d \in \mathbb{N}$ and let $\mathcal{P} = (p_j)_{j=1}^N \subset [0,1)^d$ be a sampling scheme, such that for every $\emptyset \neq u \subset [d]$ its projection on the coordinates in u is $\mathcal{D}_0^{|u|}$–$e^{\rho|u|}$-negatively dependent. Then for any weights $(\gamma_u)_{u \subset [d]\setminus\emptyset}$ and any $c > 0$, it holds*

$$D^*_{N,\gamma}(\mathcal{P}) \leq \max_{\emptyset \neq u \subset [d]} c\gamma_u \sqrt{\frac{|u|}{N}} \tag{3.17}$$

with probability at least $2 - (1 + e^{-(1.6741c^2 - 10.7042 - \rho)})^d$. Moreover, for $\theta \in (0,1)$ it holds

$$\mathbf{P}\left(D^*_{N,\gamma}(\mathcal{P}) \leq \max_{\emptyset \neq u \subset [d]} \gamma_u \sqrt{\frac{|\rho + 10.7042 + \log((2-\theta)^{\frac{1}{d}} - 1)|}{1.6741}} \sqrt{\frac{|u|}{N}} \right) \geq \theta. \tag{3.18}$$

Proof. We shall only prove the statement (3.17), the statement (3.18) follows then by simple calculations. For $\emptyset \neq u \subset [d]$ and $c > 0$, put

$$A_u = \left\{ \omega \in \Omega : D^*_N(X^u(\omega)) > c\sqrt{\frac{|u|}{N}} \right\}.$$

Here, X^u denotes the projection of X on the coordinates from u. By Theorem 3.7, it holds

$$\mathbf{P}(A_u) < e^{-(1.674c^2 - 10.7042 - \rho)|u|}.$$

Now

$$\begin{aligned}
\mathbf{P}\left(D^*_{N,\gamma}(\mathcal{P}) > \max_{\emptyset \neq u \subset [d]} c\gamma_u \sqrt{\frac{|u|}{N}} \right) &\leq \mathbf{P}\left(\bigcup_{\emptyset \neq u \subset [d]} A_u \right) \\
&< \sum_{v=1}^{d} \binom{d}{v} e^{-(1.6741c^2 - 10.7042 - \rho)v} \\
&= (1 + e^{-(1.6741c^2 - 10.7042 - \rho)})^d - 1.
\end{aligned}$$

□

3.4 Examples of negatively dependent and pairwise negatively dependent sampling schemes

Many sampling schemes, such as randomly shifted and jittered rank-1 lattices (cf. Section 3.4.2) and Latin hypercube sampling (cf. Section 3.4.3), are multidimensional generalizations of the one-dimensional *simple stratified sampling*. Simple stratified sampling is defined in the following way: let π be a uniformly chosen permutation of $\{1,\dots,N\}$ and let $(U_j)_{j=1}^N$ be independent random variables distributed uniformly on $(0,1]$. Moreover, π is independent of $(U_j)_j$. We put

$$p_j := \frac{\pi(j) - U_j}{N}, \quad j = 1,\dots,N.$$

Effectively, one is considering the partition $I_j := [\frac{j-1}{N}, \frac{j}{N}), j = 1,\dots,N$, of the unit interval and in every element of the partition putting one point, independently of all the other points. The simple lemma is a useful tool for our investigations and may be found, for example, in [33].

Lemma 3.12. *Simple stratified sampling $\mathcal{P} = (p_j)_{j=1}^N$ is pairwise negatively dependent.*

3.4.1 Negative dependence of generalized stratified sampling

We partition $[0,1)^d$ into $\beta \geq N$ sets $(B_j)_{j=1}^\beta$ with $\lambda^d(B_j) = \frac{1}{\beta}, j = 1,\dots,\beta$. Let $Y = (Y_1,\dots,Y_\beta)$ be a random vector distributed uniformly on

$$\left\{(v_1,\dots,v_\beta) \in \{0,1\}^\beta : \sum_{j=1}^{\beta} v_j = N\right\}.$$

Given the value of Y, we place one point for each $j \in [\beta]$ with $Y_j = 1$ uniformly and independently of all other points inside B_j. Symmetrizing this construction yields a sampling scheme $\mathcal{P} = (p_j)_{j=1}^N$, which we call *generalized stratified sampling* (note that every single $p \in \mathcal{P}$ is uniformly distributed in $[0,1)^d$). Here, "generalized" has to be understood in the sense that there are possibly more strata then points.

Example 3.13. There are many natural choices for the strata. The simplest one would be stripes of the form $B_j, j = 1,\dots,N$, with $B_j := [\frac{j-1}{N}, \frac{j}{N}) \times [0,1)^{d-1}$. Alternatively, one may divide $[0,1)^d$ into $N = n^d$ cubes of equal size; see, for example, [29]. However, one could also choose, for example, elementary cells (i. e., fundamental parallelepipeds) of a rank-1 lattice (cf. [16]); see Figure 3.1.

To show that generalized stratified sampling is negatively dependent, we need first a simple lemma.

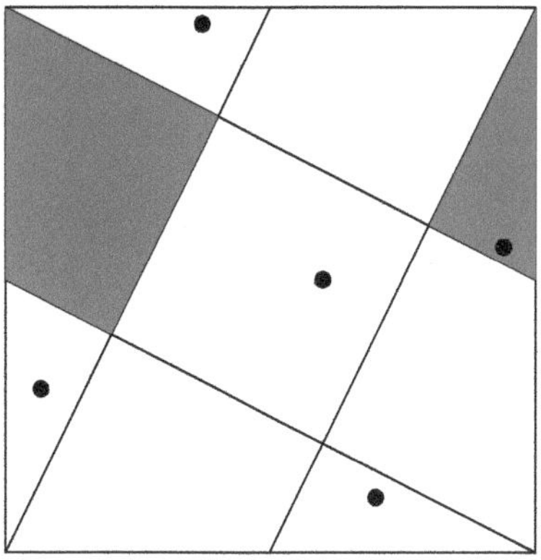

Figure 3.1: N elementary cells of a rank-1 lattice as strata, $\beta = N = 5$.

Lemma 3.14. *Let $t, N \in \mathbb{N}$, $t \le N$, $\xi \ge 0$, and let*

$$D = \left\{ x = (x_1, \dots, x_N) \in \mathbb{R}^N \,\middle|\, x \ge 0, \sum_{i=1}^{N} x_i = \xi \right\}.$$

The function

$$f : D \to \mathbb{R}, (x_1, \dots, x_N) \mapsto \sum_{J \subset [N], |J| = t} \prod_{j \in J} x_j$$

takes on its maximum in the point $(x_1, \dots, x_N) = (\frac{\xi}{N}, \dots, \frac{\xi}{N})$.

Proof. We shall prove the statement by induction on $N \ge t$. The case $N = t$ is straightforward by Lagrange multipliers theorem. Suppose we have already shown the statement for $N-1$ and we would like to prove it for N. First, let us fix the value of $x_N \in (0, \xi)$. It holds

$$\begin{aligned} \sum_{J \subset [N], |J| = t} \prod_{j \in J} x_j &= \sum_{J \subset [N], |J| = t, N \in J} \prod_{j \in J} x_j + \sum_{J \subset [N], |J| = t, N \notin J} \prod_{j \in J} x_j \\ &= x_N \sum_{J' \subset [N-1], |J'| = t-1} \prod_{j \in J'} x_j + \sum_{J' \subset [N-1], |J'| = t} \prod_{j \in J'} x_j. \end{aligned}$$

By the induction assumption for a fixed value of x_N, the last term is maximal when for $j = 1, \dots, N-1$ we have $x_j = \frac{\eta}{N-1}$, where we put $\eta = \xi - x_N$. Plugging it into the above formula, we obtain

$$\sum_{J \subset [N], |J| = t} \prod_{j \in J} x_j = (\xi - \eta) \binom{N-1}{t-1} \left(\frac{\eta}{N-1} \right)^{t-1} + \binom{N-1}{t} \left(\frac{\eta}{N-1} \right)^{t},$$

which we need to maximize with respect to η. It holds

$$\sum_{J \subset [N], |J| = t} \prod_{j \in J} x_j = Ch(\eta),$$

where $C = \frac{(N-1)!}{(t-1)!(N-t)!(N-1)^{t-1}}$ and $h(\eta) = \xi\eta^{t-1} + \left(\frac{N-t}{t(N-1)} - 1\right)\eta^t$. Now we have

$$h'(\eta) = \eta^{t-2}\left[(t-1)\xi + \left(\frac{N-t}{N-1} - t\right)\eta\right].$$

The derivative vanishes for $t \geq 3$ at $\eta_1 = 0$ and $\eta_2 = \frac{N-1}{N}\xi$. Since $h(\eta_2) > \max\{h(0), h(\xi)\}$ and η_2 is a local maximum, the claim follows. □

Theorem 3.15. *Let $\mathcal{P} = (p_j)_{j=1}^N$ be a generalized stratified sampling as described above and $A \subset [0,1)^d$ be measurable. Then for every $1 \leq t \leq N$, it holds*

$$\mathbf{P}\left(\bigcap_{j=1}^{t}\{p_j \in A\}\right) \leq \prod_{j=1}^{t}\mathbf{P}(p_j \in A).$$

In particular, generalized stratified sampling is $\mathcal{S}$-negatively dependent for any system $\mathcal{S}$ of measurable subsets of $[0,1)^d$.

Proof. Fix t as in the statement of the theorem and define

$$D_t = \{(k_1,\dots,k_t) \in [\beta]^t : \forall_{i,j\in[t]} \quad i \neq j \implies k_i \neq k_j\}.$$

Note that $|D_t| = \beta(\beta-1)\cdots(\beta-t+1)$. For $k = (k_1,\dots,k_t) \in D_t$, we have

$$\mathbf{P}\left(\bigcap_{j=1}^{t}\{Y_{k_j} = 1\}\right) = \frac{\binom{\beta-t}{N-t}}{\binom{\beta}{N}} = \frac{N(N-1)\cdots(N-t+1)}{\beta(\beta-1)\cdots(\beta-t+1)}.$$

By Lemma 3.14, it follows

$$\begin{aligned}
&\mathbf{P}\left(\bigcap_{j=1}^{t}\{p_j \in A\}\right)\\
&= \sum_{k\in D_t}\mathbf{P}\left(\bigcap_{j=1}^{t}p_j \in A \,\middle|\, \bigcap_{j=1}^{t}\{p_j \in B_{k_j}\}\right)\mathbf{P}\left(\bigcap_{j=1}^{t}\{p_j \in B_{k_j}\} \,\middle|\, \bigcap_{j=1}^{t}\{Y_{k_j} = 1\}\right)\mathbf{P}\left(\bigcap_{j=1}^{t}\{Y_{k_j} = 1\}\right)\\
&= \sum_{k\in D_t}\prod_{j=1}^{t}\frac{\lambda^d(A \cap B_{k_j})}{\lambda^d(B_{k_j})}\frac{1}{N(N-1)\cdots(N-t+1)}\frac{N(N-1)\cdots(N-t+1)}{\beta(\beta-1)\cdots(\beta-t+1)}\\
&= \frac{1}{\beta(\beta-1)\cdots(\beta-t+1)}\sum_{k\in D_t}\prod_{j=1}^{t}\frac{\lambda^d(A \cap B_{k_j})}{\lambda^d(B_{k_j})}\\
&\leq \frac{1}{\beta(\beta-1)\cdots(\beta-t+1)}\beta(\beta-1)\cdots(\beta-t+1)\left(\frac{\lambda^d(A)}{\beta}\right)^t\beta^t = (\lambda^d(A))^t.
\end{aligned}$$

□

Remark 3.16. Without further information on the strata, we cannot make any conclusions about pairwise negative dependence of generalized stratified sampling. As an

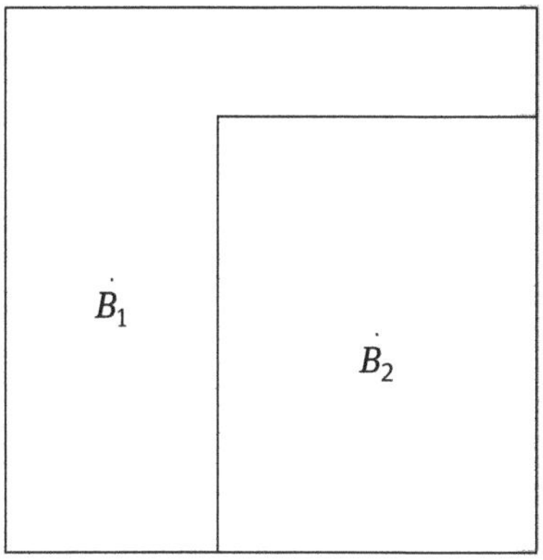

Figure 3.2: Example of strata for Remark 3.16.

example, consider a stratified sampling scheme $\mathcal{P} = (p_1, p_2)$ defined by two strata B_1, B_2 in $d \geq 2$. One may choose B_1, B_2, and $Q, R \in \mathcal{C}_1^d$ in such a way that $Q \subset B_1$, $B_2 \subset R$, and $R \neq [0,1)^d$; see Figure 3.2. In this case, however,

$$\mathbf{P}(p_2 \in R \mid p_1 \in Q) = 1,$$

and the sampling scheme is not pairwise negatively dependent.

On the other hand, if we consider strata $B_j, j = 1, \ldots, N$, with $B_j := [\frac{j-1}{N}, \frac{j}{N}) \times [0,1)^{d-1}$ then this practically boils down to the one-dimensional case and so the corresponding sampling scheme is pairwise negatively dependent; cf. Lemma 3.12.

3.4.2 Pairwise negative dependence and conditional NQD property of randomly shifted and jittered rank-1 lattices

The exposition follows closely [33]. Let N be prime. By $\mathbb{F} := \mathbb{F}_N$, we denote $\{0, 1, \ldots, N-1\}$. Moreover, $\mathbb{F}^* := \mathbb{F} \setminus \{0\}$. We also put $\widetilde{\mathbb{F}} := \frac{1}{N}\mathbb{F}$, and similarly $\widetilde{\mathbb{F}}^* := \frac{1}{N}\mathbb{F}^*$.

A discrete subgroup $\mathcal{L}$ of the d-dimensional torus $\mathbb{T}^d$ is called a lattice. A set $(y_j)_{j=1}^N$ is a rank-1 lattice if for some $g \in (\widetilde{\mathbb{F}}^*)^d$ it admits a representation

$$y_j = (j-1)g \mod 1 \quad j = 1, \ldots, N.$$

In this case, g is called a generating vector of the lattice.

Note that our definition differs from the usual one in that we allow only for generating vectors g from $(\widetilde{\mathbb{F}}^*)^d$ and not from $\widetilde{\mathbb{F}}^d$, which saves us from considering some degenerate cases.

We want now to define a sampling scheme based on rank-1 lattices which we call randomly shifted and jittered rank-1 lattice. To this end, let $(y_j)_{j=1}^N$ be a rank-1 lattice with generating vector chosen randomly uniformly from $(\widetilde{\mathbb{F}}^*)^d$. Let U be distributed uniformly on $\widetilde{\mathbb{F}}^d$, J_j, $j = 1, \dots, N$ be uniformly distributed on $[0, \frac{1}{N})^d$ and π be a uniformly chosen permutation of $\{1, \dots, N\}$. Moreover, let all of the aforementioned random variables be independent. We put

$$p_j := y_{\pi(j)} + U + J_j \bmod 1, \quad j = 1, \dots, N.$$

We call the sampling scheme $\mathcal{P} = (p_j)_{j=1}^N$ a *randomly shifted and jittered rank-1 lattice* (*RSJ rank-1 lattice*). Putting it in words: we first take a rank-1 lattice with a random generator and symmetrize it. Then we shift the lattice uniformly on the torus, where the shift has resolution $\frac{1}{N}$. In the last step, we jitter every point independently of all the other points in a cube of volume $(\frac{1}{N})^d$.

The following is Theorem 3.4 from [33].

Theorem 3.17. *Let N be prime, $d \in \mathbb{N}$. RSJ rank-1 lattice $\mathcal{P} = (p_j)_{j=1}^N$ in $[0, 1)^d$ is a coordinatewise independent NQD sampling scheme.*

In particular, RSJ rank-1 lattice is a pairwise negatively dependent and a conditionally NQD sampling scheme, which means that both alternative conditions for $\mathcal{P}$ from Theorem 3.5 hold.

In contrast to generalized stratified sampling (cf. Theorem 3.15) and Latin hypercube sampling (see Theorem 3.18), RSJ rank-1 lattice is for $d \geq 2$ and $N \geq 3$ in general not $\mathcal{C}_0^d$-negatively dependent; see Subsection 3.5.1.

3.4.3 Negative dependence, conditional NQD property, and pairwise negative dependence of Latin hypercube sampling

Let $(\pi_i)_{i=1}^d$ be independent uniformly chosen permutations of $[N]$, and $U_j^{(i)}$, $i = 1, \dots, d$, $j = 1, \dots, N$ be independent random variables distributed uniformly on $(0, 1]$ and independent also of the permutations. A sampling scheme $(p_j)_{j=1}^N$ is called a *Latin hypercube sampling* if the ith coordinate of the jth point $p_j^{(i)}$ is given by

$$p_j^{(i)} = \frac{\pi_i(j) - U_j^{(i)}}{N}, \quad i = 1, \dots, d, j = 1, \dots, N.$$

What one intuitively does is the following: one cuts $[0, 1)^d$ into slices $(S_{k,j})_{j=1}^N$, $k = 1, \dots, d$ given by

$$S_{k,j} = \prod_{j=1}^{k-1} [0, 1) \times \left[\frac{j-1}{N}, \frac{j}{N}\right) \times \prod_{j=k+1}^{d} [0, 1)$$

and puts N points in such a way that in every slice there is exactly one point.

It is worth mentioning that for $d = 1$ Latin hypercube sampling is exactly the same as RSJ rank-1 lattice (namely simple stratified sampling). For $d \geq 2$, the joint distribution of a pair of points is the same for Latin hypercube sampling as for the RSJ rank-1 lattice. But if we sample more than two points, then the joint distributions already differ; see [33].

Negative dependence of Latin hypercube sample has been studied in [11] and pairwise negative dependence has been investigated in [33].

Theorem 3.18. *Latin hypercube sample in* $[0,1)^d$ *is a sampling scheme which is:*
(i) $\mathcal{D}_0^d$–e^d*-negatively dependent,*
(ii) $\mathcal{C}_0^d$*-negatively dependent,*
(iii) *coordinatewise independent NQD.*

In the above, statements (i) and (ii) follow from Theorem 3.5. from [11], and statement (iii) is Theorem 3.4. from [33].

In particular, from (iii) it follows that LHS is pairwise negatively dependent as well as conditionally NQD.

3.4.4 Pairwise negative dependence of scrambled (0,m,s)-nets

The so-called (t, m, s)-nets belong to the most regular deterministic point sets. First defined by Niederreiter in [22], they have been subject of extensive research. For a nice introduction on (t, m, s)-nets and their randomization, see [21].

Let us fix a base $b \in \mathbb{N}_{\geq 2}$. For $j \in \mathbb{N}_0$ and $k = 0, 1, \ldots, b^j - 1$ an interval of the form

$$E_k^j = [kb^{-j}, (k+1)b^{-j})$$

is called an *elementary interval (in base b)*. Moreover, for $s \in \mathbb{N}$ and vectors $\mathbf{j} = (j_1, \ldots, j_s)$ and $\mathbf{k} = (k_1, \ldots, k_s)$ (where for every $l = 1, \ldots, s$, we require $0 \leq k_l \leq b^{j_l} - 1$) we define an s-dimensional elementary interval via

$$E_{\mathbf{k}}^{\mathbf{j}} := \prod_{l=1}^{s} E_{k_l}^{j_l}.$$

A (t, m, s)*-net* is any $P \subset [0,1]^s$ such that for any elementary interval E with $\lambda^s(E) = b^{-m+t}$ there are exactly b^t points in $P \cap E$. It is easily seen that a (t, m, s)-net consists of exactly b^m points. Specific constructions of (t, m, s)-nets are known.

Scrambling of depth m is a bijective function $S : [0,1]^s \to [0,1]^s$ such that for any elementary interval E with $\lambda^s(E) = b^{-m}$ the image $S(E)$ is again an elementary interval of volume b^{-m}.

Now let us focus on the case $t = 0$. Taking a $(0, m, s)$-net and applying to it a random scrambling of depth m, one obtains a randomized point set. Scramblings are defined in such a way that for any scrambling S of depth m and a $(0, m, s)$-net P, the point set $S(P)$ is again a $(0, m, s)$-net. By an appropriate choice of randomized scrambling $\tilde{S}$, one may make $\tilde{S}(P)$ to be a sampling scheme. In this case, we call $\tilde{S}(P)$ a scrambled $(0, m, s)$-net. Scrambling as a way of randomization of $(0, m, s)$-nets has been studied by A. B. Owen, for example, in [28].

In a recent article [20], C. Lemieux and J. Wiart have shown the following theorem (which follows from Corollary 4.10 from the aforementioned article).

Theorem 3.19. *Scrambled* $(0, m, s)$*-nets are pairwise negatively dependent sampling schemes.*

3.4.5 Mixed randomized sequences

As already mentioned, part of the success of RQMC stems from the fact that in many high-dimensional practical integration problems only a small number of coordinates is of real importance. It stands to reason that one tries to use it to his avail by constructing quadratures in which one uses RQMC on the "important" coordinates and simple (usually much cheaper) Monte Carlo for the rest of the coordinates. This method is sometimes referred to as padding and the resulting sequences of integration nodes are called hybrid-Monte Carlo sequences. Let us give a formal definition.

Definition 3.20. Let $d, d', d'' \in \mathbb{N}$ with $d = d' + d''$. Let $X = (X_k)_{k\in\mathbb{N}}$ be a sequence in $[0,1)^{d'}$, and let $Y = (Y_k)_{k\in\mathbb{N}}$ be a sequence in $[0,1)^{d''}$. The d-dimensional concatenated sequence $Z = (Z_k)_{k\in\mathbb{N}} = (X_k, Y_k)_{k\in\mathbb{N}}$ is called a *mixed sequence*. If Y is a sequence of independent uniformly distributed random points, one also says that Z results from X by *padding by Monte Carlo* and calls Z a *hybrid-Monte Carlo sequence*. If X and Y are both randomized sequences, we call Z a *mixed randomized sequence*.

Padding by Monte Carlo was introduced by Spanier in [32] to tackle problems in particle transport theory. He suggested to use a hybrid-Monte Carlo sequence resulting from padding a deterministic low-discrepancy sequence. Hybrid-Monte Carlo sequences showed a favorable performance in several numerical experiments; see, for example, [25, 26]. The latter papers also provided theoretical results on probabilistic discrepancy estimates of hybrid-Monte Carlo sequences which have been improved in [2, 10]. Favorable discrepancy bounds for padding Latin hypercube sampling (LHS) by Monte Carlo were provided in [11]. Padding a sequence by LHS (instead of by Monte Carlo) was considered earlier by Owen [27, Example 5].

A related line of research, initiated in [24], is to study the discrepancy of concatenated sequences that result from two deterministic sequences. More recent results can, for example, be found in [12, 8, 14] and the literature mentioned therein.

The following proposition shows that concatenating two mutually independent negatively dependent sampling schemes results again in a (higher dimensional) negatively dependent sampling scheme. A weaker version of the next proposition may be found in [13]; cf. Lemma 5 there.

Proposition 3.21. *Let $d, d', d'' \in \mathbb{N}$ such that $d = d' + d''$. Let $A \subseteq [0,1)^{d'}$, $B \subseteq [0,1)^{d''}$ be Borel measurable sets. Let $x_1, \dots, x_N$ be a sampling scheme in $[0,1)^{d'}$ and $y_1, \dots, y_N$ a sampling scheme in $[0,1)^{d''}$. Furthermore, let $\alpha, \beta \geq 1$.*

(i) *If the random variables $\mathbf{1}_A(x_i)$, $i = 1, \dots, N$, and $\mathbf{1}_B(y_i)$, $i = 1, \dots, N$, are upper negatively α- and β-dependent, respectively, and mutually independent, then the random variables $\mathbf{1}_{A\times B}(x_i, y_i)$, $i = 1, \dots, N$, induced by the random vectors $(x_1, y_1), \dots, (x_N, y_N)$ in $[0,1)^d$, are upper negatively $\alpha\beta$-dependent.*

(ii) *If the random variables $\mathbf{1}_A(x_i)$, $i = 1, \dots, N$, and $\mathbf{1}_B(y_i)$, $i = 1, \dots, N$, are lower negatively α- and β-dependent, respectively, and mutually independent, then the random variables $\mathbf{1}_{A\times B}(x_i, y_i)$, $i = 1, \dots, N$, induced by the random vectors $(x_1, y_1), \dots, (x_N, y_N)$ in $[0,1)^d$, are lower negatively $\alpha\beta$-dependent.*

Proof. Let us first prove statement (i). Obviously, we have for $J \subseteq [N]$,

$$\begin{aligned}
\mathbf{P}\Big(\bigcap_{j\in J}\{\mathbf{1}_{A\times B}(x_j, y_j) = 1\}\Big) &= \mathbf{P}\Big(\bigcap_{j\in J}\{x_j \in A\} \cap \bigcap_{j\in J}\{y_j \in B\}\Big) \\
&= \mathbf{P}\Big(\bigcap_{j\in J}\{x_j \in A\}\Big)\mathbf{P}\Big(\bigcap_{j\in J}\{y_j \in B\}\Big) \leq \Big(\alpha\prod_{j\in J}\mathbf{P}(x_j \in A)\Big)\Big(\beta\prod_{j\in J}\mathbf{P}(y_j \in B)\Big) \\
&= \alpha\beta\prod_{j\in J}\mathbf{P}(\mathbf{1}_{A\times B}(x_j, y_j) = 1).
\end{aligned}$$

We now prove statement (ii). Take any $\emptyset \neq J \subseteq [N]$ and set $t = |J|$. Suppose first that $((x_j, y_j))_{j=1}^N$ is a hybrid-Monte Carlo sequence, that is, $(y_j)_{j=1}^N$ is a Monte Carlo sampling scheme. Due to our assumptions in statement (ii), we obtain

$$\begin{aligned}
&\mathbf{P}\Big(\bigcap_{j\in J}\{\mathbf{1}_{A\times B}(x_j, y_j) = 0\}\Big) \\
&= \sum_{K\subseteq J}\mathbf{P}\Big(\bigcap_{j\in J}\{\mathbf{1}_{A\times B}(x_j, y_j) = 0\} \cap \bigcap_{j\in K}\{\mathbf{1}_B(y_j) = 1\} \cap \bigcap_{j\in J\setminus K}\{\mathbf{1}_B(y_j) = 0\}\Big) \\
&= \sum_{v=0}^{t}\binom{t}{v}\mathbf{P}\Big(\bigcap_{j=1}^{v}\{\mathbf{1}_A(x_j) = 0\}\Big)\mathbf{P}\Big(\bigcap_{j=1}^{v}\{\mathbf{1}_B(y_j) = 1\} \cap \bigcap_{j=v+1}^{t}\{\mathbf{1}_B(y_j) = 0\}\Big) \\
&\leq \alpha\sum_{v=0}^{t}\binom{t}{v}\mathbf{P}(\mathbf{1}_A(x_1) = 0)^v\,\mathbf{P}(\mathbf{1}_B(y_1) = 1)^v\,\mathbf{P}(\mathbf{1}_B(y_1) = 0)^{t-v} \\
&= \alpha[\mathbf{P}(\mathbf{1}_A(x_1) = 0)\,\mathbf{P}(\mathbf{1}_B(y_1) = 1) + \mathbf{P}(\mathbf{1}_B(y_1) = 0)]^t = \alpha\,\mathbf{P}(\mathbf{1}_{A\times B}(x_1, y_1) = 0)^t.
\end{aligned}$$

Now let $(y_j)_{j=1}^N$ be any sampling scheme in $[0,1)^{d''}$ such that the random variables $(\mathbf{1}_B(y_j))_{j=1}^N$ are lower β-negatively dependent and let $(\hat{y}_j)_{j=1}^N$ be a Monte Carlo sampling scheme in $[0,1)^{d''}$; we assume both sampling schemes to be mutually independent to $(x_j)_{j=1}^N$. Analogously, as in the previous case, we obtain

$$\begin{aligned}
&\mathbf{P}\Big(\bigcap_{j\in J}\{\mathbf{1}_{A\times B}(x_j,y_j)=0\}\Big)\\
&=\sum_{\nu=0}^{t}\binom{t}{\nu}\mathbf{P}\Big(\bigcap_{j=1}^{\nu}\{\mathbf{1}_A(x_j)=1\}\cap\bigcap_{j=\nu+1}^{t}\{\mathbf{1}_A(x_j)=0\}\Big)\mathbf{P}\Big(\bigcap_{j=1}^{\nu}\{\mathbf{1}_B(y_j)=0\}\Big)\\
&\le\sum_{\nu=0}^{t}\binom{t}{\nu}\mathbf{P}\Big(\bigcap_{j=1}^{\nu}\{\mathbf{1}_A(x_j)=1\}\cap\bigcap_{j=\nu+1}^{t}\{\mathbf{1}_A(x_j)=0\}\Big)\beta\,\mathbf{P}(\mathbf{1}_B(y_1)=0)^{\nu}\\
&=\beta\sum_{\nu=0}^{t}\binom{t}{\nu}\mathbf{P}\Big(\bigcap_{j=1}^{\nu}\{\mathbf{1}_A(x_j)=1\}\cap\bigcap_{j=\nu+1}^{t}\{\mathbf{1}_A(x_j)=0\}\Big)\mathbf{P}(\mathbf{1}_B(\hat{y}_1)=0)^{\nu}.
\end{aligned}$$

It follows from the case of hybrid-Monte Carlo sequences that

$$\begin{aligned}
\mathbf{P}\Big(\bigcap_{j\in J}\{\mathbf{1}_{A\times B}(x_j,y_j)=0\}\Big)&\le\beta\,\mathbf{P}\Big(\bigcap_{j\in J}\{\mathbf{1}_{A\times B}(x_j,\hat{y}_j)=0\}\Big)\\
&\le\alpha\beta\,\mathbf{P}(\mathbf{1}_{A\times B}(x_1,\hat{y}_1)=0)^t=\alpha\beta\,\mathbf{P}(\mathbf{1}_{A\times B}(x_1,y_1)=0)^t
\end{aligned}$$ □

Remark 3.22. It follows easily on closer examination of the proof that for the statement (i) of Proposition 3.21 to hold true we need only $(\mathbf{1}_A(x_j))_{j=1}^N$ and $(\mathbf{1}_B(y_j))_{j=1}^N$ to be negatively α- respectively β-upper dependent point sets, not necessarily sampling schemes. Moreover, if in (ii) we assume that $(y_j)_{j=1}^N$ is a Monte Carlo sampling scheme, we also do not need to assume that $(x_j)_{j=1}^N$ is a sampling scheme.

Remark 3.23. Let $\mathcal{S}'$, $\mathcal{S}''$ be systems of measurable sets in $[0,1)^{d'}$ and $[0,1)^{d''}$, respectively. Let $(x_j)_{j=1}^N$ be an $\mathcal{S}'$-α-negative dependent sampling scheme in $[0,1)^{d'}$ and $(y_j)_{j=1}^N$ an $\mathcal{S}''$-β-negative dependent sampling scheme in $[0,1)^{d''}$; both sampling schemes should be mutually independent. Furthermore, let $\mathcal{P} := (p_j)_{j=1}^N$ be the resulting concatenated sampling scheme in $[0,1)^d$, that is, $p_i := (x_i,y_i)$, $i=1,\dots,N$.

(i) If $\mathcal{S}'=\mathcal{C}_0^{d'}$ and $\mathcal{S}''=\mathcal{C}_0^{d''}$, we obtain from Proposition 3.21 that the mixed randomized sequence $(p_j)_{j=1}^N$ is $\mathcal{C}_0^d$-$\alpha\beta$-negatively dependent, which implies that we may directly apply Theorem 3.8 to obtain a probabilistic discrepancy bound for $\mathcal{P}$.

(ii) If $\mathcal{S}'=\mathcal{D}_0^{d'}$ and $\mathcal{S}''=\mathcal{D}_0^{d''}$, we obtain from Proposition 3.21 that $(p_j)_{j=1}^N$ is $\alpha\beta$-negatively dependent with respect to the set system

$$\mathcal{D}_0^{d'}\times\mathcal{D}_0^{d''}:=\{D'\times D''\mid D'\in\mathcal{D}_0^{d'},D''\in\mathcal{D}_0^{d''}\}\neq\mathcal{D}_0^{d}.$$

Hence Theorem 3.7 is unfortunately not directly applicable to $\mathcal{P}$. Nevertheless, one may prove a counterpart of Theorem 3.7 with slightly worse constants that relies on negative dependence with respect to $\mathcal{D}_0^{d'}\times\mathcal{D}_0^{d''}$. Namely, one may show for every

$\theta \in (0,1)$ that

$$\mathbf{P}\left(D_N^*(\mathcal{P}) \le 2 * 0.7729\sqrt{10.7042 + \rho + \frac{\ln((1-\theta)^{-1})}{d}}\sqrt{\frac{d}{N}} \right) \ge \theta. \tag{3.19}$$

The bound is based on the following simple observation: To estimate the local discrepancy of $\mathcal{P}$ in a test box $Q \in \mathcal{C}_0^d$, the strategy used in [11] (and earlier in [1]) is to decompose Q into finitely many disjoint differences of boxes $\Delta_1, \dots, \Delta_K \in \mathcal{D}_0^d$ such that $Q = \bigcup_{\nu=1}^K \Delta_\nu$. This gives

$$D_N(\mathcal{P}, Q) \le \sum_{\nu=1}^K D_N(\mathcal{P}, \Delta_\nu). \tag{3.20}$$

Now let us consider a fixed index ν. Then we find $A_\nu, B_\nu \in \mathcal{C}_0^d$ such that $A_\nu \subseteq B_\nu$ and $\Delta_\nu = B_\nu \setminus A_\nu$. Furthermore, we may write $A_\nu = A'_\nu \times A''_\nu$ and $B_\nu = B'_\nu \times B''_\nu$ with $A'_\nu, B'_\nu \in \mathcal{C}_0^{d'}$ and $A''_\nu, B''_\nu \in \mathcal{C}_0^{d''}$. Then we may represent Δ_ν as disjoint union

$$\Delta_\nu = (B'_\nu \setminus A'_\nu) \times B''_\nu \cup A'_\nu \times (B''_\nu \setminus A''_\nu) =: C_\nu^1 \cup C_\nu^2.$$

Thus

$$D_N(\mathcal{P}, \Delta_\nu) \le D_N(\mathcal{P}, C_\nu^1) + D_N(\mathcal{P}, C_\nu^2), \tag{3.21}$$

where $C_\nu^1, C_\nu^2 \in \mathcal{D}_0^{d'} \times \mathcal{D}_0^{d''}$. Now large deviation inequalities of Bernstein and Hoeffding-type can be used to obtain for each of the random variables $D_N(\mathcal{P}, C_\nu^1)$, $D_N(\mathcal{P}, C_\nu^2)$ the same upper bound as for the local discrepancy $D_N(\mathcal{P}^*, \Delta_\nu)$ of a $\mathcal{D}_0^d$-$\alpha\beta$-negative dependent sampling scheme $\mathcal{P}^*$ in the proof of [11, Theorem 4.3]. This, combined with (3.20) and (3.21), results in a probabilistic discrepancy bound for $D_N^*(\mathcal{P})$ that is as most as twice as big as the one from Theorem 3.7; for further details, see [11, Proof of Theorem 4.3].

3.5 Relations between notions of negative dependence

It may be easily seen that the coordinatewise independent NQD property implies the pairwise negative dependence property as well as the conditional NQD property. It turns out that this is the only valid implication between the considered notions of negative dependence. In this section, we give examples showing that other implications do not hold.

3.5.1 Pairwise negative dependence and negative dependence

Neither the pairwise negative dependence of a sampling scheme implies the negative dependence, nor the other way around.

Example 3.24. We first show an example of a negatively dependent sampling scheme which is not pairwise negatively dependent. To this end, consider a sampling scheme consisting of just two points (p_1, p_2) with joint CDF $F : [0,1]^2 \to [0,1]$ given by

$$F(x,y) = \min\left\{x, y, \frac{x^2+y^2}{2}\right\}.$$

It is easy to see that $F(0,0) = 0$, $F(1,1) = 1$, F is continuous, quasi-monotone, and $F(x,y) = F(y,x)$, which implies that F is a CDF of a sampling scheme. Moreover,

$$\mathbf{P}(p_1 \in [0,q), p_2 \in [0,q)) = F(q,q) = q^2,$$

so the sampling scheme is $\mathcal{C}_0^1$-negatively dependent. Notice that due to $d = 1$, it is equivalent to saying that the sampling scheme is $\mathcal{C}_1^1$-negatively dependent. However, for instance

$$\begin{aligned}\mathbf{P}\Big(p_1 \in \Big[\frac{3}{4},1\Big), p_2 \in \Big[\frac{1}{4},1\Big)\Big) &= 1 - \Big(F\Big(\frac{3}{4},1\Big) + F\Big(1,\frac{1}{4}\Big) - F\Big(\frac{3}{4},\frac{1}{4}\Big)\Big)\\ &= F\Big(\frac{3}{4},\frac{1}{4}\Big) = \frac{1}{4} > \Big(1-\frac{3}{4}\Big)\Big(1-\frac{1}{4}\Big) = \mathbf{P}\Big(p_1 \in \Big[\frac{3}{4},1\Big)\Big)\mathbf{P}\Big(p_2 \in \Big[\frac{1}{4},1\Big)\Big).\end{aligned}$$

Example 3.25. To see that even the stronger coordinatewise independent NQD property does not imply the negative dependence property, consider RSJ rank-1 lattice defined in Subsection 3.4.2. On the one hand, according to Theorem 3.17, RSJ rank-1 lattice is coordinatewise independent NQD. On the other hand, let us consider the situation for $d = 2$, and a large N to be chosen later. We put $Q = [0, \frac{3}{N})^2$. Obviously,

$$\mathbf{P}(p_1 \in Q)^3 = \left(\frac{3}{N}\right)^6.$$

We also have

$$\mathbf{P}(p_1 \in Q, p_2 \in Q, p_3 \in Q) \geq \frac{1}{\binom{N}{3}N(N-1)} = \frac{6}{N^2(N \quad 1)^2(N \quad 2)},$$

the inequality follows since for the diagonal configuration of the points (i. e., $p_j = (\frac{\pi(j)}{N}, \frac{\pi(j)}{N}) + J_j$, $j = 1, \dots, n$ for some permutation π of $\{1, \dots, N\}$, $k \in [N-1]$) there is one triple of points always lying in Q. Notice that any generating vector of the form $g = (\frac{k}{N}, \frac{k}{N})$ and any shift of the form $S = (\frac{l}{N}, \frac{l}{N})$, $l \in \{0, 1, \dots, N-1\}$, results in a diagonal configuration. Now for N large enough it holds

$$\mathbf{P}(p_1 \in Q, p_2 \in Q, p_3 \in Q) > \mathbf{P}(p_1 \in Q)^3.$$

3.5.2 Conditional NQD and pairwise negative dependence

Example 3.26. First, we show an example of a pairwise negatively dependent sampling scheme which is not conditionally NQD. Let $B_1 = [0, \frac{1}{2})^2$, $B_2 = [\frac{1}{2}, 1) \times [0, \frac{1}{2})$, $B_3 = [0, \frac{1}{2}) \times [\frac{1}{2}, 1)$, $B_4 = [\frac{1}{2}, 1)^2$ denote the slots. Now we are considering a sampling scheme $\mathcal{P} = (p_1, p_2)$ such that given the slots the points are distributed uniformly within the slots and are independent. Denote $A_{ij} := \{p_1 \in B_i, p_2 \in B_j\}$ and set

$$\begin{aligned} \mathbf{P}(A_{ii}) &= \frac{1}{16}, \quad i = 1, 2, 3, 4, \\ \mathbf{P}(A_{13}) = \mathbf{P}(A_{24}) = \mathbf{P}(A_{31}) = \mathbf{P}(A_{42}) &= \frac{1}{32}, \\ \mathbf{P}(A_{14}) = \mathbf{P}(A_{23}) = \mathbf{P}(A_{41}) = \mathbf{P}(A_{32}) &= \frac{5}{32}. \end{aligned}$$

It is easy to see that $\mathcal{P}$ is not conditionally NQD, for example,

$$\begin{aligned} &\mathbf{P}\Big(p_1^{(2)} \ge \frac{1}{2}, p_2^{(2)} \ge \frac{1}{2} \,\Big|\, p_1^{(1)} \ge \frac{1}{2}, p_2^{(1)} \ge \frac{1}{2}\Big) = \frac{1}{3} \\ &> \frac{1}{4} = \mathbf{P}\Big(p_1^{(2)} \ge \frac{1}{2} \,\Big|\, p_1^{(1)} \ge \frac{1}{2}, p_2^{(1)} \ge \frac{1}{2}\Big) \mathbf{P}\Big(p_2^{(2)} \ge \frac{1}{2} \,\Big|\, p_1^{(1)} \ge \frac{1}{2}, p_2^{(1)} \ge \frac{1}{2}\Big). \end{aligned}$$

Showing that $\mathcal{P}$ is pairwise negatively dependent requires simple but tedious calculations and as such will be omitted. Intuitively, it is clear, since the sampling scheme gives high probability to diagonal arrangements (i. e., $A_{14}, A_{23}, A_{41}, A_{32}$).

Example 3.27. Now we show an example of a sampling scheme which is conditionally NQD but not pairwise negatively dependent. To this end, let X, Y be two independent random variables distributed uniformly on $[0, 1)$. We consider a sampling scheme $\mathcal{P} = (p_1, p_2)$ given by $p_1 = (X, Y)$, $p_2 = (Y, X)$. Let $u, v \in [0, 1)^2$ and $A, B \subset [0, 1)$ be measurable. Sampling scheme $\mathcal{P}$ is conditionally NQD since

$$\begin{aligned} &\mathbf{P}(p_1^{(2)} \ge u^{(2)}, p_2^{(2)} \ge v^{(2)} \,|\, p_1^{(1)} \in A, p_2^{(1)} \in B) \\ &= \mathbf{P}(Y \ge u^{(2)}, X \ge v^{(2)} \,|\, X \in A, Y \in B) \\ &= \mathbf{P}(Y \ge u^{(2)} \,|\, X \in A, Y \in B)\, \mathbf{P}(X \ge v^{(2)} \,|\, X \in A, Y \in B) \\ &= \mathbf{P}(p_1^{(2)} \ge u^{(2)} \,|\, X \in A, Y \in B)\, \mathbf{P}(p_2^{(2)} \ge v^{(2)} \,|\, X \in A, Y \in B). \end{aligned}$$

On the other hand, $\mathcal{P}$ is not pairwise negatively dependent. To see this, note that

$$\begin{aligned} &\mathbf{P}(p_1 \ge u)\, \mathbf{P}(p_2 \ge v) \\ &= \mathbf{P}(X \ge u^{(1)}, Y \ge u^{(2)})\, \mathbf{P}(Y \ge v^{(1)}, X \ge v^{(2)}) \\ &= \mathbf{P}(X \ge u^{(1)})\, \mathbf{P}(Y \ge u^{(2)})\, \mathbf{P}(Y \ge v^{(1)})\, \mathbf{P}(X \ge v^{(2)}) \end{aligned}$$

and

$$\mathbf{P}(p_1 \ge u, p_2 \ge v) = \mathbf{P}(X \ge \max(u^{(1)}, v^{(2)}), Y \ge \max(u^{(2)}, v^{(1)})).$$

Taking for some $u^{(1)}, u^{(2)} \in (0,1)$, the point v satisfying $v^{(1)} = u^{(2)}$ and $v^{(2)} = u^{(1)}$ yields the claim.

Bibliography

[1] C. Aistleitner. Tractability results for the weighted star-discrepancy. J. Complex. **30**, 381–391 (2014).

[2] C. Aistleitner, M. T. Hofer. Probabilistic error bounds for the discrepancy of mixed sequences. Monte Carlo Methods Appl. **18**, 181–200 (2012).

[3] H. W. Block, T. H. Savits, M. Shaked. Some concepts of negative dependence. Ann. Probab. **10**, 765–772 (1982).

[4] J. Dick, F. Y. Kuo, I. H. Sloan. High dimensional integration – the quasi-Monte Carlo way. Acta Numer. **22**, 133–288 (2013).

[5] J. Dick, F. Pillichshammer. Digital Nets and Sequences. Cambridge University Press, Cambridge (2010).

[6] B. Doerr, C. Doerr, M. Gnewuch. Probabilistic lower discrepancy bounds for Latin hypercube samples. In: J. Dick, F. Y. Kuo, H. Woźniakowski (eds.) Contemporary Computational Mathematics – a Celebration of the 80th Birthday of Ian Sloan, pp. 339–350. Springer-Verlag (2018).

[7] B. Doerr, M. Gnewuch, A. Srivastav. Bounds and constructions for the star discrepancy via δ-covers. J. Complex. **21**, 691–709 (2005).

[8] M. Drmota, R. Hofer, G. Larcher. On the discrepancy of Halton-Kronecker sequences. In: Number Theory–Diophantine Problems, Uniform Distribution and Applications–Festschrift in Honour of Robert F. Tichy's 60th Birthday, pp. 219–226 (2017).

[9] M. Gnewuch. Bracketing numbers for axis-parallel boxes and applications to geometric discrepancy. J. Complex. **24**, 154–172 (2008).

[10] M. Gnewuch. On probabilistic results for the discrepancy of a hybrid-Monte Carlo sequence. J. Complex. **25**, 312–317 (2008).

[11] M. Gnewuch, N. Hebbinghaus. Discrepancy bounds for a class of negatively dependent random points including Latin hypercube samples (2018). Preprint (submitted).

[12] D. Gomez-Perez, R. Hofer, H. Niederreiter. A general discrepancy bound for hybrid sequences involving Halton sequences. Unif. Distrib. Theory **8**, 31–45 (2013).

[13] N. Hebbinghaus. Mixed sequences and application to multilevel algorithms. Master's thesis, Christ Church, University of Oxford (2012).

[14] R. Hofer. Kronecker-Halton sequences in $\mathbb{F}_p((x^{-1}))$. Finite Fields Appl. **50**, 154–177 (2018).

[15] K. Joag-Dev, F. Proschan. Negative association of random variables, with applications. Ann. Stat. **11**, 286–295 (1983).

[16] P. L'Ecuyer, C. Lemieux. Variance reduction via lattice rules. Manag. Sci. **46**, 1214–1235 (2000).

[17] E. Lehmann. Some concepts of dependence. Ann. Math. Stat. **37**, 1137–1153 (1966).

[18] C. Lemieux. Monte Carlo and Quasi-Monte Carlo Sampling. Springer, New York (2009).

[19] C. Lemieux. Negative dependence, scrambled nets, and variance bounds. Math. Oper. Res. **43**, 228–251 (2018).

[20] P. Wiart, C. Lemieux. On the dependence structure of scrambled (t,m,s)-nets (2019). arXiv:1903.09877. Preprint.

[21] J. Matoušek. Geometric Discrepancy. Springer-Verlag, Berlin Heidelberg (2010).

[22] H. Niederreiter. Point sets and sequences wit small discrepancy. Monatshefte Math. **104**, 273–337 (1987).

[23] H. Niederreiter. Random Number Generation and Quasi-Monte Carlo Methods. CBMS-NSF Regional Conference Series in Applied Mathematics, vol. 63. Society for Industrial and Applied Mathematics (SIAM), Philadelphia (1992).
[24] H. Niederreiter. On the discrepancy of some hybrid sequences. Acta Arith. **138**, 373–398 (2009).
[25] G. Ökten. A probabilistic result on the discrepancy of a hybrid-Monte Carlo sequence and applications. Monte Carlo Methods Appl. **2**, 250–270 (1996).
[26] G. Ökten, B. Tuffin, V. Burago. A central limit theorem and improved error bounds for a hybrid-Monte Carlo sequence with applications in computational finance. J. Complex. **22**, 435–458 (2006).
[27] A. B. Owen. Lattice sampling revisited: Monte Carlo variance of means over randomized orthogonal arrays. Ann. Stat. **22**, 930–945 (1994).
[28] A. B. Owen. Monte Carlo variance of scrambled net quadrature. SIAM J. Numer. Anal. **34** (1997).
[29] F. Pausinger, S. Steinerberger. On the discrepancy of jittered sampling. J. Complex. **33**, 199–216 (2016).
[30] R. Pemantle. Towards a theory of negative dependence. J. Math. Phys. **41**, 1371–1390 (2000).
[31] I. H. Sloan, H. Woźniakowski. When are quasi-Monte Carlo algorithms efficient for high dimensional integrals? J. Complex. **14**, 1–33 (1998).
[32] J. Spanier. Quasi-Monte Carlo methods for particle transport problems. In: H. Niederreiter, P. J.-S. Shiue (eds.) Monte Carlo and Quasi-Monte Carlo Methods in Scientific Computing, pp. 121–148. Springer-Verlag, Berlin (1995).
[33] M. Wnuk, M. Gnewuch. Note on pairwise negative dependence of randomized rank-1 lattices (2019). arXiv:1903.02261. Preprint (submitted).

Takashi Goda and Kosuke Suzuki

4 Recent advances in higher order quasi-Monte Carlo methods

Abstract: In this article, we review some of the recent results on higher order quasi-Monte Carlo (HoQMC) methods. After a seminal work by Dick (2007, 2008) who originally introduced the concept of HoQMC, there has been significant theoretical progress on HoQMC in terms of discrepancy as well as multivariate numerical integration. Moreover, several successful and promising applications of HoQMC to partial differential equations with random coefficients and Bayesian estimation/inversion problems have been reported recently. In this article, we start with standard quasi-Monte Carlo methods based on digital nets and sequences in the sense of Niederreiter, and then move onto their higher order version due to Dick. The Walsh analysis of smooth functions plays a crucial role in developing the theory of HoQMC, and the aim of this article is to provide a unified picture on how the Walsh analysis enables recent developments of HoQMC both for discrepancy and numerical integration.

Keywords: Higher order quasi-Monte Carlo, digital nets and sequences, Walsh analysis, discrepancy, numerical integration

MSC 2010: 11K38, 41A55, 42C10, 65C05, 65D30, 65D32

4.1 Introduction

For an integrable function $f\colon [0,1]^s \to \mathbb{R}$, we denote the integral of f by

$$I(f) = \int\limits_{[0,1]^s} f(\boldsymbol{x})\,\mathrm{d}\boldsymbol{x}.$$

Monte Carlo/Quasi-Monte Carlo (QMC) methods are a class of numerical algorithms for approximating $I(f)$ based on pointwise function evaluations. Let $P \subset [0,1]^s$ be a

Acknowledgement: This article is based on a talk which the first author gave during the discrepancy workshop of RICAM special semester "Multivariate Algorithms and their Foundations in Number Theory." He would like to thank the organizers of the workshop, Dmitriy Bilyk, Josef Dick, and Friedrich Pillichshammer, for their kind invitation. The authors sincerely acknowledge their colleague, Takehito Yoshiki. Some of the outcomes brought from the collaboration with him play a central part of this article. The work of T. G. is supported by JSPS Grant-in-Aid for Young Scientists No. 15K20964. The work of K. S. is supported by Grant-in-Aid for JSPS Fellows No. 17J00466.

Takashi Goda, School of Engineering, University of Tokyo, 7-3-1 Hongo, Bunkyo-ku, Tokyo 113-8656, Japan, e-mail: goda@frcer.t.u-tokyo.ac.jp
Kosuke Suzuki, Graduate School of Science, Hiroshima University, 1-3-1 Kagamiyama, HigashiHiroshima, 739-8526, Japan. JSPS Research Fellow

https://doi.org/10.1515/9783110652581-004

finite multiset, that is, if an element occurs multiple times, it is counted according to its multiplicity. Then $I(f)$ is approximated by

$$I(f;P) = \frac{1}{|P|}\sum_{x\in P} f(x),$$

where $|P|$ denotes the cardinality of P. It is obvious that this algorithm is exact for any choice of P if f is a constant function, but except for such a trivial case, a careful design of P and an accompanying theoretical analysis are required to show that the algorithm works well for various function classes.

One fairly easy but sensible approach is to choose each point x independently and uniformly distributed in $[0,1]^s$. This is widely known under the name of Monte Carlo methods [39]. Many fundamental results in probability theory, including the law of large numbers and the central limit theorem, apply to this approach. Looking at $I(f;P)$ as a random variable (with P being the underlying stochastic variable), we have

$$\mathbb{E}[I(f;P)] = I(f) \quad \text{and} \quad \mathbb{V}[I(f;P)] = \frac{\mathbb{V}[f]}{|P|},$$

for any function $f \in L_2([0,1]^s)$, where $\mathbb{V}[f]$ on the right-hand side of the second equality denotes the variance of f. This means, Monte Carlo methods work for any square-integrable functions, but the approximation error converges only probabilistically at the notorious "one over square root of N" rate. Thus we have a trade-off between versatility and efficiency.

In some applications where the Monte Carlo convergence is considered too slow, one needs to improve efficiency while discarding the versatility of Monte Carlo methods to some extent. QMC methods are one of the standard choices for this purpose. The classical but still central result in QMC methods is the celebrated Koksma–Hlawka inequality:

$$|I(f;P) - I(f)| \leq V_{\mathrm{HK}}(f)D^*(P), \tag{4.1}$$

where $V_{\mathrm{HK}}(f)$ denotes the total variation of f in the sense of Hardy and Krause, and $D^*(P)$ denotes the star-discrepancy of P (we shall give a precise definition of $D^*(P)$ later in Section 4.5). Although the class of functions we can deal with in this case is restricted to functions with bounded total variations (i. e., we discard versatility to some extent), through a clever design of P such that $D^*(P)$ is of order better than $|P|^{-1/2}$, the convergence rate can be improved (i. e., we improve efficiency). In fact, there are many explicit constructions of so-called *digital* (t,m,s)*-nets* and *digital* (t,s)*-sequences* achieving star-discrepancy of order $(\log N)^{s-1}/N$ and $(\log N)^s/N$, respectively;[1] see [49,

1 To be precise, for an infinite sequence of points $\mathcal{S}$, this means that there exists a constant $C_s > 0$ depending only on s such that the star-discrepancy of the first N elements of $\mathcal{S}$ is bounded by $C_s(\log N)^s/N$ uniformly for all N.

24]. Hence, it follows from the Koksma–Hlawka inequality that the integration error converges deterministically faster than the "one over square root of N" rate.

One natural question in this line is then "Can we improve efficiency further while sacrificing versatility to a greater extent?" Higher order quasi-Monte Carlo (HoQMC) methods due to Dick [7, 8] provide an affirmative solution to this question. Now let us focus on functions f having square-integrable partial mixed derivatives up to order $\alpha > 1$ in each variable, which obviously means that we discard versatility to a greater extent than standard QMC methods. In return for this drawback, however, the order of convergence of the integration error can be improved to $(\log N)^{c(\alpha,s)}/N^{\alpha}$ with some exponent $c(\alpha, s) > 0$ by employing so-called *higher order digital nets and sequences* as quadrature nodes P.[2] Hence, when the considered integrand is smooth enough, HoQMC methods can be much more efficient than standard QMC methods, not to mention Monte Carlo methods. Of course, one may ask if one can encounter such smooth functions in practice. Fortunately, there have been several successful and promising applications of HoQMC methods reported already in the literature. These include [18, 17, 20, 30, 31] on applications to partial differential equations with random coefficients (see also the review article [41]), and [11, 32, 12] on applications to Bayesian estimation/inversion problems.

Recently, there has been significant theoretical progress on HoQMC methods. The first major step was made by Dick and Pillichshammer [25]. They proved that order 5 digital sequences achieve the best possible order of L_2-discrepancy, which is $(\log N)^{s/2}/N$, uniformly for all N, and moreover, they proved that order 3 digital nets of N points achieve the best possible order of L_2-discrepancy, which is $(\log N)^{(s-1)/2}/N$ (here again, we shall give a precise definition of L_2-discrepancy later in Section 4.5). Prior to their work, there had been only one explicit construction of finite point sets (for arbitrarily fixed dimension s) with the best possible order of L_2-discrepancy due to Chen and Skriganov [5, 55]. Therefore, higher order digital nets (resp., sequences) are now recognized as the second (resp., first) explicit construction of optimal order L_2-discrepancy point sets (resp. sequences). More recently, several refined analyses for generalizing or extending the work of Dick and Pillichshammer have been conducted [10, 42, 15, 16, 4].

Another major step was made in a series of papers [36–38], where the authors refined the integration error analysis for smooth functions due to Dick [7, 8] and proved that order $(2\alpha + 1)$ digital nets and sequences achieve the best possible order of the worst-case error for a reproducing kernel Hilbert space with dominating mixed smoothness α, which is $(\log N)^{(s-1)/2}/N^{\alpha}$. Note that the original work by Dick [8] proves the worst-case error of order $(\log N)^{s\alpha}/N^{\alpha}$ for order α digital nets and sequences; see also [3]. Other than higher order digital nets and sequences, only the Frolov lattice

2 For higher order digital sequences, this order of convergence does not hold uniformly for all N, but holds for a geometric spacing of N. It is known that this cannot be improved [53].

rule in conjunction with periodization of integrands has been proven to achieve the same, best possible order of the worst-case error so far [29, 64, 45].

The results of [25] and that of [37, 38] have been obtained by using a Walsh series analysis. To introduce the concept of HoQMC methods originally, Dick managed to prove results on the decay of the Walsh coefficients of smooth functions [7, 8]. In fact, his digit interlacing construction of higher order digital nets and sequences, which shall be described in Section 4.3.2, is carefully designed to exploit the decay of the Walsh coefficients. In order to improve his seminal results, it may be sensible to attempt to exploit some further aspect of the Walsh coefficients. Both the result of [25] and that of [37, 38] rely not only on the decay but also on the sparsity of the Walsh coefficients.

In this article, we mainly focus on the papers [25, 38] and provide a unified picture on how the Walsh analysis enables recent developments of HoQMC methods both for discrepancy and numerical integration.

The rest of this article is organized as follows. In Section 4.2, we explain standard QMC methods based on digital nets and sequences in the sense of Niederreiter [49]. Although integer lattices are another important class of QMC point sets (see, for instance, [56] and [19, Section 5]), we do not cover them in this article. In Section 4.3, we introduce the definitions of higher order digital nets and sequences, and provide an explicit construction algorithm due to Dick [8]. In Section 4.4, we introduce the definition of the Walsh functions and give some key connection to digital nets. Thereafter, recent advances in HoQMC methods for discrepancy are described in Section 4.5, while those for numerical integration are in Section 4.6. We shall highlight an analogy between the approach by [25] and that of [38], where exploiting both the decay and the sparsity of the Walsh coefficients plays a crucial role. We conclude the article with some future research directions.

Notation. Throughout this article, we shall use the following notation. Let $\mathbb{N}$ be the set of positive integers and we write $\mathbb{N}_0 = \mathbb{N} \cup \{0\}$. For a prime b, let $\mathbb{F}_b$ be the finite field with b elements and we identify $\mathbb{F}_b$ with the set of integers $\{0, 1, \dots, b-1\}$ equipped with addition and multiplication modulo b. For $x \in [0, 1]$, its b-adic expansion $x = \sum_{i=1}^{\infty} \xi_i b^{-i}$ with $\xi_i \in \mathbb{F}_b$ is understood to be unique in the sense that infinitely many of the ξ_i's are different from $b-1$ if $x \neq 1$ and that all of the ξ_i's are equal to $b-1$ if $x = 1$. Note that for $k = 1 \in \mathbb{N}$ we use the b-adic expansion $1b^0$, whereas for $x = 1 \in [0, 1]$ we use $(b-1)b^{-1} + (b-1)b^{-2} + \cdots$. It will be clear from the context which expansion we use. The operator $\oplus$ denotes the digitwise addition modulo b, that is, for $x, y \in [0, 1]$ with b-adic expansions given by $x = \sum_{i=1}^{\infty} \xi_i b^{-i}$ and $y = \sum_{i=1}^{\infty} \eta_i b^{-i}$, respectively, $\oplus$ is defined as

$$x \oplus y = \sum_{i=1}^{\infty} \zeta_i b^{-i}, \quad \text{where } \zeta_i = \xi_i + \eta_i \pmod{b}.$$

Similarly, we use $\oplus$ for digitwise addition for nonnegative integers based on the b-adic expansions. In case of vectors in $[0,1]^s$ or $\mathbb{N}_0^s$, the operator $\oplus$ is applied componentwise.

4.2 Standard quasi-Monte Carlo

4.2.1 Digital nets and sequences

We start with a general construction scheme for a class of QMC point sets called *digital nets* due to Niederreiter [49]. Note that both, digital (t,m,s)-nets and higher order digital nets can be regarded as special subclasses of QMC point sets.

Definition 4.1 (Digital nets). Let $m,n \in \mathbb{N}$, and let $C_1,\dots,C_s$ be $n \times m$ matrices over $\mathbb{F}_b$. For an integer $0 \le h < b^m$ with b-adic expansion $h = \eta_0 + \eta_1 b + \cdots + \eta_{m-1}b^{m-1}$, define the point $\boldsymbol{x}_h = (x_{h,1},\dots,x_{h,s}) \in [0,1]^s$ by

$$x_{h,j} = \frac{\xi_{1,h,j}}{b} + \frac{\xi_{2,h,j}}{b^2} + \cdots + \frac{\xi_{n,h,j}}{b^n},$$

where

$$(\xi_{1,h,j},\xi_{2,h,j},\dots,\xi_{n,h,j}) = (\eta_0,\eta_1,\dots,\eta_{m-1}) \cdot C_j^{\top}.$$

The set $P = \{\boldsymbol{x}_h \mid 0 \le h < b^m\} \subset [0,1]^s$ is called a digital net over $\mathbb{F}_b$ (with generating matrices $C_1,\dots,C_s$).

It is obvious from the definition that the parameter m determines the total number of points, which is b^m, while the parameter n determines the precision of points. We can extend this definition to construct infinite sequences of points called *digital sequences*. Again, both, digital (t,s)-sequences and higher order digital sequences can be regarded as special subclasses of digital sequences.

Definition 4.2 (Digital sequences). Let $C_1,\dots,C_s$ be $\mathbb{N} \times \mathbb{N}$ matrices over $\mathbb{F}_b$. For an integer $h \in \mathbb{N}_0$ with b-adic expansion $h = \eta_0 + \eta_1 b + \cdots$, where all but a finite number of η_i are 0, define the point $\boldsymbol{x}_h = (x_{h,1},\dots,x_{h,s}) \in [0,1]^s$ by

$$x_{h,j} = \frac{\xi_{1,h,j}}{b} + \frac{\xi_{2,h,j}}{b^2} + \cdots,$$

where

$$(\xi_{1,h,j},\xi_{2,h,j},\dots) = (\eta_0,\eta_1,\dots) \cdot C_j^{\top}.$$

The sequence of points $\mathcal{S} = \{\boldsymbol{x}_h \mid h \in \mathbb{N}_0\} \subset [0,1]^s$ is called a digital sequence over $\mathbb{F}_b$ (with generating matrices $C_1,\dots,C_s$).

Remark 4.1. Assume that for each $1 \le j \le s$ there exists a function $K_j \colon \mathbb{N} \to \mathbb{N}$ such that $C_j = (c_{k,l}^{(j)})_{k,l \in \mathbb{N}}$ satisfies $c_{k,l}^{(j)} = 0$ whenever $k > K_j(l)$. Then for any $m \in \mathbb{N}$ the first b^m elements of a digital sequence over $\mathbb{F}_b$ with generating matrices $C_1, \ldots, C_s$ can be identified with a digital net over $\mathbb{F}_b$ with generating matrices $C_1^{[n\times m]}, \ldots, C_s^{[n\times m]}$ with the precision parameter

$$n = \max_{1\le j\le s} \max_{1\le l\le m} K_j(l),$$

where $C_j^{[n\times m]}$ denotes the upper-left $n \times m$ submatrix of C_j.

4.2.2 Quality measure

In order to generate point sets or sequences from the above construction scheme such that the star-discrepancy is small, we need to design generating matrices $C_1, \ldots, C_s$. In this subsection, we introduce the widely-used quality measure called *t-value*, which is based on the Niederreiter–Rosenbloom–Tsfasman (NRT) weight function [46, 54].

Definition 4.3 (Dual nets). Let $m, n \in \mathbb{N}$ and let P be a digital net over $\mathbb{F}_b$ with generating matrices $C_1, \ldots, C_s \in \mathbb{F}_b^{n\times m}$. The dual net of P, denoted by $P^\perp$, is defined by

$$P^\perp := \{\mathbf{k} = (k_1, \ldots, k_s) \in \mathbb{N}_0^s \mid C_1^\top v_n(k_1) \oplus \cdots \oplus C_s^\top v_n(k_s) = \mathbf{0} \in \mathbb{F}_b^m\},$$

where

$$v_n(k) = (\kappa_0, \ldots, \kappa_{n-1})^\top \in \mathbb{F}_b^n$$

for $k \in \mathbb{N}_0$ with b-adic expansion $k = \kappa_0 + \kappa_1 b + \cdots$, where all but a finite number of κ_i are 0.

Remark 4.2. Let $\mathcal{S}$ be a digital sequence over $\mathbb{F}_b$ for which there exist functions $K_j \colon \mathbb{N} \to \mathbb{N}$ such that $C_j = (c_{k,l}^{(j)})_{k,l \in \mathbb{N}}$ satisfies $c_{k,l}^{(j)} = 0$ whenever $k > K_j(l)$. Then the dual net can be defined for the first b^m elements of $\mathcal{S}$ for any $m \in \mathbb{N}$, since they can be identified with a digital net as discussed in Remark 4.1.

Definition 4.4 (NRT weight function). For $k \in \mathbb{N}$, we denote the b-adic expansion of k by

$$k = \kappa_1 b^{c_1 - 1} + \kappa_2 b^{c_2 - 1} + \cdots + \kappa_v b^{c_v - 1}$$

with $\kappa_1, \ldots, \kappa_v \in \mathbb{F}_b \setminus \{0\}$ and $c_1 > \cdots > c_v > 0$. Then the NRT weight function $\mu_1 \colon \mathbb{N}_0 \to \mathbb{N}_0$ is defined by $\mu_1(0) = 0$ and $\mu_1(k) = c_1$. In case of vectors in $\mathbb{N}_0^s$, we define

$$\mu_1(k_1, \ldots, k_s) = \sum_{j=1}^{s} \mu_1(k_j).$$

We are ready to introduce the definition of t-value.

Definition 4.5 (t-value). Let $m, n \in \mathbb{N}$, and let P be a digital net over $\mathbb{F}_b$ with generating matrices $C_1, \ldots, C_s \in \mathbb{F}_b^{n\times m}$. We write

$$\mu_1(P^\perp) := \min_{\boldsymbol{k}\in P^\perp\setminus\{\boldsymbol{0}\}} \mu_1(\boldsymbol{k}).$$

Then P is called a *digital* (t, m, s)*-net* over $\mathbb{F}_b$ with $t \in \mathbb{N}_0$ which satisfies

$$m - \mu_1(P^\perp) + 1 \le t \le m.$$

Such parameter t is called the t-value of P and is said to be strict if $t = m - \mu_1(P^\perp) + 1$.

This definition of the t-value is based on the concept of duality theory of digital nets as originally studied in [50]. There is another but equivalent definition of t-value: let ρ be an integer such that, for any choice $d_1, \ldots, d_s \in \mathbb{N}_0$ with $d_1 + \cdots + d_s = \rho$,

$$\begin{gathered}\text{the first } d_1 \text{ row vectors of } C_1\\ \text{the first } d_2 \text{ row vectors of } C_2\\ \vdots\\ \text{the first } d_s \text{ row vectors of } C_s\end{gathered}$$

are linearly independent over $\mathbb{F}_b$. Then the t-value can be also defined by $m - \rho$.

Because of the linear independence of the row vectors of generating matrices, any digital (t, m, s)-net over $\mathbb{F}_b$ has the following equi-distribution property: every b-adic elementary box of the form

$$E = \prod_{j=1}^{s}\left[\frac{a_j}{b^{c_j}}, \frac{a_j+1}{b^{c_j}}\right)$$

with $c_1, \ldots, c_s \ge 0$, $c_1 + \cdots + c_s = m - t$ and $0 \le a_j < b^{c_j}$ for all j, whose volume is b^{t-m}, contains exactly b^t points. Hence, as the t-value is smaller, digital nets are more equi-distributed over $[0, 1]^s$. This is why the t-value works as a quality measure of digital nets.

For digital sequences, the t-value is defined as follows.

Definition 4.6. Let $\mathcal{S}$ be a digital sequence over $\mathbb{F}_b$. $\mathcal{S}$ is called a *digital* (t, s)*-sequence* over $\mathbb{F}_b$ if there exists a $t \in \mathbb{N}_0$ such that the first b^m points of $\mathcal{S}$ are a digital (t, m, s)-net for any $m \ge t$.

The following result states that the star-discrepancy of digital (t, m, s)-nets and digital (t, s)-sequences are of order $(\log N)^{s-1}/N$ and $(\log N)^s/N$, respectively, as mentioned in the first section. We refer to [49, Theorems 4.10 and 4.17] for the proof.

Theorem 4.1. *The following holds true:*

1. *Let P be a digital (t,m,s)-net over $\mathbb{F}_b$. There exists a constant $B^{(1)}_{s,b,t}$ such that the star-discrepancy of P is bounded by*

$$D^*(P) \le B^{(1)}_{s,b,t} \frac{m^{s-1}}{b^m}.$$

2. *Let $\mathcal{S} = \{\boldsymbol{x}_h \mid h \in \mathbb{N}_0\}$ be a digital (t,s)-sequence over $\mathbb{F}_b$. There exists a constant $B^{(2)}_{s,b,t}$ such that the star-discrepancy of the first N points of $\mathcal{S}$ is bounded by*

$$D^*(\{\boldsymbol{x}_0,\dots,\boldsymbol{x}_{N-1}\}) \le B^{(2)}_{s,b,t} \frac{(\log N)^s}{N},$$

for any $N \ge 2$.

We end this subsection by providing one useful result in analyzing the integration error of QMC rules using digital (t,m,s)-nets.

Lemma 4.1. *Let P be a digital (t,m,s)-net over $\mathbb{F}_b$. The following holds true:*

1. *For $z \in \mathbb{N}_0$,*

$$\begin{aligned} &\left|\{\boldsymbol{k} \in P^\perp \setminus \{\boldsymbol{0}\} \mid \mu_1(\boldsymbol{k}) = z\}\right| \\ &\quad \le \begin{cases} 0 & \text{if } z < \mu_1(P^\perp), \\ b^{z-\mu_1(P^\perp)+1}(z+1)^{s-1} & \text{otherwise.} \end{cases} \end{aligned}$$

2. *For any real $\lambda > 1$,*

$$\sum_{\boldsymbol{k} \in P^\perp \setminus \{\boldsymbol{0}\}} b^{-\lambda\mu_1(\boldsymbol{k})} \le 2^{s-1} b^\lambda \frac{(\mu_1(P^\perp))^{s-1}}{b^{\lambda\mu_1(P^\perp)}} \sum_{z=1}^\infty b^{(1-\lambda)z} z^{s-1}.$$

Proof. In this proof, we put $A_z = |\{\boldsymbol{k} \in P^\perp \setminus \{\boldsymbol{0}\} \mid \mu_1(\boldsymbol{k}) = z\}|$. It holds that

$$A_z = \sum_{\substack{z_1,\dots,z_s \in \mathbb{N}_0 \\ z_1+\dots+z_s = z}} \left|\{\boldsymbol{k} \in P^\perp \setminus \{\boldsymbol{0}\} \mid \mu_1(k_1) = z_1, \dots, \mu_1(k_s) = z_s\}\right|.$$

Following [55, Lemma 2.2], the summand is bounded above by

$$\begin{aligned} &\left|\{\boldsymbol{k} \in P^\perp \setminus \{\boldsymbol{0}\} \mid \mu_1(k_1) = z_1, \dots, \mu_1(k_s) = z_s\}\right| \\ &\quad \le \begin{cases} 0 & \text{if } z < \mu_1(P^\perp), \\ b^{z-\mu_1(P^\perp)+1} & \text{otherwise.} \end{cases} \end{aligned}$$

Thus we have $A_z = 0$ if $z < \mu_1(P^\perp)$, since each summand is 0. For $z \ge \mu_1(P^\perp)$, this bound gives

$$A_z \le b^{z-\mu_1(P^\perp)+1} \sum_{\substack{z_1,\dots,z_s \in \mathbb{N}_0 \\ z_1+\dots+z_s = z}} 1 = b^{z-\mu_1(P^\perp)+1} \binom{z+s-1}{s-1}$$

$$= b^{z-\mu_1(P^\perp)+1} \prod_{j=1}^{s-1} \frac{z+j}{j} \le b^{z-\mu_1(P^\perp)+1}(z+1)^{s-1},$$

which proves the first assertion of the lemma.

Using the result of the first assertion and then applying the change of variables $z \mapsto z + \mu_1(P) - 1$, we have

$$\begin{aligned}
\sum_{k\in P^\perp\setminus\{\mathbf{0}\}} b^{-\lambda\mu_1(k)} &= \sum_{z=\mu_1(P^\perp)}^{\infty} b^{-\lambda z} A_z \le \sum_{z=\mu_1(P^\perp)}^{\infty} b^{(1-\lambda)z-\mu_1(P^\perp)+1}(z+1)^{s-1} \\
&= b^{-\lambda(\mu_1(P^\perp)-1)} \sum_{z=1}^{\infty} b^{(1-\lambda)z}(z+\mu_1(P^\perp))^{s-1} \\
&\le 2^{s-1}(\mu_1(P^\perp))^{s-1} b^{-\lambda(\mu_1(P^\perp)-1)} \sum_{z=1}^{\infty} b^{(1-\lambda)z} z^{s-1},
\end{aligned}$$

where the last sum over z is finite since $\lambda > 1$. Hence we complete the proof. □

4.2.3 Explicit constructions

Theorem 4.1 together with the Koksma–Hlawka inequality (4.1) gives a motivation to construct digital (t, m, s)-nets or digital (t, s)-sequences with small t-value. In fact, many explicit constructions of digital (t, s)-sequences with small t-value are already known. Examples are given by Sobol' [58], Faure [26], Niederreiter [47], Tezuka [63], Niederreiter and Xing [51], as well as many others. Here, we give one example from [47, 63].

Let $p_1, p_2, \ldots \in \mathbb{F}_b[x]$ be a sequence of distinct monic irreducible polynomials over $\mathbb{F}_b$ with $\deg(p_1) \le \deg(p_2) \le \cdots$. For each $j \in \mathbb{N}$, let $e_j = \deg(p_j)$ and consider the following Laurent series expansion

$$\frac{x^{e_j-z-1}}{(p_j(x))^i} = \sum_{l=1}^{\infty} \frac{a^{(j)}(i,z,l)}{x^l} \in \mathbb{F}_b((x^{-1})) \tag{4.2}$$

for integers $i \ge 1$ and $0 \le z < e_j$. Define the matrix $C_j = (c_{k,l}^{(j)})_{k,l\in\mathbb{N}}$ by

$$c_{k,l}^{(j)} = a^{(j)}\left(\left\lfloor \frac{k-1}{e_j} \right\rfloor + 1, (k-1) \bmod e_j, l\right).$$

Here, we see that the rows of C_j (from upper to lower) correspond to the Laurent series expansions of

$$\frac{x^{e_j-1}}{p_j(x)}, \ldots, \frac{1}{p_j(x)}, \frac{x^{e_j-1}}{(p_j(x))^2}, \ldots, \frac{1}{(p_j(x))^2}, \ldots.$$

Hence, we have $K_j(l) = l$ for any j, l in light of Remark 4.1, that is, $c_{k,l}^{(j)} = 0$ whenever $k > l$, meaning that C_j is an upper triangular matrix.

It is straightforward from the definition that this explicit construction of digital sequences is extensible in the dimension. The first s matrices $C_1, \dots, C_s$ generate a digital (t, s)-sequence over $\mathbb{F}_b$ with

$$t \leq \sum_{j=1}^{s} (e_j - 1).$$

We refer to [24, Theorem 8.2] for the proof of this fact.

Remark 4.3. Some comments are in order:
1. If we replace the numerator $x^{e_j - z - 1}$ of (4.2) by x^z, then we obtain the construction algorithm originally introduced in [47], which is nowadays known as *Niederreiter sequences*. For the original Niederreiter sequence, the strict t-value is equal to $\sum_{j=1}^{s}(e_j - 1)$ [21].
2. A generalization of Niederreiter sequences by Tezuka [63] is to use the set of linearly independent polynomials $\{y_{j,i,z}(x) \mid 0 \leq z < e_j\}$ over $\mathbb{F}_b$ for the numerator of (4.2) rather than the simplest set $\{x^z \mid 0 \leq z < e_j\}$.
3. The Sobol' sequences due to [58] are a subclass of the generalized Niederreiter sequences over $\mathbb{F}_2$, where the primitive polynomials are used for $p_1, p_2, \dots$. Recently, Faure and Lemieux [27] gave some precise connections between the Sobol' sequences and the generalized Niederreiter sequences.

4.2.4 Polynomial lattice point sets

We end this section by introducing another important class of digital nets called *polynomial lattice point sets* introduced by Niederreiter [48].

Definition 4.7. Let $m \in \mathbb{N}$, and let $p \in \mathbb{F}_b[x]$ and $\boldsymbol{q} = (q_1, \dots, q_s) \in (\mathbb{F}_b[x])^s$ such that $\deg(p) = m$ and $\deg(q_j) < m$. For $1 \leq j \leq s$, consider the Laurent series expansion

$$\frac{q_j(x)}{p(x)} = \sum_{l=1}^{\infty} \frac{a_l^{(j)}}{x^l} \in \mathbb{F}_b((x^{-1}))$$

and define the Hankel matrix $C_j = (c_{k,l}^{(j)})_{1 \leq k,l \leq m} \in \mathbb{F}_b^{m \times m}$ by

$$c_{k,l}^{(j)} = a_{k+l-1}^{(j)}.$$

Then a digital net over $\mathbb{F}_b$ with these generating matrices $C_1, \dots, C_s$ is called a *polynomial lattice point set* (with modulus p and generating vector $\boldsymbol{q}$).

Indeed, polynomial lattice point sets can be constructed without using generating matrices explicitly. Define the map $v_m \colon \mathbb{F}_b((x^{-1})) \to [0, 1]$ by

$$v_m\left(\sum_{i=w}^{\infty} a_i x^{-i}\right) := \sum_{i=\max\{1,w\}}^{m} a_i b^{-i}.$$

We identify $h \in \mathbb{N}_0$, whose finite b-adic expansion is given by $h = \eta_0 + \eta_1 b + \cdots$, with the polynomial over $\mathbb{F}_b$ given by $h(x) = \eta_0 + \eta_1 x + \cdots$. Put

$$\boldsymbol{x}_h = \left(v_m\left(\frac{h(x)q_1(x)}{p(x)} \right), \ldots, v_m\left(\frac{h(x)q_s(x)}{p(x)} \right) \right) \in [0,1]^s.$$

Then a point set $\{\boldsymbol{x}_h \mid 0 \le h < b^m\}$ is nothing but the polynomial lattice point set as defined above.

The modulus p is often chosen to be either the monomial $p(x) = x^m$ or irreducible. The difficulty is in how to choose the generating vector $\boldsymbol{q}$. In particular, for $s \ge 3$, no explicit way for this choice has been known yet. Currently, one of the most standard approaches is to recursively choose one component q_j from the set $\{q \in \mathbb{F}_b[x] \mid \deg(q) < m\}$ which minimizes a chosen criterion while the earlier ones $q_1, \ldots, q_{j-1}$ are kept unchanged. This greedy algorithm is known as *component-by-component construction* [57]. Another well-known approach is to restrict ourselves to vectors of the form

$$\boldsymbol{q} = (1, q, \ldots, q^{s-1}) \in (\mathbb{F}_b[x])^s$$

for $q \in \mathbb{F}_b[x]$ with $\deg(q) < m$, and then to choose one optimal q with respect to a chosen criterion. This algorithm is known as *Korobov construction*. Compared to Korobov construction, the component-by-component construction has the advantage that it is extensible in the dimension and that one can use the fast Fourier transform to find good generating vectors [52]. However, neither of both is extensible in the number of points.

4.3 Higher order quasi-Monte Carlo

4.3.1 Quality measure

To introduce the definitions of higher order digital nets and sequences, we start with generalizing the NRT weight function.

Definition 4.8 (Dick weight function). Let $\alpha \in \mathbb{N}$. For $k \in \mathbb{N}$, we denote the b-adic expansion of k by

$$k = \kappa_1 b^{c_1-1} + \kappa_2 b^{c_2-1} + \cdots + \kappa_v b^{c_v-1}$$

with $\kappa_1, \ldots, \kappa_v \in \mathbb{F}_b \setminus \{0\}$ and $c_1 > \cdots > c_v > 0$. Then the Dick weight function $\mu_\alpha \colon \mathbb{N}_0 \to \mathbb{N}_0$ is defined by $\mu_\alpha(0) = 0$ and

$$\mu_\alpha(k) = \sum_{i=1}^{\min(\alpha,v)} c_i.$$

In case of vectors in $\mathbb{N}_0^s$, we define

$$\mu_\alpha(k_1, \dots, k_s) = \sum_{j=1}^{s} \mu_\alpha(k_j).$$

It is obvious that the Dick weight function coincides with the NRT weight function when $\alpha = 1$. As a natural generalization of digital (t, m, s)-nets and (t, s)-sequences based on the NRT weight function, higher order digital nets and sequences due to Dick [7, 8] are defined by using the Dick weight function as follows.

Definition 4.9 (Higher order digital nets). Let $\alpha \in \mathbb{N}$. Let $m, n \in \mathbb{N}$, and let P be a digital net over $\mathbb{F}_b$ with generating matrices $C_1, \dots, C_s \in \mathbb{F}_b^{n\times m}$. We write

$$\mu_\alpha(P^\perp) := \min_{\boldsymbol{k}\in P^\perp\setminus\{\boldsymbol{0}\}} \mu_\alpha(\boldsymbol{k}).$$

Then P is called an *order α digital (t_α, m, s)-net* over $\mathbb{F}_b$ with $t_\alpha \in \mathbb{N}_0$ which satisfies

$$\alpha m - \mu_\alpha(P^\perp) + 1 \leq t_\alpha \leq \alpha m.$$

The parameter t_α is said to be strict if $t_\alpha = \alpha m - \mu_\alpha(P^\perp) + 1$.

Definition 4.10 (Higher order digital sequences). Let $\mathcal{S}$ be a digital sequence over $\mathbb{F}_b$. $\mathcal{S}$ is called an *order α digital (t_α, s)-sequence* over $\mathbb{F}_b$ if there exists $t_\alpha \in \mathbb{N}_0$ such that the first b^m points of $\mathcal{S}$ are an order α digital (t_α, m, s)-net for any $m \geq t_\alpha/\alpha$.

Obviously, for fixed α, the t_α-value works as a quality measure of order α digital nets and sequences.

Our definition of higher order digital nets is again based on the concept of duality theory of digital nets, and looks different from the original definition by Dick which has an additional parameter β. When $n \geq \alpha m$, however, by setting $\beta = 1$, our definition becomes equivalent to his original definition based on the linear independence of row vectors of generating matrices described below: let ρ be an integer such that, for any choice $1 \leq d_{j,v_j} < \cdots < d_{j,1} \leq n$, where $0 \leq v_j \leq m$ for all $1 \leq j \leq s$ with

$$\sum_{j=1}^{s} \sum_{i=1}^{\min(v_j,\alpha)} d_{j,i} = \rho,$$

$$\begin{gathered}
\text{the } d_{1,v_1}, \dots, d_{1,1}\text{-th row vectors of } C_1 \\
\text{the } d_{2,v_2}, \dots, d_{2,1}\text{-th row vectors of } C_2 \\
\vdots \\
\text{the } d_{s,v_s}, \dots, d_{s,1}\text{-th row vectors of } C_s
\end{gathered}$$

are linearly independent over $\mathbb{F}_b$. Then a digital net with generating matrices $C_1, \dots, C_s$ is an order α digital (t_α, m, s)-net over $\mathbb{F}_b$ with $t_\alpha = \alpha m - \rho$. This linear independence of the row vectors of generating matrices ensures that a higher order digital net has a similar geometric equi-distribution property to what is described in Subsection 4.2.2; see [24, Chapter 15.3] for more details.

Here, we provide one useful property called *propagation rule* of higher order digital nets and sequences shown in [7, Theorem 3.3]. We give a different proof which does not rely on the linear independence of generating matrices.

Lemma 4.2. *For $\beta \in \mathbb{N}, \beta \geq 2$, let P and $\mathcal{S}$ be an order β digital (t_β, m, s)-net and an order β digital (t_β, s)-sequence over $\mathbb{F}_b$, respectively. Then, for any $1 \leq \alpha < \beta$, P and $\mathcal{S}$ are also an order α digital (t_α, m, s)-net over $\mathbb{F}_b$ and an order α digital (t_α, s)-sequence over $\mathbb{F}_b$, respectively, both with $t_\alpha \leq \lceil t_\beta \alpha/\beta \rceil$.*

Proof. First, we prove that the inequality

$$\frac{\mu_\alpha(\boldsymbol{k})}{\alpha} \geq \frac{\mu_\beta(\boldsymbol{k})}{\beta}$$

holds for any $\boldsymbol{k} \in \mathbb{N}_0^s$ and $1 \leq \alpha \leq \beta$. Since the weight function for vector $\boldsymbol{k}$ is defined as the sum of the weight function for each coordinate, it suffices to prove the one-dimensional case. Since the result for $k = 0$ follows trivially, let us consider $k > 0$. Denote the b-adic expansion of k by

$$k = \kappa_1 b^{c_1 - 1} + \kappa_2 b^{c_2 - 1} + \cdots + \kappa_v b^{c_v - 1}$$

with $\kappa_1, \ldots, \kappa_v \in \mathbb{F}_b \setminus \{0\}$ and $c_1 > \cdots > c_v > 0$. Then we have

$$\frac{\mu_\alpha(k)}{\alpha} = \frac{1}{\alpha} \sum_{i=1}^{\min(\alpha,v)} c_i \geq \frac{1}{\beta} \sum_{i=1}^{\min(\beta,v)} c_i = \frac{\mu_\beta(k)}{\beta},$$

which proves the assertion.

Now let us consider an order β digital (t_β, m, s)-net over $\mathbb{F}_b$. Using the above inequality, we obtain

$$\mu_\alpha(P^\perp) = \min_{\boldsymbol{k} \in P^\perp \setminus \{\boldsymbol{0}\}} \mu_\alpha(\boldsymbol{k}) \geq \min_{\boldsymbol{k} \in P^\perp \setminus \{\boldsymbol{0}\}} \frac{\alpha}{\beta} \mu_\beta(\boldsymbol{k}) = \frac{\alpha}{\beta} \mu_\beta(P^\perp).$$

Thus P is an order α digital (t_α, m, s)-net over $\mathbb{F}_b$ with

$$\begin{aligned} t_\alpha &= \alpha m - \mu_\alpha(P^\perp) + 1 \leq \alpha m - \frac{\alpha}{\beta} \mu_\beta(P^\perp) + 1 \\ &= \frac{\alpha}{\beta}(\beta m - \mu_\beta(P^\perp) + 1) + \frac{\beta - \alpha}{\beta} = \frac{\alpha}{\beta} t_\beta + \frac{\beta - \alpha}{\beta}. \end{aligned}$$

Given that t_α is a nonnegative integer and that the fraction $(\beta - \alpha)/\beta$ is less than 1, the t_α-value can be bounded above by $\lceil t_\beta \alpha/\beta \rceil$. The result for an order β digital (t_β, s)-sequence follows immediately. □

Importantly, this result implies that any higher order digital nets and sequences are also digital (t, m, s)-nets and digital (t, s)-sequences, respectively.

We end this subsection by providing two useful results in analyzing the integration error of QMC rules using higher order digital nets. The first lemma is a higher order version of Lemma 4.1.

Lemma 4.3. *For $\alpha \geq 2$, let P be an order α digital (t_α, m, s)-net over $\mathbb{F}_b$. The following holds true:*

1. *For $z \in \mathbb{N}_0$,*

$$\begin{aligned} &|\{\boldsymbol{k} \in P^\perp \setminus \{\boldsymbol{0}\} \mid \mu_\alpha(\boldsymbol{k}) = z\}| \\ &\leq \begin{cases} 0 & \text{if } z < \mu_\alpha(P^\perp), \\ (b-1)^{s\alpha} b^{(z-\mu_\alpha(P^\perp))/\alpha} (z+2)^{s\alpha-1} & \text{otherwise.} \end{cases} \end{aligned}$$

2. *For any real $\lambda > 1/\alpha$,*

$$\sum_{\boldsymbol{k} \in P^\perp \setminus \{\boldsymbol{0}\}} b^{-\lambda \mu_\alpha(\boldsymbol{k})} \leq 2^{s\alpha-1}(b-1)^{s\alpha} \frac{(\mu_\alpha(P^\perp))^{s\alpha-1}}{b^{\lambda \mu_\alpha(P^\perp)}} \sum_{z=0}^{\infty} b^{(1/\alpha-\lambda)z}(z+2)^{s\alpha-1}.$$

Proof. Similar to the proof of Lemma 4.1, we put $A_{\alpha,z} = |\{\boldsymbol{k} \in P^\perp \setminus \{\boldsymbol{0}\} \mid \mu_\alpha(\boldsymbol{k}) = z\}|$. For $1 \leq j \leq s$, denote the b-adic expansion of $k_j \in \mathbb{N}_0$ by

$$k_j = \kappa_{1,j} b^{c_{1,j}-1} + \kappa_{2,j} b^{c_{2,j}-1} + \cdots + \kappa_{v_j,j} b^{c_{v_j,j}-1}$$

with $\kappa_{1,j}, \ldots, \kappa_{v_j,j} \in \mathbb{F}_b \setminus \{0\}$ and $c_{1,j} > \cdots > c_{v_j,j} > 0$, and write $c_{v_j+1,j} = c_{v_j+2,j} = \cdots = 0$. If $k_j = 0$, let $c_{1,j} = c_{2,j} = \cdots = 0$. Then we have

$$A_{\alpha,z} = \sum_{\substack{z_{1,j} \geq \cdots \geq z_{\alpha,j} \in \mathbb{N}_0, \forall j=1,\ldots,s \\ \sum_{j=1}^s \sum_{i=1}^\alpha z_{i,j} = z}} |\{\boldsymbol{k} \in P^\perp \setminus \{\boldsymbol{0}\} \mid c_{i,j} = z_{i,j}, 1 \leq i \leq \alpha, 1 \leq j \leq s\}|.$$

It can be inferred from [24, Proof of Lemma 15.20] that the summand is bounded above by

$$\begin{aligned} &|\{\boldsymbol{k} \in P^\perp \setminus \{\boldsymbol{0}\} \mid c_{i,j} = z_{i,j}, 1 \leq i \leq \alpha, 1 \leq j \leq s\}| \\ &\leq \begin{cases} 0 & \text{if } z < \mu_\alpha(P^\perp), \\ (b-1)^{s\alpha} & \text{if } z \geq \mu_\alpha(P^\perp) \text{ and} \\ & \quad z_{\alpha,1} + \cdots + z_{\alpha,s} < \mu_\alpha(P^\perp)/\alpha, \\ (b-1)^{s\alpha} b^{z_{\alpha,1}+\cdots+z_{\alpha,s}-\mu_\alpha(P^\perp)/\alpha} & \text{if } z \geq \mu_\alpha(P^\perp) \text{ and} \\ & \quad z_{\alpha,1} + \cdots + z_{\alpha,s} \geq \mu_\alpha(P^\perp)/\alpha. \end{cases} \end{aligned}$$

Thus, we have $A_{\alpha,z} = 0$ if $z < \mu_\alpha(P^\perp)$, since each summand is 0. For $z \geq \mu_\alpha(P^\perp)$, this bound gives

$$A_{\alpha,z} \leq (b-1)^{s\alpha} \sum_{\substack{z_{1,j} \geq \cdots \geq z_{\alpha,j} \in \mathbb{N}_0, \forall j=1,\ldots,s \\ z_{1,1}+\cdots+z_{\alpha,1}+\cdots+z_{1,s}+\cdots+z_{\alpha,s}=z}} \max(1, b^{z_{\alpha,1}+\cdots+z_{\alpha,s}-\mu_\alpha(P^\perp)/\alpha})$$

$$\begin{aligned}
&\leq (b-1)^{s\alpha} \sum_{i=0}^{\lfloor z/\alpha \rfloor} \binom{i+s-1}{s-1} \binom{z-i+s(\alpha-1)-1}{s(\alpha-1)-1} \max(1, b^{i-\mu_\alpha(P^\perp)/\alpha}) \\
&\leq (b-1)^{s\alpha} \sum_{i=0}^{\lfloor z/\alpha \rfloor} (i+1)^{s-1}(z-i+1)^{s(\alpha-1)-1} \max(1, b^{i-\mu_\alpha(P^\perp)/\alpha}) \\
&\leq (b-1)^{s\alpha}(z+2)^{s\alpha-2} \sum_{i=0}^{\lfloor z/\alpha \rfloor} \max(1, b^{i-\mu_\alpha(P^\perp)/\alpha}) \\
&\leq (b-1)^{s\alpha}(z+2)^{s\alpha-2}(z/\alpha+1) b^{z/\alpha-\mu_\alpha(P^\perp)/\alpha},
\end{aligned}$$

which proves the first assertion of the lemma.

Using the result of the first assertion and then applying the change of variables $z \mapsto z + \mu_\alpha(P)$, we have

$$\begin{aligned}
\sum_{\boldsymbol{k} \in P^\perp \setminus \{\boldsymbol{0}\}} b^{-\lambda \mu_\alpha(\boldsymbol{k})} &= \sum_{z=\mu_\alpha(P^\perp)}^{\infty} b^{-\lambda z} A_{\alpha,z} \\
&\leq (b-1)^{s\alpha} \sum_{z=\mu_\alpha(P^\perp)}^{\infty} b^{-\lambda z + (z-\mu_\alpha(P^\perp))/\alpha} (z+2)^{s\alpha-1} \\
&\leq (b-1)^{s\alpha} b^{-\lambda \mu_\alpha(P^\perp)} \sum_{z=0}^{\infty} b^{(1/\alpha-\lambda)z} \left(z+\mu_\alpha(P^\perp)+2\right)^{s\alpha-1} \\
&\leq 2^{s\alpha-1}(b-1)^{s\alpha} \left(\mu_\alpha(P^\perp)\right)^{s\alpha-1} b^{-\lambda\mu_\alpha(P^\perp)} \sum_{z=0}^{\infty} b^{(1/\alpha-\lambda)z}(z+2)^{s\alpha-1},
\end{aligned}$$

where the last sum over z is finite since $\lambda > 1/\alpha$. Hence we complete the proof. □

Before stating the second useful lemma, we need to recall the notion of "type (p, q)" introduced in [38].

Definition 4.11. For $k, l \in \mathbb{N}_0$, we denote the b-adic expansions of k and l by

$$k = \sum_{i=1}^{v} \kappa_i b^{c_i-1} \quad \text{and} \quad l = \sum_{i=1}^{w} \lambda_i b^{d_i-1},$$

respectively, where $\kappa_1, \ldots, \kappa_v, \lambda_1, \ldots, \lambda_w \in \{1, \ldots, b-1\}$, $c_1 > c_2 > \cdots > c_v > 0$ and $d_1 > d_2 > \cdots > d_w > 0$. For $k = 0$ ($l = 0$, resp.), we assume that $v = 0$ and $\kappa_0 b^{c_0-1} = 0$ ($w = 0$ and $\lambda_0 b^{d_0-1} = 0$, resp.). For $p, q \in \mathbb{N}_0$, we write

$$k^{(p)} = \sum_{i=p+1}^{v} \kappa_i b^{c_i-1} \quad \text{and} \quad l^{(q)} = \sum_{i=q+1}^{w} \lambda_i b^{d_i-1},$$

where the empty sum equals 0. Then we say that (k, l) is of type (p, q) if $k^{(p)} = l^{(q)}$ and $\kappa_p b^{c_p-1} \neq \lambda_q b^{d_q-1}$, where we set $\kappa_0 b^{c_0-1} = \lambda_0 b^{d_0-1} = 0$, except the case $k = l$ where we say that (k, l) is of type $(0, 0)$.

In what follows, we write $(k,l) \in T_{\geq\alpha}$ if (k,l) is of type (p,q) with $p+q \geq \alpha$. In case of vectors in $\mathbb{N}_0^s$, we write $(\boldsymbol{k},\boldsymbol{l}) \in T_{\geq\alpha}$ if there exists at least one index $1 \leq j \leq s$ such that $(k_j,l_j) \in T_{\geq\alpha}$.

Now the following result, which can be regarded as a generalization of the result shown in [25, Lemma 3.7], is proven in [38, Lemma 8]. Here, we state the result in a slightly more general form.

Lemma 4.4. *For $\alpha \in \mathbb{N}$, let P be an order α digital (t_α,m,s)-net over $\mathbb{F}_b$. For $z \in \mathbb{N}_0$, there exists a constant $B_{\alpha,b,s,t_\alpha} > 0$ such that the following holds:*

$$\begin{aligned}&\left|\{(\boldsymbol{k},\boldsymbol{l}) \in (P^\perp \setminus \{\boldsymbol{0}\})^2 \mid \mu_1(\boldsymbol{k})+\mu_1(\boldsymbol{l}) = z, (\boldsymbol{k},\boldsymbol{l}) \notin T_{\geq\alpha}\}\right| \\ &\quad \leq \begin{cases} 0 & \text{if } z < 2\mu_1(P^\perp), \\ B_{\alpha,b,s,t_\alpha}(z-2\mu_1(P^\perp))^{s(\alpha-1)+1} z^{s-1} b^{(z-2\mu_1(P^\perp))/2} & \text{otherwise.}\end{cases}\end{aligned}$$

4.3.2 Digit interlacing construction

Here, we give an explicit construction of higher order digital nets and sequences based on the *digit interlacing function* due to Dick [7, 8]:

Definition 4.12. Let $\alpha \in \mathbb{N}$ and let $\boldsymbol{x} = (x_1,\dots,x_\alpha) \in [0,1]^\alpha$. For $1 \leq j \leq \alpha$, we denote the b-adic expansion of x_j by $x_j = \sum_{i=1}^\infty \xi_{i,j} b^{-i}$. The digit interlacing function (of factor α) $\mathcal{D}_\alpha : [0,1]^\alpha \to [0,1]$ is defined by

$$\mathcal{D}_\alpha(x_1,\dots,x_\alpha) := \sum_{i=1}^\infty \sum_{j=1}^\alpha \frac{\xi_{i,j}}{b^{\alpha(i-1)+j}}.$$

In case of vectors in $[0,1]^{\alpha s}$, we apply $\mathcal{D}_\alpha$ to every nonoverlapping consecutive α components, that is,

$$\mathcal{D}_\alpha(x_1,\dots,x_{\alpha s}) := (\mathcal{D}_\alpha(x_1,\dots,x_\alpha),\dots,\mathcal{D}_\alpha(x_{\alpha(s-1)+1},\dots,x_{\alpha s})) \in [0,1]^s.$$

Lemma 4.5. *The following holds true:*

1. *Let P be a digital $(t,m,\alpha s)$-net over $\mathbb{F}_b$. The set $\mathcal{D}_\alpha(P) = \{\mathcal{D}_\alpha(\boldsymbol{x}) \mid \boldsymbol{x} \in P\}$ is an order α digital (t_α,m,s)-net over $\mathbb{F}_b$ with*
$$t_\alpha \leq \alpha \min\left(m, t+\left\lfloor \frac{s(\alpha-1)}{2} \right\rfloor\right).$$
2. *Let $\mathcal{S}$ be a digital $(t,\alpha s)$-sequence over $\mathbb{F}_b$. The sequence $\mathcal{D}_\alpha(\mathcal{S}) = \{\mathcal{D}_\alpha(\boldsymbol{x}) \mid \boldsymbol{x} \in \mathcal{S}\}$ is an order α digital (t_α,s)-sequence over $\mathbb{F}_b$ with*
$$t_\alpha \leq \alpha t + \frac{s\alpha(\alpha-1)}{2}.$$

Since there are many explicit constructions of digital (t, m, s)-nets and (t, s)-sequences with small t-value for arbitrarily dimension s, as described in Subsection 4.2.3, the above lemma from [8, Theorems 4.11 and 4.12] directly implies that higher order digital nets and sequences can be explicitly constructed.

Remark 4.4. Let P be a digital $(t, m, \alpha s)$-net over $\mathbb{F}_b$ with generating matrices $C_1, \dots, C_{\alpha s} \in \mathbb{F}_b^{m\times m}$. Let $\boldsymbol{c}_i^{(j)}$ denote the ith row vector of C_j. For each $1 \le j \le s$, construct the matrix $D_j \in \mathbb{F}_b^{\alpha m\times m}$, whose ith row vector is denoted by $\boldsymbol{d}_i^{(j)}$, from the matrices $C_{\alpha(j-1)+1}, \dots, C_{\alpha j}$ as

$$\boldsymbol{d}_{\alpha(h-1)+i}^{(j)} = \boldsymbol{c}_h^{(\alpha(j-1)+i)}, \tag{4.3}$$

for $1 \le h \le m$ and $1 \le i \le \alpha$. Then the set $\mathcal{D}_\alpha(P)$ is a digital net over $\mathbb{F}_b$ with generating matrices $D_1, \dots, D_s$.

Similarly, the sequence $\mathcal{D}_\alpha(\mathcal{S})$ can be identified with a digital sequence over $\mathbb{F}_b$ with generating matrices $D_1, \dots, D_s \in \mathbb{F}_b^{\mathbb{N}\times\mathbb{N}}$ which are constructed from the generating matrices $C_1, \dots, C_{\alpha s} \in \mathbb{F}_b^{\mathbb{N}\times\mathbb{N}}$ of $\mathcal{S}$, where the row vectors of D_j are given by (4.3) for $h \ge 1$ and $1 \le i \le \alpha$.

To construct an interlaced finite point set $\mathcal{D}_\alpha(P)$, one can use polynomial lattice point sets in dimension αs instead of digital $(t, m, \alpha s)$-nets. The resulting point set $\mathcal{D}_\alpha(P)$ is called an *interlaced polynomial lattice point set*, and has been often used in applications of HoQMC methods; see [18, 17, 20, 11, 30–32, 12]. Here, we need to find good generating vectors $\boldsymbol{q} = (q_1, \dots, q_{\alpha s})$, but the digit interlacing composition makes it nontrivial whether each component q_j can be searched for one-by-one or consecutive α components $q_{\alpha(j-1)+1}, \dots, q_{\alpha j}$ should be searched for simultaneously. The papers [35, 33] originally gave a justification for employing the former approach, that is, a component-by-component construction.

4.4 Walsh functions

4.4.1 Definitions

The Walsh functions were originally introduced by Walsh [66] and have been studied thereafter, for instance, in [28, 6]. In what follows, let ω_b denote the primitive bth root of unity $\exp(2\pi\sqrt{-1}/b)$. The one-dimensional Walsh functions are defined as follows.

Definition 4.13. Let $k \in \mathbb{N}_0$ with b-adic expansion $k = \sum_{i=0}^{\infty} \kappa_i b^i$, where all but a finite number of κ_i are 0. The kth b-adic Walsh function ${}_b\mathrm{wal}_k\colon [0,1] \to \{1, \omega_b, \dots, \omega_b^{b-1}\}$ is defined by

$${}_b\mathrm{wal}_k(x) := \omega_b^{\kappa_0\xi_1+\kappa_1\xi_2+\cdots},$$

where the b-adic expansion of $x \in [0,1]$ is denoted by $x = \sum_{i=1}^{\infty} \xi_i b^{-i}$, unique in the sense that infinitely many of the ξ_i are different from $b-1$ if $x \neq 1$.

It is clear from the definition that every Walsh function is piecewise constant since it depends only on some finite number of the digits ξ_i. Multidimensional Walsh functions are given by generalizing the one-dimensional Walsh functions.

Definition 4.14. Let $s \geq 1$. For $\boldsymbol{k} = (k_1, \ldots, k_s) \in \mathbb{N}_0^s$, the $\boldsymbol{k}$th b-adic Walsh function ${}_b\mathrm{wal}_{\boldsymbol{k}}\colon [0,1]^s \to \{1, \omega_b, \ldots, \omega_b^{b-1}\}$ is defined by

$${}_b\mathrm{wal}_{\boldsymbol{k}}(\boldsymbol{x}) := \prod_{j=1}^{s} {}_b\mathrm{wal}_{k_j}(x_j).$$

Several important properties of the Walsh functions are listed below; see [24, Appendix A.2] for the proof.

Lemma 4.6. *The following holds true:*
1. *For $\boldsymbol{k}, \boldsymbol{l} \in \mathbb{N}_0^s$ and $\boldsymbol{x}, \boldsymbol{y} \in [0,1]^s$,*

$${}_b\mathrm{wal}_{\boldsymbol{k}}(\boldsymbol{x})\, {}_b\mathrm{wal}_{\boldsymbol{l}}(\boldsymbol{x}) = {}_b\mathrm{wal}_{\boldsymbol{k} \oplus \boldsymbol{l}}(\boldsymbol{x}) \quad \text{and} \quad {}_b\mathrm{wal}_{\boldsymbol{k}}(\boldsymbol{x})\, {}_b\mathrm{wal}_{\boldsymbol{k}}(\boldsymbol{y}) = {}_b\mathrm{wal}_{\boldsymbol{k}}(\boldsymbol{x} \oplus \boldsymbol{y}).$$

2. *For $\boldsymbol{k} \in \mathbb{N}_0^s$,*

$$\int_{[0,1]^s} {}_b\mathrm{wal}_{\boldsymbol{k}}(\boldsymbol{x})\, \mathrm{d}\boldsymbol{x} = \begin{cases} 1 & \text{if } \boldsymbol{k} = \boldsymbol{0}, \\ 0 & \text{otherwise.} \end{cases}$$

3. *For $\boldsymbol{k}, \boldsymbol{l} \in \mathbb{N}_0^s$,*

$$\int_{[0,1]^s} {}_b\mathrm{wal}_{\boldsymbol{k}}(\boldsymbol{x}) \overline{{}_b\mathrm{wal}_{\boldsymbol{l}}(\boldsymbol{x})}\, \mathrm{d}\boldsymbol{x} = \begin{cases} 1 & \text{if } \boldsymbol{k} = \boldsymbol{l}, \\ 0 & \text{otherwise.} \end{cases}$$

4. *For any $s \in \mathbb{N}$, the Walsh system $\{{}_b\mathrm{wal}_{\boldsymbol{k}} \mid \boldsymbol{k} \in \mathbb{N}_0^s\}$ is a complete orthonormal system in $L_2([0,1]^s)$.*

It follows from the fourth assertion of Lemma 4.6 that we can define the Walsh series of $f \in L_2([0,1]^s)$:

$$\sum_{\boldsymbol{k} \in \mathbb{N}_0^s} \hat{f}(\boldsymbol{k})\, {}_b\mathrm{wal}_{\boldsymbol{k}}(\boldsymbol{x}),$$

where $\hat{f}(\boldsymbol{k})$ denotes the $\boldsymbol{k}$th Walsh coefficient of f defined by

$$\hat{f}(\boldsymbol{k}) := \int_{[0,1]^s} f(\boldsymbol{x}) \overline{{}_b\mathrm{wal}_{\boldsymbol{k}}(\boldsymbol{x})}\, \mathrm{d}\boldsymbol{x}.$$

For any continuous function f which satisfies $\sum_{\boldsymbol{k} \in \mathbb{N}_0^s} |\hat{f}(\boldsymbol{k})| < \infty$, the above Walsh series of f equals pointwise to f itself; see [24, Theorem A.20].

4.4.2 Connection to digital nets

It follows from the first assertion of Lemma 4.6 that the Walsh functions hold the following important character property. The proof can be found, for instance, in [24, Lemma 4.75].

Lemma 4.7. *Let $P \subset [0,1]^s$ be a digital net over $\mathbb{F}_b$. For $\boldsymbol{k} \in \mathbb{N}_0^s$ we have*

$$\sum_{\boldsymbol{x}\in P} {}_b\mathrm{wal}_{\boldsymbol{k}}(\boldsymbol{x}) = \begin{cases} |P| & \text{if } \boldsymbol{k} \in P^\perp, \\ 0 & \text{otherwise.} \end{cases}$$

In what follows, by using this lemma, we show how the Walsh functions play a crucial role in analyzing the QMC integration error.

As preparation, let us consider a reproducing kernel Hilbert space H equipped with reproducing kernel $K\colon [0,1]^s \times [0,1]^s \to \mathbb{R}$ and inner product $\langle \cdot, \cdot \rangle_K$. The norm of $f \in H$ is simply given by $\|f\|_K = \sqrt{\langle f, f\rangle_K}$. The worst-case error of QMC integration using a point set P is defined by

$$e^{\mathrm{wor}}(H,P) := \sup_{\substack{f\in H\\ \|f\|_K\le 1}} \left|I(f) - I(f;P)\right|,$$

while the initial error is defined as reference by

$$e^{\mathrm{wor}}(H,0) := \sup_{\substack{f\in H\\ \|f\|_K\le 1}} \left|I(f)\right|.$$

Both the initial error and the worst-case error have explicit formulas relying only on K and P as follows; see [24, Chapter 2.3.3] for the proof.

Proposition 4.1. *For a reproducing kernel Hilbert space H whose reproducing kernel satisfies $\int_{[0,1]^s} \sqrt{K(\boldsymbol{x},\boldsymbol{x})}\,\mathrm{d}\boldsymbol{x} < \infty$, the squared initial error is given by*

$$\left(e^{\mathrm{wor}}(H,0)\right)^2 = \int_{[0,1]^{2s}} K(\boldsymbol{x},\boldsymbol{y})\,\mathrm{d}\boldsymbol{x}\,\mathrm{d}\boldsymbol{y}.$$

The squared worst-case error of QMC integration using a point set P is given by

$$\begin{aligned}\left(e^{\mathrm{wor}}(H,P)\right)^2 = {} & \int_{[0,1]^{2s}} K(\boldsymbol{x},\boldsymbol{y})\,\mathrm{d}\boldsymbol{x}\,\mathrm{d}\boldsymbol{y} - \frac{2}{|P|}\sum_{\boldsymbol{x}\in P}\int_{[0,1]^s} K(\boldsymbol{x},\boldsymbol{y})\,\mathrm{d}\boldsymbol{y} \\ & + \frac{1}{|P|^2}\sum_{\boldsymbol{x},\boldsymbol{y}\in P} K(\boldsymbol{x},\boldsymbol{y}). \end{aligned} \tag{4.4}$$

Now let us consider the Walsh series of a reproducing kernel K:

$$\sum_{\boldsymbol{k},\boldsymbol{l}\in\mathbb{N}_0^s} \hat{K}(\boldsymbol{k},\boldsymbol{l})\,{}_b\mathrm{wal}_{\boldsymbol{k}}(\boldsymbol{x})\overline{{}_b\mathrm{wal}_{\boldsymbol{l}}(\boldsymbol{y})},$$

where the $(\boldsymbol{k},\boldsymbol{l})$-th Walsh coefficient $\hat{K}(\boldsymbol{k},\boldsymbol{l})$ is defined by

$$\hat{K}(\boldsymbol{k},\boldsymbol{l}) = \int_{[0,1]^{2s}} K(\boldsymbol{x},\boldsymbol{y})\overline{{}_b\mathrm{wal}_{\boldsymbol{k}}(\boldsymbol{x})}\,{}_b\mathrm{wal}_{\boldsymbol{l}}(\boldsymbol{y})\,\mathrm{d}\boldsymbol{x}\,\mathrm{d}\boldsymbol{y}.$$

Again the pointwise equality holds between the above Walsh series and K itself if K is continuous and $\sum_{\boldsymbol{k},\boldsymbol{l}\in\mathbb{N}_0^s}|\hat{K}(\boldsymbol{k},\boldsymbol{l})| < \infty$.

The following proposition provides a simple expression of the squared worst-case error when the point set is a *digitally shifted digital net*

$$P \oplus \boldsymbol{\delta} = \{\boldsymbol{x} \oplus \boldsymbol{\delta} \mid \boldsymbol{x} \in P\},$$

where P is a digital net over $\mathbb{F}_b$ and $\boldsymbol{\delta} \in [0,1]^s$. The following result is well understood, but as far as the authors know, this result is not available in the literature in this full generality, so that we provide a proof for the sake of completeness.

Proposition 4.2. *Let P be a digital net over $\mathbb{F}_b$ and $\boldsymbol{\delta} \in [0,1]^s$. For a reproducing kernel Hilbert space H whose reproducing kernel K is continuous and satisfies $\int_{[0,1]^s}\sqrt{K(\boldsymbol{x},\boldsymbol{x})}\,\mathrm{d}\boldsymbol{x} < \infty$ and $\sum_{\boldsymbol{k},\boldsymbol{l}\in\mathbb{N}_0^s}|\hat{K}(\boldsymbol{k},\boldsymbol{l})| < \infty$, we have*

$$(e^{\mathrm{wor}}(H, P\oplus\boldsymbol{\delta}))^2 = \sum_{\boldsymbol{k},\boldsymbol{l}\in P^\perp\setminus\{\boldsymbol{0}\}} \hat{K}(\boldsymbol{k},\boldsymbol{l})\,{}_b\mathrm{wal}_{\boldsymbol{k}}(\boldsymbol{\delta})\overline{{}_b\mathrm{wal}_{\boldsymbol{l}}(\boldsymbol{\delta})}.$$

Proof. It is trivial from the definition of the Walsh functions that the first term on the right-hand side of (4.4) is equal to $\hat{K}(\boldsymbol{0},\boldsymbol{0})$. For the second term on the right-hand side of (4.4), by using the symmetry of K, the first and second assertions of Lemma 4.6 and Lemma 4.7, we have

$$\begin{aligned}
&\frac{2}{|P|}\sum_{\boldsymbol{x}\in P}\int_{[0,1]^s} K(\boldsymbol{x}\oplus\boldsymbol{\delta},\boldsymbol{y})\,\mathrm{d}\boldsymbol{y}\\
&= \frac{1}{|P|}\sum_{\boldsymbol{x}\in P}\int_{[0,1]^s} K(\boldsymbol{x}\oplus\boldsymbol{\delta},\boldsymbol{y})\,\mathrm{d}\boldsymbol{y} + \frac{1}{|P|}\sum_{\boldsymbol{x}\in P}\int_{[0,1]^s} K(\boldsymbol{y},\boldsymbol{x}\oplus\boldsymbol{\delta})\,\mathrm{d}\boldsymbol{y}\\
&= \frac{1}{|P|}\sum_{\boldsymbol{x}\in P}\sum_{\boldsymbol{k},\boldsymbol{l}\in\mathbb{N}_0^s} \hat{K}(\boldsymbol{k},\boldsymbol{l})\,{}_b\mathrm{wal}_{\boldsymbol{k}}(\boldsymbol{x}\oplus\boldsymbol{\delta})\int_{[0,1]^s}\overline{{}_b\mathrm{wal}_{\boldsymbol{l}}(\boldsymbol{y})}\,\mathrm{d}\boldsymbol{y}\\
&\quad + \frac{1}{|P|}\sum_{\boldsymbol{x}\in P}\sum_{\boldsymbol{k},\boldsymbol{l}\in\mathbb{N}_0^s} \hat{K}(\boldsymbol{k},\boldsymbol{l})\overline{{}_b\mathrm{wal}_{\boldsymbol{l}}(\boldsymbol{x}\oplus\boldsymbol{\delta})}\int_{[0,1]^s}{}_b\mathrm{wal}_{\boldsymbol{k}}(\boldsymbol{y})\,\mathrm{d}\boldsymbol{y}\\
&= \sum_{\boldsymbol{k}\in\mathbb{N}_0^s}\hat{K}(\boldsymbol{k},\boldsymbol{0})\,{}_b\mathrm{wal}_{\boldsymbol{k}}(\boldsymbol{\delta})\frac{1}{|P|}\sum_{\boldsymbol{x}\in P}{}_b\mathrm{wal}_{\boldsymbol{k}}(\boldsymbol{x})\\
&\quad + \sum_{\boldsymbol{l}\in\mathbb{N}_0^s}\hat{K}(\boldsymbol{0},\boldsymbol{l})\overline{{}_b\mathrm{wal}_{\boldsymbol{l}}(\boldsymbol{\delta})}\frac{1}{|P|}\sum_{\boldsymbol{x}\in P}\overline{{}_b\mathrm{wal}_{\boldsymbol{l}}(\boldsymbol{x})}\\
&= \sum_{\boldsymbol{k}\in P^\perp}\hat{K}(\boldsymbol{k},\boldsymbol{0})\,{}_b\mathrm{wal}_{\boldsymbol{k}}(\boldsymbol{\delta}) + \sum_{\boldsymbol{l}\in P^\perp}\hat{K}(\boldsymbol{0},\boldsymbol{l})\overline{{}_b\mathrm{wal}_{\boldsymbol{l}}(\boldsymbol{\delta})}.
\end{aligned}$$

Finally, for the third term on the right-hand side of (4.4), by using the first assertion of Lemma 4.6 and Lemma 4.7, we have

$$\begin{aligned}
&\frac{1}{|P|^2}\sum_{\boldsymbol{x},\boldsymbol{y}\in P} K(\boldsymbol{x}\oplus\boldsymbol{\delta},\boldsymbol{y}\oplus\boldsymbol{\delta})\\
&=\frac{1}{|P|^2}\sum_{\boldsymbol{x},\boldsymbol{y}\in P}\sum_{\boldsymbol{k},\boldsymbol{l}\in\mathbb{N}_0^s}\hat{K}(\boldsymbol{k},\boldsymbol{l})\,{}_b\mathrm{wal}_{\boldsymbol{k}}(\boldsymbol{x}\oplus\boldsymbol{\delta})\overline{{}_b\mathrm{wal}_{\boldsymbol{l}}(\boldsymbol{y}\oplus\boldsymbol{\delta})}\\
&=\sum_{\boldsymbol{k},\boldsymbol{l}\in\mathbb{N}_0^s}\hat{K}(\boldsymbol{k},\boldsymbol{l})\,{}_b\mathrm{wal}_{\boldsymbol{k}}(\boldsymbol{\delta})\overline{{}_b\mathrm{wal}_{\boldsymbol{l}}(\boldsymbol{\delta})}\frac{1}{|P|}\sum_{\boldsymbol{x}\in P}{}_b\mathrm{wal}_{\boldsymbol{k}}(\boldsymbol{x})\frac{1}{|P|}\sum_{\boldsymbol{y}\in P}\overline{{}_b\mathrm{wal}_{\boldsymbol{l}}(\boldsymbol{y})}\\
&=\sum_{\boldsymbol{k},\boldsymbol{l}\in P^\perp}\hat{K}(\boldsymbol{k},\boldsymbol{l})\,{}_b\mathrm{wal}_{\boldsymbol{k}}(\boldsymbol{\delta})\overline{{}_b\mathrm{wal}_{\boldsymbol{l}}(\boldsymbol{\delta})}.
\end{aligned}$$

Altogether we obtain

$$\begin{aligned}
(e^{\mathrm{wor}}(H,P\oplus\boldsymbol{\delta}))^2 &= \hat{K}(\boldsymbol{0},\boldsymbol{0})-\sum_{\boldsymbol{k}\in P^\perp}\hat{K}(\boldsymbol{k},\boldsymbol{0})\,{}_b\mathrm{wal}_{\boldsymbol{k}}(\boldsymbol{\delta})-\sum_{\boldsymbol{l}\in P^\perp}\hat{K}(\boldsymbol{0},\boldsymbol{l})\overline{{}_b\mathrm{wal}_{\boldsymbol{l}}(\boldsymbol{\delta})}\\
&\quad+\sum_{\boldsymbol{k},\boldsymbol{l}\in P^\perp}\hat{K}(\boldsymbol{k},\boldsymbol{l})\,{}_b\mathrm{wal}_{\boldsymbol{k}}(\boldsymbol{\delta})\overline{{}_b\mathrm{wal}_{\boldsymbol{l}}(\boldsymbol{\delta})}\\
&=\sum_{\boldsymbol{k},\boldsymbol{l}\in P^\perp\setminus\{\boldsymbol{0}\}}\hat{K}(\boldsymbol{k},\boldsymbol{l})\,{}_b\mathrm{wal}_{\boldsymbol{k}}(\boldsymbol{\delta})\overline{{}_b\mathrm{wal}_{\boldsymbol{l}}(\boldsymbol{\delta})},
\end{aligned}$$

which completes the proof. □

Using Proposition 4.2 we obtain the following result.

Corollary 4.1. *Let P be a digital net over $\mathbb{F}_b$ and H be a reproducing kernel Hilbert space whose reproducing kernel K is continuous and satisfies $\int_{[0,1]^s}\sqrt{K(\boldsymbol{x},\boldsymbol{x})}\,\mathrm{d}\boldsymbol{x}<\infty$ and $\sum_{\boldsymbol{k},\boldsymbol{l}\in\mathbb{N}_0^s}|\hat{K}(\boldsymbol{k},\boldsymbol{l})|<\infty$. Then we have*

$$(e^{\mathrm{wor}}(H,P))^2=\sum_{\boldsymbol{k},\boldsymbol{l}\in P^\perp\setminus\{\boldsymbol{0}\}}\hat{K}(\boldsymbol{k},\boldsymbol{l}),$$

and

$$\int_{[0,1]^s}(e^{\mathrm{wor}}(H,P\oplus\boldsymbol{\delta}))^2\,\mathrm{d}\boldsymbol{\delta}=\sum_{\boldsymbol{k}\in P^\perp\setminus\{\boldsymbol{0}\}}\hat{K}(\boldsymbol{k},\boldsymbol{k}).$$

Proof. The first assertion follows immediately from Proposition 4.2 by considering the case $\boldsymbol{\delta}=\boldsymbol{0}$. For the second assertion, it follows from Proposition 4.2 and the third assertion of Lemma 4.6 that

$$\begin{aligned}
\int_{[0,1]^s}(e^{\mathrm{wor}}(H,P\oplus\boldsymbol{\delta}))^2\,\mathrm{d}\boldsymbol{\delta}&=\sum_{\boldsymbol{k},\boldsymbol{l}\in P^\perp\setminus\{\boldsymbol{0}\}}\hat{K}(\boldsymbol{k},\boldsymbol{l})\int_{[0,1]^s}{}_b\mathrm{wal}_{\boldsymbol{k}}(\boldsymbol{\delta})\overline{{}_b\mathrm{wal}_{\boldsymbol{l}}(\boldsymbol{\delta})}\,\mathrm{d}\boldsymbol{\delta}\\
&=\sum_{\boldsymbol{k}\in P^\perp\setminus\{\boldsymbol{0}\}}\hat{K}(\boldsymbol{k},\boldsymbol{k}).
\end{aligned}$$

□

We would like to emphasize that the worst-case error of a QMC rule using a digital net is given by the double sum of the Walsh coefficients of reproducing kernel, whereas the shift-averaged worst-case error is simply given by the single sum of their diagonal elements. This way applying a random digital shift has an effect on vanishing all nondiagonal terms, which sometimes makes the error analysis much easier as we shall see in the subsequent sections.

4.5 Discrepancy

4.5.1 Definitions

As represented by the Koksma–Hlawka inequality (4.1), the star-discrepancy, or more generally speaking, the L_p-discrepancy is an extremely important quantitative measure of how uniformly a point set is distributed over $[0,1]^s$.

Definition 4.15. Let P be a point set in $[0,1]^s$. For $\boldsymbol{y} = (y_1,\dots,y_s) \in [0,1]^s$, we write $[\boldsymbol{0},\boldsymbol{y}) = [0,y_1) \times \cdots \times [0,y_s)$ and define the so-called *local discrepancy function*

$$\Delta_P(\boldsymbol{y}) := \frac{1}{|P|} \sum_{\boldsymbol{x} \in P} \mathbf{1}_{[\boldsymbol{0},\boldsymbol{y})}(\boldsymbol{x}) - \prod_{j=1}^{s} y_j,$$

where $\mathbf{1}_{[\boldsymbol{0},\boldsymbol{y})}$ denotes the indicator function which is equal to 1 if $\boldsymbol{x} \in [\boldsymbol{0},\boldsymbol{y})$, and 0 otherwise. For $1 \le p \le \infty$, the L_p-discrepancy of P is defined as the L_p-norm of $\Delta_P(\boldsymbol{y})$, that is,

$$L_p(P) := \left(\int_{[0,1]^s} |\Delta_P(\boldsymbol{y})|^p \, \mathrm{d}\boldsymbol{y} \right)^{1/p},$$

with the obvious modification for $p = \infty$.

We speak of the star-discrepancy if $p = \infty$ and the different notation such as $D^*(P)$ have been often used in the literature.

4.5.2 L_2-discrepancy and worst-case error

In this section, we are particularly interested in the L_2-discrepancy. Here, we follow the exposition of [24, Chapter 2.4] to give a connection between the L_2-discrepancy and the worst-case error for some reproducing kernel Hilbert space.

Let $H^{\downarrow}$ denote the reproducing kernel Hilbert space whose reproducing kernel is given by

$$K^{\downarrow}(\boldsymbol{x},\boldsymbol{y}) = \prod_{j=1}^{s} \min(1-x_j, 1-y_j).$$

It is known from [1, Section 8] that the function space $H^{\downarrow}$ is the s-fold tensor product of the univariate reproducing kernel Hilbert space with reproducing kernel $K^{\downarrow}(x,y) = \min(1-x, 1-y)$. Moreover, $H^{\downarrow}$ contains the completion with respect to the L_2-norm of the derivatives of all products of absolutely continuous functions which satisfy

$$\frac{\partial^{|u|} f}{\partial \boldsymbol{x}_u}(\boldsymbol{x}_u, \boldsymbol{1}) = 0 \quad \text{for all } \emptyset = u \subsetneq \{1, \dots, s\},$$

where we write $\boldsymbol{x}_u = (x_j)_{j \in u}$ and $(\boldsymbol{x}_u, \boldsymbol{1})$ is the vector whose jth component is equal to x_j if $j \in u$, and 1 otherwise. Note that the symbol $\downarrow$ is used to represent this "anchor" property of the space. The inner product is given by

$$\langle f, g \rangle_{K^{\downarrow}} = \int_{[0,1]^s} \frac{\partial^s f}{\partial \boldsymbol{x}}(\boldsymbol{x}) \frac{\partial^s g}{\partial \boldsymbol{x}}(\boldsymbol{x}) \, \mathrm{d}\boldsymbol{x},$$

for $f, g \in H^{\downarrow}$. Then the following result is known.

Lemma 4.8. *For any point set $P \subset [0,1]^s$, we have*

$$L_2(P) = e^{\mathrm{wor}}(H^{\downarrow}, P).$$

Therefore, when P is a digital net over $\mathbb{F}_b$, combining this identity with Corollary 4.1 gives an expression for the squared L_2-discrepancy:

$$(L_2(P))^2 = \sum_{\boldsymbol{k}, \boldsymbol{l} \in P^{\perp} \setminus \{\boldsymbol{0}\}} \widehat{K^{\downarrow}}(\boldsymbol{k}, \boldsymbol{l}),$$

and also an expression for the shift-averaged squared L_2-discrepancy:

$$\int_{[0,1]^s} (L_2(P \oplus \boldsymbol{\delta}))^2 \, \mathrm{d}\boldsymbol{\delta} = \sum_{\boldsymbol{k} \in P^{\perp} \setminus \{\boldsymbol{0}\}} \widehat{K^{\downarrow}}(\boldsymbol{k}, \boldsymbol{k}).$$

In [25, Lemma 2.2], Dick and Pillichshammer conducted an exact evaluation of the Walsh coefficients $\widehat{K^{\downarrow}}(\boldsymbol{k}, \boldsymbol{l})$ for all $\boldsymbol{k}, \boldsymbol{l} \in \mathbb{N}_0^s$ when $b = 2$. Here, we simplify their result in a way that will be sufficient for readers to understand the proof of optimal order L_2-discrepancy bounds.

Lemma 4.9. *For $\boldsymbol{k}, \boldsymbol{l} \in \mathbb{N}_0^s$, we have:*
1. $|\widehat{K^{\downarrow}}(\boldsymbol{k}, \boldsymbol{l})| \le 3^{-s} \cdot 2^{-\mu_1(\boldsymbol{k}) - \mu_1(\boldsymbol{l})}$.
2. $\widehat{K^{\downarrow}}(\boldsymbol{k}, \boldsymbol{l}) = 0$ *if* $(\boldsymbol{k}, \boldsymbol{l}) \in T_{\ge 3}$.

Here, we recall that the notion $T_{\ge \alpha}$ was introduced in Subsection 4.3.1. The first assertion of the lemma shows the decay of the Walsh coefficients, while the second assertion shows the sparsity of the Walsh coefficients.

4.5.3 Optimal order L_2-discrepancy bounds

Before stating the result by Dick and Pillichshammer [25] on the L_2-discrepancy bound for higher order digital nets, we start from another result by the same authors shown in [23] by using Lemma 4.1.

Theorem 4.2. *Let P be a digital (t, m, s)-net over $\mathbb{F}_2$. Then there exists a constant $D^{(1)}_{s,t}$ such that the following holds:*

$$\int_{[0,1]^s} (L_2(P \oplus \boldsymbol{\delta}))^2 \,\mathrm{d}\boldsymbol{\delta} \le D^{(1)}_{s,t} \frac{m^{s-1}}{2^{2m}}.$$

Proof. Using Lemma 4.9 and the second assertion of Lemma 4.1 (with $\lambda = 2$), we have

$$\begin{aligned}\int_{[0,1]^s} (L_2(P \oplus \boldsymbol{\delta}))^2 \,\mathrm{d}\boldsymbol{\delta} &= \sum_{\boldsymbol{k} \in P^\perp \setminus \{\boldsymbol{0}\}} \widehat{K^\downarrow}(\boldsymbol{k}, \boldsymbol{k}) \le \frac{1}{3^s} \sum_{\boldsymbol{k} \in P^\perp \setminus \{\boldsymbol{0}\}} 2^{-2\mu_1(\boldsymbol{k})} \\ &\le \frac{2^{s+1}}{3^s} \cdot \frac{(\mu_1(P^\perp))^{s-1}}{2^{2\mu_1(P^\perp)}} \sum_{z=1}^{\infty} 2^{-z} z^{s-1}.\end{aligned}$$

Since we have $\mu_1(P^\perp) = m - t + 1$, the result of the theorem follows. □

This theorem directly implies the existence of a digital shift $\boldsymbol{\delta} \in [0,1]^s$ such that the digitally shifted digital (t, m, s)-net satisfies the best possible order of L_2-discrepancy:

$$L_2(P \oplus \boldsymbol{\delta}) \le \sqrt{D^{(1)}_{s,t}} \frac{m^{(s-1)/2}}{2^m} = \sqrt{D^{(1)}_{s,t}} \frac{(\log_2 N)^{(s-1)/2}}{N}.$$

However, this is a probabilistic result since we do not know how to find such $\boldsymbol{\delta}$ explicitly.

We note that the optimal exponent $(s - 1)$ of the numerator in Theorem 4.2 comes from the inequality

$$\left|\{\boldsymbol{k} \in P^\perp \setminus \{\boldsymbol{0}\} \mid \mu_1(\boldsymbol{k}) = z\}\right| \le b^{z-\mu_1(P^\perp)+1}(z+1)^{s-1}$$

given in the first assertion of Lemma 4.1. As is clear from the proof, since the expression of the shift-averaged squared L_2-discrepancy has only the diagonal terms of the Walsh coefficients, there is no necessity of exploiting the sparsity of the Walsh coefficients. In order to obtain a deterministic counterpart of Theorem 4.2, however, it seems insufficient to exploit the decay of the Walsh coefficients only, and one approach is to exploit both the decay and the sparsity of the Walsh coefficients simultaneously. To do this, we rely on Lemma 4.4. The following theorem is from [25, Theorem 4.1].

Theorem 4.3. *Let P be an order 3 digital (t_3, m, s)-net over $\mathbb{F}_2$. Then there exists a constant $D^{(2)}_{s,t_3}$ such that the following holds:*

$$(L_2(P))^2 \le D^{(2)}_{s,t_3} \frac{m^{s-1}}{2^{2m}}.$$

Proof. Using Lemmas 4.9 and 4.4 (with $\alpha = 3$) and then applying the change of variables $z \to z + 2\mu_1(P^\perp)$, we have

$$\begin{aligned}
(L_2(P))^2 &= \sum_{\boldsymbol{k},\boldsymbol{l} \in P^\perp \setminus \{\boldsymbol{0}\}} \widehat{K^\downarrow}(\boldsymbol{k},\boldsymbol{l}) \le \frac{1}{3^s} \sum_{\substack{\boldsymbol{k},\boldsymbol{l} \in P^\perp \setminus \{\boldsymbol{0}\} \\ (\boldsymbol{k},\boldsymbol{l}) \notin T_{\ge 3}}} 2^{-\mu_1(\boldsymbol{k}) - \mu_1(\boldsymbol{l})} \\
&= \frac{1}{3^s} \sum_{z=2\mu_1(P^\perp)}^{\infty} 2^{-z} \left| \{(\boldsymbol{k},\boldsymbol{l}) \in (P^\perp \setminus \{\boldsymbol{0}\})^2 \mid \mu_1(\boldsymbol{k}) + \mu_1(\boldsymbol{l}) = z, (\boldsymbol{k},\boldsymbol{l}) \notin T_{\ge 3}\} \right| \\
&\le \frac{B_{3,2,s,t_3}}{3^s} \sum_{z=2\mu_1(P^\perp)}^{\infty} 2^{-z} (z - 2\mu_1(P^\perp))^{2s+1} z^{s-1} 2^{(z-2\mu_1(P^\perp))/2} \\
&= \frac{B_{3,2,s,t_3}}{3^s} 2^{-2\mu_1(P^\perp)} \sum_{z=0}^{\infty} 2^{-z/2} z^{2s+1} (z + 2\mu_1(P^\perp))^{s-1} \\
&\le \frac{B_{3,2,s,t_3}}{3^s} \cdot \frac{2^{2(s-1)} (\mu_1(P^\perp))^{s-1}}{2^{2\mu_1(P^\perp)}} \sum_{z=0}^{\infty} 2^{-z/2} z^{3s}.
\end{aligned}$$

The last sum over z is trivially finite. It follows from Lemma 4.2 that an order 3 digital (t_3, m, s)-net over $\mathbb{F}_2$ is also a digital (t, m, s)-net over $\mathbb{F}_2$ with $t \le \lceil t_3/3 \rceil$, that is, $\mu_1(P^\perp) \ge m - \lceil t_3/3 \rceil + 1$. Thus, the result of the theorem follows. □

Remark 4.5. If we do not take the sparsity of the Walsh coefficients into account, we might proceed like

$$\begin{aligned}
(L_2(P))^2 &= \sum_{\boldsymbol{k},\boldsymbol{l} \in P^\perp \setminus \{\boldsymbol{0}\}} \widehat{K^\downarrow}(\boldsymbol{k},\boldsymbol{l}) \le \frac{1}{3^s} \sum_{\boldsymbol{k},\boldsymbol{l} \in P^\perp \setminus \{\boldsymbol{0}\}} 2^{-\mu_1(\boldsymbol{k}) - \mu_1(\boldsymbol{l})} \\
&= \frac{1}{3^s} \left(\sum_{\boldsymbol{k} \in P^\perp \setminus \{\boldsymbol{0}\}} 2^{-\mu_1(\boldsymbol{k})} \right)^2 \\
&= \frac{1}{3^s} \left(\sum_{z=\mu_1(P^\perp)}^{\infty} 2^{-z} \left| \{\boldsymbol{k} \in P^\perp \setminus \{\boldsymbol{0}\} \mid \mu_1(\boldsymbol{k}) = z\} \right| \right)^2 \\
&\le \frac{2^2}{3^s \cdot 2^{\mu_1(P^\perp)}} \left(\sum_{z=\mu_1(P^\perp)}^{\infty} (z+1)^{s-1} \right)^2,
\end{aligned}$$

where we used Lemma 4.1 in the last inequality. Since the last sum over z obviously diverges, this argument ends up with a trivial upper bound.

We recall that Theorem 4.3 is for order 3 digital nets over $\mathbb{F}_2$, a class of finite point sets. By considering order 5 digital sequences, as the other main result of the paper [25], Dick and Pillichshammer proved the optimal order L_2-discrepancy bound which holds uniformly for all N.

4.6 Numerical integration

4.6.1 Sobolev spaces

Let us move on to multivariate numerical integration. The function space of our interest in this section is defined as follows. First, let us consider the one-dimensional case. For $\alpha \in \mathbb{N}$, the Sobolev space with smoothness α is

$$H_\alpha := \{f: [0,1] \to \mathbb{R} \mid \\ f^{(r)}: \text{absolutely continuous for } r = 0, \dots, \alpha - 1, f^{(\alpha)} \in L^2([0,1])\},$$

where $f^{(r)}$ denotes the rth derivative of f. According to [65, Chapter 10.2], the space H_α is a reproducing kernel Hilbert space with the reproducing kernel

$$K_\alpha(x,y) = \sum_{r=0}^{\alpha} \frac{B_r(x)B_r(y)}{(r!)^2} + (-1)^{\alpha+1}\frac{B_{2\alpha}(|x-y|)}{(2\alpha)!},$$

for $x, y \in [0,1]$, where B_r denotes the Bernoulli polynomial of degree r, and with the inner product

$$\langle f, g\rangle_{K_\alpha} = \sum_{r=0}^{\alpha-1} \int_0^1 f^{(r)}(x)\,\mathrm{d}x \int_0^1 g^{(r)}(x)\,\mathrm{d}x + \int_0^1 f^{(\alpha)}(x) g^{(\alpha)}(x)\,\mathrm{d}x,$$

for $f, g \in H_\alpha$.

For the s-dimensional case, we consider the s-fold tensor product space of the one-dimensional space introduced above. Thus the Sobolev space $H_{\alpha,s}$ which we consider is simply given by $H_{\alpha,s} = \bigotimes_{j=1}^{s} H_\alpha$. Again it is known from [1, Section 8] that the reproducing kernel of the space $H_{\alpha,s}$ is the product of the reproducing kernels for the one-dimensional space H_α. Therefore, $H_{\alpha,s}$ is the reproducing kernel Hilbert space with the reproducing kernel

$$K_{\alpha,s}(\boldsymbol{x},\boldsymbol{y}) = \prod_{j=1}^{s} K_\alpha(x_j, y_j),$$

and with the inner product

$$\langle f, g\rangle_{K_{\alpha,s}} = \sum_{u \subseteq \{1,\dots,s\}} \sum_{\boldsymbol{r}_u \in \{0,\dots,\alpha-1\}^{|u|}} \int_{[0,1]^{s-|u|}} \\ \left(\int_{[0,1]^{|u|}} f^{(\boldsymbol{r}_u, \boldsymbol{\alpha})}(\boldsymbol{x})\,\mathrm{d}\boldsymbol{x}_u \right)\left(\int_{[0,1]^{|u|}} g^{(\boldsymbol{r}_u, \boldsymbol{\alpha})}(\boldsymbol{x})\,\mathrm{d}\boldsymbol{x}_u \right) \mathrm{d}\boldsymbol{x}_{\{1,\dots,s\}\setminus u},$$

for $f, g \in H_{\alpha,s}$. In the above, we used the following notation: For $u \subseteq \{1,\dots,s\}$ and $\boldsymbol{x} \in [0,1]^s$, we write $\boldsymbol{x}_u = (x_j)_{j\in u}$. Moreover, for $\boldsymbol{r}_u = (r_j)_{j \in u} \in \{0,\dots,\alpha-1\}^{|u|}$, $(\boldsymbol{r}_u, \boldsymbol{\alpha})$ denotes the s-dimensional vector whose jth component equals r_j if $j \in u$, and α otherwise. Note that an integral and sum over the empty set is defined to be the identity operator. Since the dimension s is fixed, we shall simply write H_α and K_α instead of $H_{\alpha,s}$ and $K_{\alpha,s}$, respectively.

If P is a digital net over $\mathbb{F}_b$, Corollary 4.1 gives

$$(e^{\mathrm{wor}}(H_\alpha, P))^2 = \sum_{\boldsymbol{k},\boldsymbol{l} \in P^\perp \setminus \{\boldsymbol{0}\}} \widehat{K_\alpha}(\boldsymbol{k}, \boldsymbol{l}),$$

and

$$\int_{[0,1]^s} (e^{\mathrm{wor}}(H_\alpha, P \oplus \boldsymbol{\delta}))^2 \, \mathrm{d}\boldsymbol{\delta} = \sum_{\boldsymbol{k} \in P^\perp \setminus \{\boldsymbol{0}\}} \widehat{K_\alpha}(\boldsymbol{k}, \boldsymbol{k}).$$

Similar to Lemma 4.9, the following result is known.

Lemma 4.10. *For $\boldsymbol{k}, \boldsymbol{l} \in \mathbb{N}_0^s$, we have:*
1. $|\widehat{K_\alpha}(\boldsymbol{k}, \boldsymbol{l})| \le C_{\alpha,b}^s b^{-\mu_\alpha(\boldsymbol{k})-\mu_\alpha(\boldsymbol{l})}$ *with*

$$C_{\alpha,b} = \max_{1 \le v \le \alpha} \left\{ \sum_{\tau=v}^{\alpha} \frac{(C'_{\tau,b})^2}{b^{2(\tau-v)}} + \frac{2C'_{2\alpha,b}}{b^{2(\alpha-v)}} \right\},$$

where

$$C'_{1,b} = \frac{1}{2\sin(\pi/b)} \quad \textit{and} \quad C'_{\tau,b} = \frac{(1 + 1/b + 1/(b(b+1)))^{\tau-2}}{(2\sin(\pi/b))^\tau} \quad \textit{for } \tau \ge 2.$$

2. $\widehat{K_\alpha}(\boldsymbol{k}, \boldsymbol{l}) = 0$ *if* $(\boldsymbol{k}, \boldsymbol{l}) \in T_{\ge 2\alpha+1}$.

The first assertion on the decay of the Walsh coefficients was shown in [3], while the second assertion of the sparsity of the Walsh coefficients was shown in [38]. Regarding the first assertion, we also refer to more recent works [62, 67] which introduce different approaches from the one by Dick [7–9] for evaluating the Walsh coefficients. In many cases, one may obtain smaller constants $C_{\alpha,b}$.

Remark 4.6. A lower bound on the worst-case error of order $(\log N)^{(s-1)/2}/N^\alpha$, which holds for any (nonlinear/adaptive) quadrature rule based on N function evaluations, can be proven by adapting the bump function technique from [2]. We also refer to [22, Theorem 4 and Appendix] whose result directly applies to the present problem. As we shall show, higher order digital nets and sequences achieve this best possible order exactly.

4.6.2 Optimal order error bounds

First, we discuss what happens if we exploit only the decay of the Walsh coefficients. A similar result has been already proven in [3, Theorem 30] but with a slightly worse exponent of the $\log N$ term.

Theorem 4.4. *Let $\alpha \geq 2$. Let P be an order α digital (t_α, m, s)-net over $\mathbb{F}_b$. Then there exists a constant $E^{(1)}_{\alpha,b,s,t_\alpha} > 0$ such that the following holds:*

$$\int_{[0,1]^s} (e^{\mathrm{wor}}(H_\alpha, P \oplus \boldsymbol{\delta}))^2 \, \mathrm{d}\boldsymbol{\delta} \leq E^{(1)}_{\alpha,b,s,t_\alpha} \frac{m^{s\alpha-1}}{b^{2\alpha m}}.$$

Proof. Using the first assertion of Lemma 4.10 and the second assertion of Lemma 4.3 (with $\lambda = 2$), we have

$$\begin{aligned}
&\int_{[0,1]^s} (e^{\mathrm{wor}}(H_\alpha, P \oplus \boldsymbol{\delta}))^2 \, \mathrm{d}\boldsymbol{\delta} \\
&= \sum_{\boldsymbol{k} \in P^\perp \setminus \{\boldsymbol{0}\}} \widehat{K_\alpha}(\boldsymbol{k}, \boldsymbol{k}) \leq C^s_{\alpha,b} \sum_{\boldsymbol{k} \in P^\perp \setminus \{\boldsymbol{0}\}} b^{-2\mu_\alpha(\boldsymbol{k})} \\
&\leq C^s_{\alpha,b} 2^{s\alpha-1} (b-1)^{s\alpha} \frac{(\mu_\alpha(P^\perp))^{s\alpha-1}}{b^{2\mu_\alpha(P^\perp)}} \sum_{z=0}^{\infty} b^{(1/\alpha-2)z} (z+2)^{s\alpha-1}.
\end{aligned}$$

By considering the equality $\mu_\alpha(P^\perp) = \alpha m - t_\alpha + 1$, the theorem follows. □

Again the exponent $(s\alpha - 1)$ of the numerator of the theorem stems from the inequality

$$\left|\{\boldsymbol{k} \in P^\perp \setminus \{\boldsymbol{0}\} \mid \mu_\alpha(\boldsymbol{k}) = z\}\right| \leq (b-1)^{s\alpha} b^{(z-\mu_\alpha(P^\perp))/\alpha} (z+2)^{s\alpha-1}$$

given in the first assertion of Lemma 4.3. It seems hard to fundamentally improve this bound. Thus, even before exploiting the sparsity of the Walsh coefficients, there is a difficulty in obtaining the best possible order of the shift-averaged worst-case error.

To overcome this issue, let us go back to the bound given in the first assertion of Lemma 4.1:

$$\left|\{\boldsymbol{k} \in P^\perp \setminus \{\boldsymbol{0}\} \mid \mu_1(\boldsymbol{k}) = z\}\right| \leq b^{z-\mu_1(P^\perp)+1} (z+1)^{s-1}.$$

As we discussed in Subsection 4.5.3, this gives the optimal order of the shift-averaged L_2-discrepancy. Considering that the best possible exponent of the $\log N$ term for the present integration problem is the same as that of the L_2-discrepancy, which is $(s-1)/2$, one idea is to switch the weight function from μ_α to μ_1 in the error analysis. The following interpolation inequality was shown in [36] to realize this.

Lemma 4.11. *Let $\alpha, \beta \in \mathbb{N}$ with $1 < \alpha < \beta$. For any $\boldsymbol{k} \in \mathbb{N}_0^s$, we have*

$$\mu_\alpha(\boldsymbol{k}) \geq \frac{\alpha-1}{\beta-1} \mu_\beta(\boldsymbol{k}) + \frac{\beta-\alpha}{\beta-1} \mu_1(\boldsymbol{k}).$$

By using this inequality together with the propagation rule of higher order digital nets (Lemma 4.2), we can prove the optimal order of the shift-averaged worst-case error by using order 2α digital nets rather than order α digital nets. The following result is from [36].

Theorem 4.5. *Let $\alpha \geq 2$. Let P be an order 2α digital $(t_{2\alpha}, m, s)$-net over $\mathbb{F}_b$. Then there exists a constant $E^{(2)}_{\alpha,b,s,t_{2\alpha}} > 0$ such that the following holds:*

$$\int_{[0,1]^s} (e^{\mathrm{wor}}(H_\alpha, P \oplus \boldsymbol{\delta}))^2 \, \mathrm{d}\boldsymbol{\delta} \leq E^{(2)}_{\alpha,b,s,t_{2\alpha}} \frac{m^{s-1}}{b^{2\alpha m}}.$$

Proof. We write $A = (\alpha - 1)/(2\alpha - 1)$ and $B = \alpha/(2\alpha - 1)$. Using the first assertion of Lemma 4.10, Lemma 4.11 (with $\beta = 2\alpha$), and the second assertion of Lemma 4.1 (with $\lambda = 2B$), we have

$$\begin{aligned}
&\int_{[0,1]^s} (e^{\mathrm{wor}}(H_\alpha, P \oplus \boldsymbol{\delta}))^2 \, \mathrm{d}\boldsymbol{\delta} \\
&= \sum_{\boldsymbol{k} \in P^\perp \setminus \{\boldsymbol{0}\}} \widehat{K_\alpha}(\boldsymbol{k}, \boldsymbol{k}) \leq C^s_{\alpha,b} \sum_{\boldsymbol{k} \in P^\perp \setminus \{\boldsymbol{0}\}} b^{-2\mu_\alpha(\boldsymbol{k})} \\
&\leq C^s_{\alpha,b} \sum_{\boldsymbol{k} \in P^\perp \setminus \{\boldsymbol{0}\}} b^{-2A\mu_{2\alpha}(\boldsymbol{k}) - 2B\mu_1(\boldsymbol{k})} \\
&\leq C^s_{\alpha,b} b^{-2A\mu_{2\alpha}(P^\perp)} \sum_{\boldsymbol{k} \in P^\perp \setminus \{\boldsymbol{0}\}} b^{-2B\mu_1(\boldsymbol{k})} \\
&\leq C^s_{\alpha,b} 2^{s-1} b^{2B} \frac{(\mu_1(P^\perp))^{s-1}}{b^{2A\mu_{2\alpha}(P^\perp) + 2B\mu_1(P^\perp)}} \sum_{z=1}^{\infty} b^{(1-2B)z} z^{s-1}.
\end{aligned}$$

Since $2B - 1 = 1/(2\alpha - 1) > 0$, the last sum over z is finite. Using Lemma 4.2, we have

$$\mu_1(P^\perp) \geq m - \lceil t_{2\alpha}/(2\alpha) \rceil + 1,$$

and so

$$\begin{aligned}
2A\mu_{2\alpha}(P^\perp) + 2B\mu_1(P^\perp) &\geq 2A(2\alpha m - t_{2\alpha} + 1) + 2B(m - \lceil t_{2\alpha}/(2\alpha) \rceil + 1) \\
&= 2\alpha m - 2A(t_{2\alpha} - 1) - 2B(\lceil t_{2\alpha}/(2\alpha) \rceil - 1),
\end{aligned}$$

from which the result of the theorem follows. □

It is important to recall that we consider order 2α digital nets in this theorem, instead of order α digital nets as in Theorem 4.4. The order 2α is required here to ensure the finiteness of the sum

$$\sum_{z=1}^{\infty} b^{(1-2B)z} z^{s-1}.$$

Finally, combining the idea of switching the weight function (Theorem 4.5) with the idea of exploiting both the decay and the sparsity of the Walsh coefficients simultaneously (Theorem 4.3), we arrive at a deterministic counterpart of Theorem 4.5, proven in [38].

Theorem 4.6. *Let $\alpha \geq 2$. Let P be an order $(2\alpha+1)$ digital $(t_{2\alpha+1}, m, s)$-net over $\mathbb{F}_b$. Then there exists a constant $E^{(3)}_{\alpha,b,s,t_{2\alpha+1}} > 0$ such that the following holds:*

$$(e^{\mathrm{wor}}(H_\alpha, P))^2 \leq E^{(3)}_{\alpha,b,s,t_{2\alpha+1}} \frac{m^{s-1}}{b^{2\alpha m}}.$$

Proof. We write $A = (\alpha-1)/(2\alpha)$ and $B = (\alpha+1)/(2\alpha)$. Using Lemmas 4.10, 4.11 (with $\beta = 2\alpha+1$), and 4.4 (with α replaced by $2\alpha+1$) in order and then applying the change of variables $z \to z + 2\mu_1(P^\perp)$, we have

$$\begin{aligned}
(e^{\mathrm{wor}}(H_\alpha, P))^2 &= \sum_{\boldsymbol{k},\boldsymbol{l} \in P^\perp \setminus \{\boldsymbol{0}\}} \widehat{K_\alpha}(\boldsymbol{k},\boldsymbol{l}) \leq C^s_{\alpha,b} \sum_{\substack{\boldsymbol{k},\boldsymbol{l} \in P^\perp \setminus \{\boldsymbol{0}\} \\ (\boldsymbol{k},\boldsymbol{l}) \notin T_{\geq 2\alpha+1}}} b^{-\mu_\alpha(\boldsymbol{k}) - \mu_\alpha(\boldsymbol{l})} \\
&\leq C^s_{\alpha,b} b^{-2A\mu_{2\alpha+1}(P^\perp)} \sum_{\substack{\boldsymbol{k},\boldsymbol{l} \in P^\perp \setminus \{\boldsymbol{0}\} \\ (\boldsymbol{k},\boldsymbol{l}) \notin T_{\geq 2\alpha+1}}} b^{-B(\mu_1(\boldsymbol{k}) + \mu_1(\boldsymbol{l}))} \\
&= C^s_{\alpha,b} b^{-2A\mu_{2\alpha+1}(P^\perp)} \sum_{z=2\mu_1(P^\perp)}^{\infty} b^{-Bz} \\
&\quad \times \left|\{(\boldsymbol{k},\boldsymbol{l}) \in (P^\perp \setminus \{\boldsymbol{0}\})^2 \mid \mu_1(\boldsymbol{k}) + \mu_1(\boldsymbol{l}) = z, (\boldsymbol{k},\boldsymbol{l}) \notin T_{\geq 2\alpha+1}\}\right| \\
&\leq B_{2\alpha+1,b,s,t_{2\alpha+1}} C^s_{\alpha,b} b^{-2A\mu_{2\alpha+1}(P^\perp)} \\
&\quad \times \sum_{z=2\mu_1(P^\perp)}^{\infty} b^{-Bz} (z - 2\mu_1(P^\perp))^{2s\alpha+1} z^{s-1} b^{(z-2\mu_1(P^\perp))/2} \\
&\leq B_{2\alpha+1,b,s,t_{2\alpha+1}} 2^{s-1} C^s_{\alpha,b} \frac{(\mu_1(P^\perp))^{s-1}}{b^{2A\mu_{2\alpha+1}(P^\perp) + 2B\mu_1(P^\perp)}} \sum_{z=0}^{\infty} b^{(1/2-B)z} z^{(2\alpha+1)s}.
\end{aligned}$$

Since $B - 1/2 = 1/(2\alpha) > 0$, the last sum over z is finite. Using Lemma 4.2, we have

$$\mu_1(P^\perp) \geq m - \lceil t_{2\alpha+1}/(2\alpha+1) \rceil + 1,$$

and so

$$\begin{aligned}
2A\mu_{2\alpha+1}(P^\perp) + 2B\mu_1(P^\perp) &\geq 2A((2\alpha+1)m - t_{2\alpha+1} + 1) + 2B(m - \lceil t_{2\alpha+1}/(2\alpha+1) \rceil + 1) \\
&= 2\alpha m - 2A(t_{2\alpha+1} - 1) - 2B(\lceil t_{2\alpha+1}/(2\alpha+1) \rceil - 1),
\end{aligned}$$

from which the result of the theorem follows. □

The error bound shown in Theorem 4.6 also applies to the first b^m points of an order $(2\alpha+1)$ digital $(t_{2\alpha+1}, s)$-sequence over $\mathbb{F}_b$ if $m \geq t_{2\alpha+1}/(2\alpha+1)$, since they can be identified with an order $(2\alpha+1)$ digital $(t_{2\alpha+1}, m, s)$-net over $\mathbb{F}_b$.

4.7 Conclusions and outlook

In this article, we have reviewed some of recent results on HoQMC methods with the particular aim to provide a unified picture on how the Walsh analysis enables these developments. The challenge in analyzing either the L_2-discrepancy or the worst-case error in a reproducing kernel Hilbert space is that we have to deal with the double sum over the Walsh coefficients. Considering the shift-averaged worst-case error instead, the problem becomes much easier in many cases, but of course, the outcome will remain probabilistic. To obtain a deterministic counterpart of such a probabilistic result, exploiting the decay and the sparsity of the Walsh coefficients simultaneously seems to be a reasonable strategy to attack the problem. In fact, as we have seen, both the optimal order L_2-discrepancy bound in [25] and the optimal order quadrature error bound in [38] are obtained by employing this strategy.

Looking into the future, there are some possible directions for further research as raised below.

1. **Choice of an orthonormal basis:** The system of Walsh functions fits quite well with digital nets as emphasized in this article, but is not the only choice. Indeed, recent papers [42, 15, 16, 4] use the system of *Haar functions* and succeed in generalizing or extending the result of [25]. For instance, it was proven in [15] that order 2 (instead of order 5) digital sequences achieve the best possible order of L_2-discrepancy. Also, prior to the works of [36–38], Hinrichs et al. used the system of *Faber functions* to analyze the worst-case error of order 2 digital nets for different function spaces [40]. As natural questions from the current status, one may ask "Can we get better results in numerical integration problems by using the system of Haar/Faber functions? Can we lower the necessary order of digital nets and sequences from $2\alpha+1$ to achieve the best possible error rate?" We do not have any progress on these questions so far.
2. **Alternative construction scheme:** Higher order digital nets and sequences can be explicitly constructed through the digit interlacing function (Definition 4.12 and Lemma 4.5). Quite recently, it has been shown in [14] that Richardson extrapolation can be used as an alternative to the digit interlacing when the class of underlying point sets are restricted to polynomial lattice point sets. Also, the paper [34] proposed a different usage of Richardson extrapolation in the context of HoQMC methods. It is desirable to have more different options for explicit construction of higher order digital nets and sequences.
3. **Universality for various function classes:** One major drawback of higher order digital nets and sequences is that we need to construct point sets or sequences depending on dominating mixed smoothness of the considered function space. Propagation rule (Lemma 4.2) says that, once we construct point sets or sequences which work for a certain smoothness β, they also work for any smaller smoothness $1 \le \alpha < \beta$ but not for larger smoothness. Ideally what we want in practice is point sets or sequences which work for all ranges of smoothness. One straightforward idea is to construct *infinite order* digital nets and sequences and then study their propagation rule. In this line of research, we refer to [62, 67] for the Walsh analysis of infinitely many times differentiable functions, and furthermore, to [44, 59–61, 13, 43] for the relevant literature.

Bibliography

[1] N. Aronszajn. Theory of reproducing kernels. Trans. Am. Math. Soc. **68**, 337–404 (1950).

[2] N. S. Bakhvalov. Approximate computation of multiple integrals. Vestn. Mosk. Univ., Ser. Mat. Meh. Astron. Fiz. Him. **4**, 3–18 (1959) (in Russian). English translation in: J. Complex. **31**, 502–516 (2015).

[3] J. Baldeaux, J. Dick. QMC rules of arbitrary high order: Reproducing kernel Hilbert space approach. Constr. Approx. **30**, 495–527 (2009).

[4] D. Bilyk, L. Markhasin. BMO and exponential Orlicz space estimates of the discrepancy function in arbitrary dimension. J. Anal. Math. **135**, 249–269 (2018).

[5] W. W. L. Chen, M. M. Skriganov. Explicit constructions in the classical mean squares problem in irregularities of point distribution. J. Reine Angew. Math. **545**, 67–95 (2002).

[6] H. E. Chrestenson. A class of generalised Walsh functions. Pac. J. Math. **5**, 17–31 (1955).

[7] J. Dick. Explicit constructions of quasi-Monte Carlo rules for the numerical integration of high-dimensional periodic functions. SIAM J. Numer. Anal. **45**, 2141–2176 (2007).

[8] J. Dick. Walsh spaces containing smooth functions and quasi-Monte Carlo rules of arbitrary high order. SIAM J. Numer. Anal. **46**, 1519–1553 (2008).

[9] J. Dick. The decay of the Walsh coefficients of smooth functions. Bull. Aust. Math. Soc. **80**, 430–453 (2009).

[10] J. Dick. Discrepancy bounds for infinite-dimensional order two digital sequences over $\mathbb{F}_2$. J. Number Theory **136**, 204–232 (2014).

[11] J. Dick, R. N. Gantner, Q. T. Le Gia, C. Schwab. Multilevel higher-order quasi-Monte Carlo Bayesian estimation. Math. Models Methods Appl. Sci. **27**, 953–995 (2017).

[12] J. Dick, R. N. Gantner, Q. T. Le Gia, C. Schwab. Higher order Quasi-Monte Carlo integration for Bayesian PDE Inversion. Comput. Math. Appl. **77**, 144–172 (2019).

[13] J. Dick, T. Goda, K. Suzuki, T. Yoshiki. Construction of interlaced polynomial lattice rules for infinitely differentiable functions. Numer. Math. **137**, 257–288 (2017).

[14] J. Dick, T. Goda, T. Yoshiki. Richardson extrapolation of polynomial lattice rules. SIAM J. Numer. Anal. **57**, 44–69 (2019).

[15] J. Dick, A. Hinrichs, L. Markhasin, F. Pillichshammer. Optimal L_p-discrepancy bounds for second order digital sequences. Isr. J. Math. **221**, 489–510 (2017).

[16] J. Dick, A. Hinrichs, L. Markhasin, F. Pillichshammer. Discrepancy of second order digital sequences in function spaces with dominating mixed smoothness. Mathematika **63**, 863–894 (2017).

[17] J. Dick, F. Y. Kuo, Q. T. Le Gia, C. Schwab. Multilevel higher order QMC Petrov-Galerkin discretization for affine parametric operator equations. SIAM J. Numer. Anal. **54**, 2541–2568 (2016).

[18] J. Dick, F. Y. Kuo, Q. T. Le Gia, D. Nuyens, C. Schwab. Higher order QMC Petrov-Galerkin discretization for affine parametric operator equations with random field inputs. SIAM J. Numer. Anal. **52**, 2676–2702 (2014).

[19] J. Dick, F. Y. Kuo, I. H. Sloan. High-dimensional integration: The quasi-Monte Carlo way. Acta Numer. **22**, 133–288 (2013).

[20] J. Dick, Q. T. Le Gia, C. Schwab. Higher order quasi-Monte Carlo integration for holomorphic, parametric operator equations. SIAM/ASA J. Uncertain. Quantificat. **4**, 48–79 (2016).

[21] J. Dick, H. Niederreiter. On the exact t-value of Niederreiter and Sobol' sequences. J. Complex. **24**, 572–581 (2008).

[22] J. Dick, D. Nuyens, F. Pillichshammer. Lattice rules for nonperiodic smooth integrands. Numer. Math. **126**, 259–291 (2014).

[23] J. Dick, F. Pillichshammer. On the mean square weighted $\mathcal{L}_2$ discrepancy of randomized digital (t, m, s)-nets over $\mathbb{Z}_2$. Acta Arith. **117**, 371–403 (2005).

[24] J. Dick, F. Pillichshammer. Digital Nets and Sequences: Discrepancy Theory and Quasi-Monte Carlo Integration. Cambridge University Press, Cambridge (2010).
[25] J. Dick, F. Pillichshammer. Optimal $\mathcal{L}_2$ discrepancy bounds for higher order digital sequences over the finite field $\mathbb{F}_2$. Acta Arith. **162**, 65–99 (2014).
[26] H. Faure. Discrépances de suites associées à un système de numération (en dimension s). Acta Arith. **41**, 337–351 (1982).
[27] H. Faure, C. Lemieux. Irreducible Sobol' sequences in prime power bases. Acta Arith. **173**, 59–80 (2016).
[28] N. J. Fine. On the Walsh functions. Trans. Am. Math. Soc. **65**, 372–414 (1949).
[29] K. K. Frolov. Upper error bounds for quadrature formulas on function classes. Dokl. Akad. Nauk SSSR **231**, 818–821 (1976).
[30] R. N. Gantner, L. Herrmann, C. Schwab. Quasi-Monte Carlo integration for affine-parametric, elliptic PDEs: local supports and product weights. SIAM J. Numer. Anal. **56**, 111–135 (2018).
[31] R. N. Gantner, L. Herrmann, C. Schwab. Multilevel QMC with product weights for affine-parametric, elliptic PDEs. In: J. Dick, F. Y. Kuo, H. Wózniakowski (eds.) Contemporary Computational Mathematics – A Celebration of the 80th Birthday of Ian Sloan. Springer, Cham (2018).
[32] R. N. Gantner, M. D. Peters. Higher-order quasi-Monte Carlo for Bayesian shape inversion. SIAM/ASA J. Uncertain. Quantificat. **6**, 707–736 (2018).
[33] T. Goda. Good interlaced polynomial lattice rules for numerical integration in weighted Walsh spaces. J. Comput. Appl. Math. **285**, 279–294 (2015).
[34] T. Goda. Richardson extrapolation allows truncation of higher order digital nets and sequences. IMA J. Numer. Anal. (appeared online) DOI:10.1093/imanum/drz017.
[35] T. Goda, J. Dick. Construction of interlaced scrambled polynomial lattice rules of arbitrary high order. Found. Comput. Math. **15**, 1245–1278 (2015).
[36] T. Goda, K. Suzuki, T. Yoshiki. Optimal order quasi-Monte Carlo integration in weighted Sobolev spaces of arbitrary smoothness. IMA J. Numer. Anal. **37**, 505–518 (2017).
[37] T. Goda, K. Suzuki, T. Yoshiki. An explicit construction of optimal order quasi-Monte Carlo rules for smooth integrands. SIAM J. Numer. Anal. **54**, 2664–2683 (2016).
[38] T. Goda, K. Suzuki, T. Yoshiki. Optimal order quadrature error bounds for infinite-dimensional higher-order digital sequences. Found. Comput. Math. **18**, 433–458 (2018).
[39] J. M. Hammersley, D. C. Handscomb. Monte Carlo Methods. Chapman & Hall, London (1964).
[40] A. Hinrichs, L. Markhasin, J. Oettershagen, T. Ullrich. Optimal quasi-Monte Carlo rules on order 2 digital nets for the numerical integration of multivariate periodic functions. Numer. Math. **134**, 163–196 (2016).
[41] F. Y. Kuo, D. Nuyens. Application of quasi-Monte Carlo methods to elliptic PDEs with random diffusion coefficients: a survey of analysis and implementation. Found. Comput. Math. **16**, 1631–1696 (2016).
[42] L. Markhasin. L_p- and $S^r_{p,q}B$-discrepancy of (order 2) digital nets. Acta Arith. **168**, 139–159 (2015).
[43] M. Matsumoto, R. Ohori, T. Yoshiki. Approximation of Quasi-Monte Carlo worst case error in weighted spaces of infinitely times smooth functions. J. Comput. Appl. Math. **330**, 155–164 (2018).
[44] M. Matsumoto, M. Saito, K. Matoba. A computable figure of merit for quasi-Monte Carlo point sets. Math. Comput. **83**, 1233–1250 (2014).
[45] V. K. Nguyen, M. Ullrich, T. Ullrich. Change of variable in spaces of mixed smoothness and numerical integration of multivariate functions on the unit cube. Constr. Approx. **46**, 69–108 (2017).
[46] H. Niederreiter. Low-discrepancy point sets. Monatshefte Math. **102**, 155–167 (1986).

[47] H. Niederreiter. Low-discrepancy and low-dispersion sequences. J. Number Theory **30**, 51–70 (1988).
[48] H. Niederreiter. Low-discrepancy point sets obtained by digital constructions over finite fields. Czechoslov. Math. J. **42**, 143–166 (1992).
[49] H. Niederreiter. Random Number Generation and Quasi-Monte Carlo Methods. CBMS-NSF Regional Conference Series in Applied Mathematics, vol. 63. SIAM, Philadelphia (1992).
[50] H. Niederreiter, G. Pirsic. Duality for digital nets and its applications. Acta Arith. **97**, 173–182 (2001).
[51] H. Niederreiter, C. P. Xing. Rational Points on Curves over Finite Fields: Theory and Applications. London Mathematical Society Lecture Note Series, vol. 285. Cambridge University Press, Cambridge (2001).
[52] D. Nuyens, R. Cools. Fast component-by-component construction, a reprise for different kernels. In: Monte Carlo and quasi-Monte Carlo methods 2004, pp. 373–387. Springer, Berlin (2006).
[53] A. B. Owen. A constraint on extensible quadrature rules. Numer. Math. **132**, 511–518 (2016).
[54] M.Yu. Rosenbloom, M. A. Tsfasman. Codes for the m-metric. Probl. Inf. Transm. **33**, 55–63 (1997).
[55] M. M. Skriganov. Harmonic analysis on totally disconnected groups and irregularities of point distributions. J. Reine Angew. Math. **600**, 25–49 (2006).
[56] I. H. Sloan, S. Joe. Lattice Methods for Multiple Integration. Oxford University Press, Oxford (1994).
[57] I. H. Sloan, A. V. Reztsov. Component-by-component construction of good lattice rules. Math. Comput. **71**, 263–273 (2002).
[58] I. M. Sobol'. The distribution of points in a cube and approximate evaluation of integrals. Zh. Vycisl. Mat. i Mat. Fiz. **7**, 784–802 (1967).
[59] K. Suzuki. An explicit construction of point sets with large minimum Dick weight. J. Complex. **30**, 347–354 (2014).
[60] K. Suzuki. WAFOM on abelian groups for quasi-Monte Carlo point sets. Hiroshima Math. J. **45**, 341–364 (2015).
[61] K. Suzuki. Super-polynomial convergence and tractability of multivariate integration for infinitely times differentiable functions. J. Complex. **39**, 51–68 (2017).
[62] K. Suzuki, T. Yoshiki. Formulas for the Walsh coefficients of smooth functions and their application to bounds on the Walsh coefficients. J. Approx. Theory **205**, 1–24 (2016).
[63] S. Tezuka. Polynomial arithmetic analogue of Halton sequences. ACM Trans. Model. Comput. Simul. **3**, 99–107 (1993).
[64] M. Ullrich, T. Ullrich. The role of Frolov's cubature formula for functions with bounded mixed derivative. SIAM J. Numer. Anal. **54**, 969–993 (2016).
[65] G. Wahba. Spline Models for Observational Data. CBMS-NSF Regional Conference Series in Applied Mathematics, vol. 59. SIAM, Philadelphia (1990).
[66] J. L. Walsh. A closed set of normal orthogonal functions. Am. J. Math. **45**, 5–24 (1923).
[67] T. Yoshiki. Bounds on Walsh coefficients by dyadic difference and a new Koksma-Hlawka type inequality for Quasi-Monte Carlo integration. Hiroshima Math. J. **47**, 155–179 (2017).

Sigrid Grepstad, Lisa Kaltenböck, and Mario Neumüller

5 On the asymptotic behavior of the sine product $\prod_{r=1}^{n} |2 \sin \pi r\alpha|$

Abstract: In this paper, we review recently established results on the asymptotic behavior of the trigonometric product $P_n(\alpha) = \prod_{r=1}^{n} |2\sin \pi r\alpha|$ as $n \to \infty$. We focus on irrationals α whose continued fraction coefficients are bounded. Our main goal is to illustrate that when discussing the regularity of $P_n(\alpha)$, not only the boundedness of the coefficients plays a role; also their size, as well as the structure of the continued fraction expansion of α, is important.

Keywords: Trigonometric product, Ostrowski representation, Kronecker sequence, golden ratio

MSC 2010: Primary 26D05, 41A60, 11B39, Secondary 11L15, 11K31

5.1 Introduction

The trigonometric product

$$P_n(\alpha) = \prod_{r=1}^{n} |2\sin \pi r\alpha|$$

has been subject to mathematical investigations for more than 50 years. It arises naturally in a number of mathematical fields, such as partition theory, Padé approximation, and discrepancy theory. Of particular interest is the asymptotic behavior of $P_n(\alpha)$ as $n \to \infty$, which has proven surprisingly difficult to determine. In Figure 5.1, we have plotted $P_n(\alpha)$ for $n = 1, \dots, 250$ and different values of irrational α. These plots illustrate the chaotic nature of the product sequence $P_n(\alpha)$. Yet we see that for certain values of α, there is some self-similarity in the behavior of $P_n(\alpha)$ with increasing n.

In this paper, we review known bounds on the growth and decay of $P_n(\alpha)$, focusing on breakthroughs in the last 5 years. These recent developments deal mainly with

Acknowledgement: Sigrid Grepstad is supported in part by Grant 275113 of the Research Council of Norway. Lisa Kaltenböck and Mario Neumüller are funded by the Austrian Science Fund (FWF): Project F5507-N26 and Project F5509-N26, which are both part of the Special Research Program "Quasi-Monte Carlo Methods: Theory and Applications."

Sigrid Grepstad, Department of Mathematical Sciences, Norwegian University of Science and Technology, 7491 Trondheim, Norway, e-mail: sigrid.grepstad@ntnu.no
Lisa Kaltenböck, Mario Neumüller, Department of Financial Mathematics and Applied Number Theory, Johannes Kepler University, Altenbergerstraße 69, 4040 Linz, Austria, e-mails: lisa.kaltenboeck@jku.at, mario.neumueller@jku.at

https://doi.org/10.1515/9783110652581-005

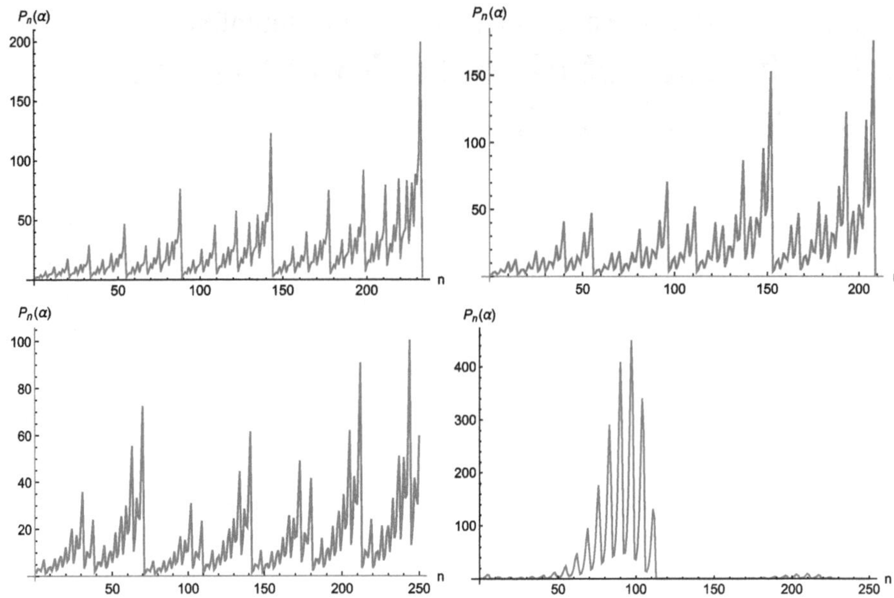

Figure 5.1: Values of $P_n(\alpha)$ for $\alpha = (\sqrt{5}-1)/2$ (upper left), $\alpha = \sqrt{3}$ (upper right), $\alpha = e$ (lower left) and $\alpha = \pi$ (lower right).

the case when α has bounded continued fraction coefficients. As shown by Lubinsky 20 years ago, this is a case in which the behavior of $P_n(\alpha)$ is exceptionally regular (see Section 5.1.2). What recent results have come to reveal, is that also the structure of the continued fraction expansion of α affects regularity. For instance, certain limit phenomena appear only for very structured expansions (see Section 5.3). Moreover, and perhaps more surprisingly, also the specific sizes of the continued fraction coefficients play a role. This is evident when discussing the long-standing open question (now resolved) of whether $\liminf_{n\to\infty} P_n(\alpha) = 0$ for all irrationals α.

5.1.1 Growth of $P_n(\alpha)$

Let us briefly review what is known about the growth of $P_n(\alpha)$ as $n \to \infty$. Note first that if $\alpha = p/q$ is rational, then $P_n(\alpha) = 0$ for all $n \geq q$. Moreover, we have that $P_n(\alpha) = P_n(\{\alpha\})$, where $\{\cdot\}$ denotes the fractional part, so we may safely restrict our attention to irrationals α in the unit interval.

It was established by Sudler [11] and Wright [13] in the 1960s that the norm $\|P_n(\alpha)\| = \sup_{0<\alpha<1} |P_n(\alpha)|$ grows exponentially as $n \to \infty$, and

$$\lim_{n\to\infty} \|P_n(\alpha)\|^{1/n} = C \approx 1.22. \tag{5.1}$$

(See also [3] for an alternative approach and the exact value of C.) In light of (5.1), one might expect that also the pointwise growth of $P_n(\alpha)$ is exponential, but this is not the case. It was shown by Lubinsky and Saff in [9] that for almost every $\alpha \in (0,1)$, we have

$$\lim_{n\to\infty} P_n(\alpha)^{1/n} = 1.$$

In later work, Lubinsky provides a more precise growth bound on $P_n(\alpha)$, namely

$$|\log P_n(\alpha)| = O(\log n(\log\log n)^{1+\varepsilon}) \tag{5.2}$$

for any $\varepsilon > 0$, and this holds for almost every α [8]. In the opposite direction, $P_n(\alpha)$ grows almost linearly for infinitely many n. We have that

$$\limsup_{n\to\infty} \frac{\log P_n(\alpha)}{\log n} \geq 1$$

for all irrationals $\alpha \in (0,1)$.

5.1.2 Significance of the continued fraction expansion

In his 1999 paper [8], Lubinsky illustrates a significant difference in nature of $P_n(\alpha)$ depending on whether or not the continued fraction expansion of α has bounded coefficients. If this is the case, then there exist positive constants C_1 and C_2 such that

$$n^{-C_2} \leq P_n(\alpha) \leq n^{C_1}, \tag{5.3}$$

that is, $P_n(\alpha)$ can be polynomially bounded (see [8, Theorem 1.3]).

When α has *unbounded* continued fraction coefficients, the upper bound in (5.2) (valid for almost all such α) has yet to be improved upon. Moreover, Lubinsky showed that

$$\liminf_{n\to\infty} P_n(\alpha) = 0 \tag{5.4}$$

in this case, and that for almost all α the decay to 0 is faster than any negative power of n for infinitely many n.

The focus of this paper will be on the more regular case when α has bounded continued fraction coefficients, and on two closely related questions raised by Lubinsky in [8], namely:

1. Does (5.4) still hold in the case of bounded continued fraction coefficients?
2. What is the smallest value that we can choose for C_2 in (5.3)?

Our interest in these questions was sparked by a recent paper by Mestel and Verschueren [12], where the special case $\alpha = (\sqrt{5}-1)/2$ is studied in great detail. We review key results from this paper in Section 5.3. Using these key results, we argue in Section 5.4 that for $\alpha = (\sqrt{5}-1)/2$, equality (5.4) does *not* hold. We will see in Section 5.5 that, in fact, it appears one may choose $C_2 = 0$ for this specific α. In the same section, we explain why this simplest choice of C_2 cannot possibly be valid for all α with bounded continued fraction coefficients; this was also alluded to by Lubinsky in [8].

A third question natural to raise is: what is the smallest value we may choose for C_1 in (5.3)? We firmly believe that for the special case $\alpha = (\sqrt{5}-1)/2$, the answer to this question is $C_1 = 1$ (see Figure 5.2). More precisely, we believe that $P_n(\alpha) < cn$ for some constant $c > 0$ independent of n. Upper bounds on $P_n(\alpha)$ will not be the focus of this paper. Nevertheless, we will briefly return to this question for the special case $\alpha = (\sqrt{5}-1)/2$ in Section 5.4.

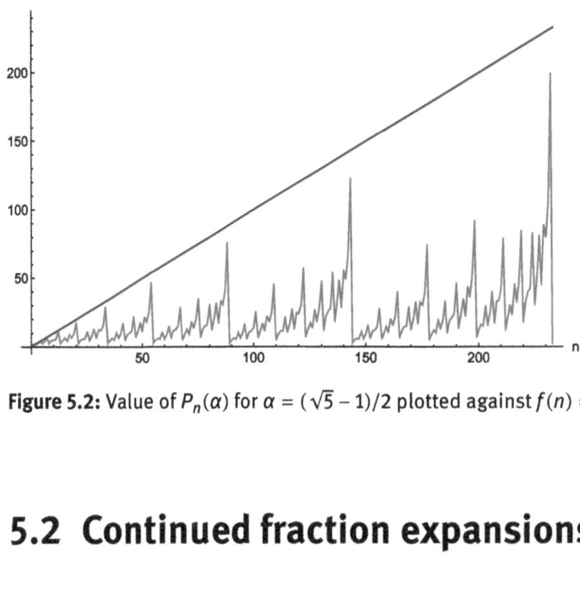

Figure 5.2: Value of $P_n(\alpha)$ for $\alpha = (\sqrt{5}-1)/2$ plotted against $f(n) = n$.

5.2 Continued fraction expansions

In order to set the notation for the remainder of the paper, we briefly review some facts about continued fraction expansions. Any irrational $\alpha \in (0,1)$ has a unique and infinite continued fraction expansion

$$\alpha = \cfrac{1}{a_1 + \cfrac{1}{a_2 + \cfrac{1}{a_3 + \cdots}}} = [0; a_1, a_2, a_3, \ldots],$$

where $a_i \in \mathbb{N}$ for all $i \in \mathbb{N}$. A best rational approximation of α is given by p_n/q_n, where p_n and q_n are defined recursively by

$$q_0 = 0, \quad q_1 = 1, \quad q_{n+1} = a_n q_n + q_{n-1};$$
$$p_0 = 1, \quad p_1 = 0, \quad p_{n+1} = a_n p_n + p_{n-1}.$$

This approximation is best possible in the sense that for no $q < q_n$ can we find $p \in \mathbb{N}$ such that

$$\left|\alpha - \frac{p}{q}\right| < \left|\alpha - \frac{p_n}{q_n}\right|.$$

We call p_n and q_n the best approximation numerator and denominator of α, respectively. The fraction p_n/q_n is called the nth convergent of α, and it is well known that

$$\left|\alpha - \frac{p_n}{q_n}\right| \leq \frac{1}{q_{n+1}q_n}. \tag{5.5}$$

Finally, we recall that given a sequence of best approximation denominators $\{q_0, q_1, q_2, \ldots\}$ corresponding to some irrational α, any natural number N has a unique *Ostrowski expansion* in terms of this sequence.

Theorem 5.1 (Ostrowski representation). *Let $\alpha \in (0,1)$ be an irrational with continued fraction expansion $[0; a_1, a_2, \ldots]$ and best approximation denominators $(q_n)_{n\geq 1}$. Then every natural number N has a unique expansion*

$$N = \sum_{j=1}^{z} b_j q_j, \tag{5.6}$$

where:
1. $0 \leq b_1 \leq a_1 - 1$ *and* $0 \leq b_j \leq a_j$ *for* $j > 1$.
2. *If* $b_j = a_j$ *for some* j, *then* $b_{j-1} = 0$.
3. $z = z(N) = O(\log N)$.

We refer to (5.6) *as the Ostrowski representation of N in base α.*

A proof of Theorem 5.1 can be found in [7, p. 126]. For further reading on the Ostrowski expansion, see [1] or [10].

5.3 Convergence along subsequences

In a recent paper by Mestel and Verschueren [12], the authors give a detailed exposition on the product $P_n(\alpha)$ in the special case when $\alpha = \varphi = (\sqrt{5} - 1)/2$ is the (fractional part of the) golden mean. The irrational number φ has the simplest possible continued fraction expansion

$$\varphi = \cfrac{1}{1 + \cfrac{1}{1 + \cfrac{1}{1 + \cdots}}} = [0; \overline{1}],$$

and the sequence of best approximation denominators of φ is the well-known Fibonacci sequence

$$(F_n)_{n\geq 0} = (0, 1, 1, 2, 3, 5, 8, 13, \ldots). \tag{5.7}$$

Mestel and Verschueren give a rigorous proof of an intriguing fact which was observed experimentally in [6] by Knill and Tangerman, namely that the subsequence $P_{F_n}(\varphi)$ converges to a positive constant as $n \to \infty$.

Theorem 5.2 ([12, Theorem 3.1]). *Let* $\varphi = (\sqrt{5} - 1)/2$ *and let* $(F_n)_{n\geq 0}$ *be the Fibonacci sequence in* (5.7). *The subsequence* $(P_{F_n}(\varphi))_{n\geq 1}$ *is convergent, and*

$$\lim_{n\to\infty} P_{F_n}(\varphi) = \lim_{n\to\infty} \prod_{r=1}^{F_n} |2\sin \pi r\varphi| > 0.$$

Numerical calculations suggest that the limiting value of $P_{F_n}(\varphi)$ is approximately 2.4 (see Figure 5.3).

It turns out that the convergence of the subsequence $P_{F_n}(\varphi)$ is not a property specific to the golden mean. The same property can be established for any irrational α with continued fraction expansion $\alpha = [0; \overline{a}]$, and a similar phenomenon is observed for any irrational with a periodic continued fraction expansion.

Theorem 5.3 ([5, Theorem 1.2]). *Suppose* α *has a periodic continued fraction expansion of the form* $\alpha = [0; \overline{a_1, a_2, \ldots, a_\ell}]$ *with period* ℓ, *and let* $(q_n)_{n\geq 0}$ *be its sequence of best approximation denominators. Then there exist positive constants* $C_0, C_1, \ldots, C_{\ell-1}$ *such that*

$$\lim_{m\to\infty} P_{q_{\ell m+k}}(\alpha) = \lim_{m\to\infty} \prod_{r=1}^{q_{\ell m+k}} |2\sin \pi r\alpha| = C_k$$

for each $k = 0, 1, \ldots, \ell - 1$.

Adding a preperiod to the continued fraction expansion of α in Theorem 5.3 does not alter the conclusion, and accordingly this result extends to all quadratic irrationals α. See [5] for further details.

In Figure 5.3 below, we have plotted the subsequences $P_{q_n}(\alpha)$ for $\alpha = \varphi$ and $\alpha = \sqrt{3}$. In the latter case, the continued fraction expansion of α has period $\ell = 2$, and accordingly we observe that the two subsequences $P_{q_{2m}}(\alpha)$ and $P_{q_{2m+1}}(\alpha)$ converge rapidly to two different positive constants.

5.4 A positive lower bound for $P_n(\alpha)$

The limit phenomenon observed in Theorem 5.2 sheds new light on the old and longstanding open problem of whether

$$\liminf_{n\to\infty} P_n(\alpha) = 0 \tag{5.8}$$

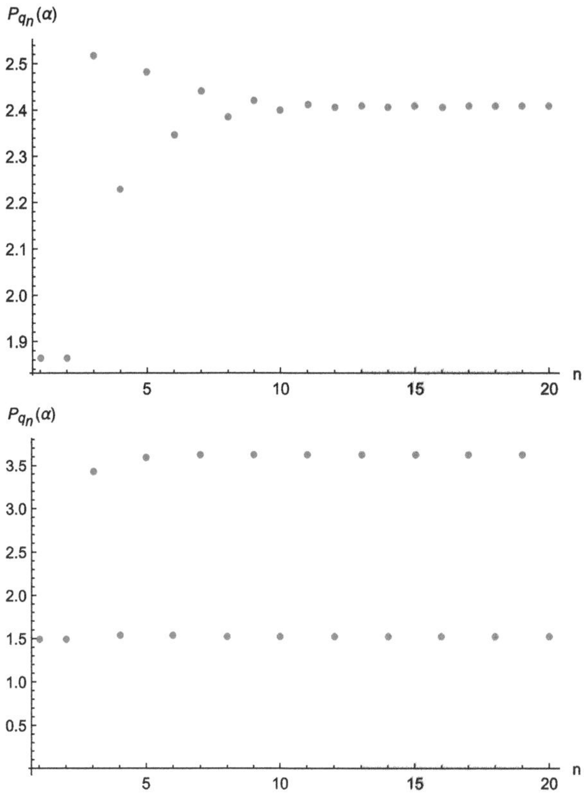

Figure 5.3: Values of $P_{q_n}(\alpha)$ for $\alpha = \varphi$ (above) and $\alpha = \sqrt{3} = [1; \overline{1, 2}]$ (below), where q_n is the nth best approximation denominator of α.

for all irrationals α. As mentioned in Section 5.1.2, this question was raised by Lubinsky in [8], but the problem goes back much further; also Erdős and Szekeres asked this question already in the 1950s [2]. Lubinsky showed that (5.8) indeed holds for all α with unbounded continued fraction coefficients, and suggested it is likely that (5.8) holds in general.

However, when $\alpha = \varphi$ is the golden mean, numerics indicate that it is precisely along the subsequence $(F_n)_{n\geq 1}$ of Fibonacci numbers that $P_n(\varphi)$ takes on its minimum values. On the other hand, peaks of $P_n(\varphi)$ appear to be occurring along the subsequence $(F_n - 1)_{n\geq 1}$. Specifically, numerical calculations are suggesting that

$$P_{F_{n-1}}(\varphi) \leq P_N(\varphi) \leq P_{F_n-1}(\varphi) \tag{5.9}$$

for $n \geq 3$ and $N \in \{F_{n-1}, \ldots, F_n - 1\}$. This is illustrated in Figure 5.4.

The inequalities in (5.9) have two immediate and important consequences. First of all, should the upper bound in (5.9) hold, then it would follow that the growth of $P_n(\varphi)$ is at most linear. Using the convergence of the subsequence $P_{F_n}(\varphi)$, it is derived

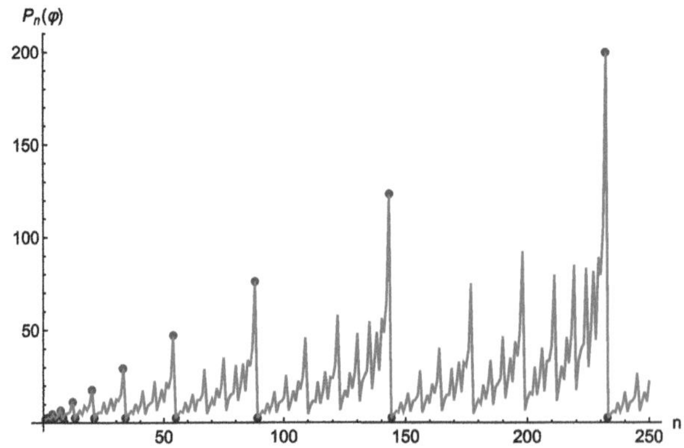

Figure 5.4: Value of $P_n(\varphi)$, with the two subsequences $P_{F_n-1}(\varphi)$ (peaks) and $P_{F_n}(\varphi)$ (dips) highlighted.

in [12] that $P_{F_n-1}(\varphi) \leq cF_n$, and combining this with (5.9) we get

$$P_N(\varphi) \leq cF_n \leq 2cN.$$

Second, should the lower bound in (5.9) hold, then it would follow immediately from Theorem 5.2 that

$$\liminf_{n\to\infty} P_n(\varphi) \geq \lim_{n\to\infty} P_{F_n}(\varphi) > 0. \tag{5.10}$$

To the best of our knowledge, the inequalities in (5.9) have not been proven rigorously. Nevertheless, it turns out that (5.10) can be deduced from Theorem 5.2 by a slightly extended argument.

Theorem 5.4 ([4, Theorem 1.1]). *If* $\varphi = (\sqrt{5}-1)/2$, *then*

$$\liminf_{n\to\infty} P_n(\varphi) = \liminf_{n\to\infty} \prod_{r=1}^{n} |2\sin \pi r\varphi| > 0.$$

The main idea in the proof of Theorem 5.4 is rather simple: For any $N \in \mathbb{N}$, let $N = \sum_{j=1}^{m} F_{n_j}$ be its Ostrowski representation in base φ (also known as its Zeckendorf representation [14]). We may then express $P_N(\varphi)$ as the double product

$$P_N(\varphi) = \prod_{r=1}^{N} |2\sin \pi r\varphi| = \prod_{j=1}^{m} \prod_{r=1}^{F_{n_j}} \big|2\sin \pi(r\varphi + k_j\varphi)\big|, \tag{5.11}$$

where $k_j = \sum_{s=j+1}^{m} F_{n_s}$ for $1 \leq j \leq m-1$ and $k_m = 0$. Observe that the inner product on the right-hand side in (5.11) is a perturbed version of $P_{F_j}(\varphi)$. It was shown in [12, pp. 220–221] that for these perturbed products, there exist constants $0 < K_1 \leq 1 \leq K_2$

such that

$$K_1 \le \prod_{r=1}^{F_{n_j}} |2 \sin \pi(r\varphi + k_j\varphi)| \le K_2 \tag{5.12}$$

for all $1 \le j \le m$. Now notice that the fractional part of the perturbation $k_j\varphi$ is tending to zero with increasing values of j. This is a consequence of the identity

$$F_n\varphi = F_{n-1} - (-\varphi)^n.$$

We know from Theorem 5.2 that the unperturbed sequence $P_{F_{n_j}}(\varphi)$ tends to a constant $c \approx 2.4$ as j increases, and it is thus tempting to suggest that the lower bound $K_1 \le 1$ in (5.12) can be raised to some value greater than 1 if j is chosen sufficiently large. Indeed it turns out that

$$\prod_{r=1}^{F_{n_j}} |2 \sin \pi(r\varphi + k_j\varphi)| \ge 1,$$

for all j greater than some threshold value $J \in \mathbb{N}$ (independent of N), and accordingly it follows from (5.11) and (5.12) that

$$P_N(\varphi) \ge K_1^J$$

for all $N \in \mathbb{N}$.

For a detailed exposition of the proof of Theorem 5.4, see [4].

5.5 Possible extensions of Theorem 5.4

We have now seen that $\liminf_{n\to\infty} P_n(\alpha) = 0$ fails for the golden mean $\alpha = \varphi$, and it is natural to ask whether

$$\liminf_{n\to\infty} P_n(\alpha) > 0$$

also for other irrationals. Since the fact that $\liminf_{n\to\infty} P_n(\varphi) > 0$ is deduced from Theorem 5.2, and Theorem 5.2 has a natural extension to quadratic irrationals (Theorem 5.3), one is led to guess that Theorem 5.4 might be generalized to all quadratic irrationals. Unfortunately, this is too much to hope for.

Theorem 5.5. *Let $\alpha = [0; a_1, a_2, \ldots]$ have bounded continued fraction coefficients, and let $M = \max_{j\in\mathbb{N}} a_j$. Provided M is sufficiently large, there exists some threshold value $K = K(M)$ such that if $a_j \ge K$ infinitely often, then*

$$\liminf_{n\to\infty} P_n(\alpha) = 0. \tag{5.13}$$

Remark. Theorem 5.5 is a consequence of a result by Lubinsky (Proposition 5.6 below). Lubinsky himself claims in [8] that Theorem 5.5 is true for a general threshold K independent of M. However, this is not rigorously proven, and we have not managed to verify it. Basing our argument on Proposition 5.6 below, we do not see that the dependency on M can be omitted.

Proposition 5.6 ([8, Proposition 5.1]). *Let $\alpha = [0; a_1, a_2, \ldots]$, and for $n \in \mathbb{N}$ let $n = \sum_{j=1}^{z} b_j q_j$ be its Ostrowski expansion in base α. Denote by $z^\#$ the length of this expansion*

$$z^\# = z^\#(n) = \#\{j : 1 \le j \le z \text{ and } b_j \ne 0\}.$$

We then have

$$\begin{aligned} \log P_n(\alpha) \le 800 z^\# + 151 \sum_{j=1}^{z} \frac{b_j}{a_j} \max_{k<j} \log a_k + \frac{3}{2} \sum_{j=1}^{z} \log^+ b_j \\ + \sum_{j=1}^{z} b_j \log\left(\frac{2\pi b_j q_j |q_j \alpha - p_j|}{e} \right), \end{aligned} \tag{5.14}$$

where $\log^+ x = \max\{\log x, 0\}$.

Remark. The fact that $\liminf_{n\to\infty} P_n(\alpha) = 0$ whenever $\alpha = [0; a_1, a_2, \ldots]$ has *unbounded* continued fraction coefficients is a straightforward consequence of this proposition (as illustrated by Lubinsky in [8]). To see this, simply construct a strictly increasing subsequence of coefficients a_{n_j} where

$$a_{n_j} > a_k \quad \text{for all } k < n_j.$$

Then putting $n = N_j = q_{n_j}$ in (5.14), it is easily verified that this inequality reduces to

$$\log P_{N_j}(\alpha) \le C - \log a_{n_j}$$

for some absolute constant C, and since $a_{n_j} \to \infty$ as $j \to \infty$ it follows that

$$\lim_{j\to\infty} P_{N_j}(\alpha) = 0.$$

Let us now see how Theorem 5.5 is deduced from Proposition 5.6.

Proof of Theorem 5.5. Let $\alpha = [0; a_1, a_2, \ldots]$ with $M = \max_j a_j$, and suppose $a_j \ge K$ infinitely often for some natural number $K \le M$. Denote by $(n_i)_{i\in\mathbb{N}}$ a sequence of indices such that $a_{n_i} \ge K$ for every i. We may choose this sequence so that

$$n_i - n_{i-1} > 1 \quad \text{for all } i > 1.$$

Now construct a sequence of integers N_m by letting $N_m = \sum_{i=1}^{m} q_{n_i}$. We have then given N_m in its Ostrowski representation to base α, as

$$N_m = \sum_{i=1}^{m} q_{n_i} = \sum_{j=1}^{n_m} b_j q_j,$$

where

$$b_j = \begin{cases} 1, & \text{if } j \in (n_i)_{i \in \mathbb{N}} \\ 0, & \text{otherwise,} \end{cases}$$

and where no two consecutive coefficients b_j are both nonzero.

We now use Proposition 5.6 to estimate $\log P_{N_m}(\alpha)$. Since $b_j \in \{0, 1\}$, it is clear that the third term on the right-hand side in (5.14) is zero. For the second term on the right-hand side in (5.14), we have the upper bound

$$151 \sum_{j=1}^{n_m} \frac{b_j}{a_j} \max_{k<j} \log a_k \leq 151 \log M \sum_{i=1}^{m} \frac{1}{a_{n_i}} \leq \frac{151 \log M}{K} m. \tag{5.15}$$

Finally, for the fourth term on the right-hand side in (5.14), we observe that if $b_j = 1$, then

$$b_j \log\left(\frac{2\pi b_j q_j |q_j \alpha - p_j|}{e}\right) \leq \log\left(\frac{2\pi q_j}{e q_{j+1}}\right) \leq \log\left(\frac{\pi q_j}{a_j q_j}\right) = \log \pi - \log a_j,$$

where for the first inequality we have used (5.5). It follows that

$$\sum_{j=1}^{n_m} b_j \log\left(\frac{2\pi b_j q_j |q_j \alpha - p_j|}{e}\right) \leq (\log \pi - \log K) m, \tag{5.16}$$

and inserting (5.15) and (5.16) in (5.14), we arrive at

$$\log P_{N_m}(\alpha) \leq \left(802 + \frac{151 \log M}{K} - \log K\right) m. \tag{5.17}$$

If M is sufficiently small, then the right-hand side in (5.17) is positive regardless of the size of $K \leq M$. However, once M is sufficiently large, one can find $K = K(M)$ such that

$$802 + \frac{151 \log M}{K} - \log K < 0.$$

In this case, it is clear from (5.17) that

$$\log P_{N_m}(\alpha) \to -\infty$$

as $m \to \infty$, and accordingly

$$\lim_{m \to \infty} P_{N_m}(\alpha) = 0.$$

This concludes the proof of Theorem 5.5. □

5.5.1 Irrationals of the form $\alpha = [0; \overline{a}]$

Let us finally have an extra look at irrationals of the form

$$\alpha = [0; \overline{a}].$$

For this special case, we have $M = K = a$ in Theorem 5.5, and it is clear from the proof that $\liminf_{n\to\infty} P_n(\alpha) = 0$ if

$$802 + \frac{151 \log a}{a} - \log a < 0,$$

or equivalently if $a \geq e^{802+\varepsilon}$ for some small $\varepsilon > 0$.

Studying the product $P_n(\alpha)$ numerically, it appears that the true lower bound on a for when $\liminf_{n\to\infty} P_n(\alpha) = 0$ might actually be significantly lower. In Table 5.1, we have listed the evolution of minima of $P_n(\alpha)$ for $\alpha = [0; \overline{a}]$, $a = 1, 2, \ldots, 8$, determined numerically.

Table 5.1: Evolution of minima of $P_n(\alpha)$ for $n = 1, \ldots, 50\,000$.

α	Evolution of minima $(P_n(\alpha), n)$
$[0; \overline{1}]$	(1.865, 1)
$[0; \overline{2}]$	(1.928, 1)
$[0; \overline{3}]$	(1.333, 1)
$[0; \overline{4}]$	(1.351, 1)
$[0; \overline{5}]$	(1.138, 1)
$[0; \overline{6}]$	(0.977, 1), (0.907, 7), (0.849, 44), (0.794, 272), (0.742, 1 677), (0.693, 10 335)
$[0; \overline{7}]$	(0.852, 1), (0.708, 8), (0.589, 58), (0.491, 415), (0.408, 2 964), (0.340, 21 164)
$[0; \overline{8}]$	(0.755, 1), (0.564, 9), (0.422, 74), (0.316, 602), (0.236, 4 891), (0.177, 39 731)

It is curious that for $a \leq 5$, we have

$$\min_{1 \leq n \leq 50\,000} P_n(\alpha) = P_1(\alpha),$$

whereas for $a > 5$, the minimal value of $P_n(\alpha)$ is decreasing slowly with increasing n. The apparent change in behavior at the cutoff $a = 5$ leads us to close by posing the following conjecture.

Conjecture 5.7. *Let $\alpha = [0; \overline{a}]$. If $a \leq 5$, then*

$$\liminf_{n\to\infty} P_n(\alpha) \geq P_1(\alpha) > 0.$$

If $a > 5$, then

$$\liminf_{n\to\infty} P_n(\alpha) = 0.$$

Bibliography

[1] J. Allouche, J. Shallit. Automatic Sequences: Theory, Applications, Generalizations. Cambridge University Press, Cambridge (2003).

[2] P. Erdős, G. Szekeres. On the product $\prod_{k=1}^{n}(1-z^{a_k})$. Acad. Serbe Sci. Publ. Inst. Math. **13**, 29–34 (1959).

[3] G. Freiman, H. Halberstam. On a product of sines. Acta Arith. **49**(4), 377–385 (1988).

[4] S. Grepstad, L. Kaltenböck, M. Neumüller. A positive lower bound for $\liminf_{N\to\infty} \prod_{r=1}^{N} |2 \sin \pi r\varphi|$. Proc. Amer. Math. Soc. **147**(11), 4863–4876 (2019).

[5] S. Grepstad, M. Neumüller. Asymptotic behaviour of the Sudler product of sines for quadratic irrationals. J. Math. Anal. Appl. **465**(2), 928–960 (2018).

[6] O. Knill, F. Tangerman. Self-similarity and growth in the Birkhoff sum for the golden rotation. Nonlinearity **24**(11), 3115–3127 (2011).

[7] L. Kuipers, H. Niederreiter. Uniform Distribution of Sequences. John Wiley, New York (1974).

[8] D. Lubinsky. The size of $(q;q)_n$ for q on the unit circle. J. Number Theory **76**(2), 217–247 (1999).

[9] D. Lubinsky, E. B. Saff. Convergence of Padé approximants of partial theta functions and the Rogers-Szegő polynomials. Constr. Approx. **3**, 331–361 (1987).

[10] A. Rockett, P. Szüsz. Continued Fractions. World Scientific Publishing Co., Inc., River Edge, NJ (1992).

[11] C. Jr. Sudler. An estimate for a restricted partition function. Quart. J. Math. Oxf. Ser. **15**, 1–10 (1964).

[12] P. Verschueren, B. Mestel. Growth of the Sudler product of sines at the golden rotation number. J. Math. Anal. Appl. **433**, 200–226 (2016).

[13] E. M. Wright. Proof of a conjecture of Sudler's. Quart. J. Math. Oxf. Ser. **15**(2), 11–15 (1964).

[14] E. Zeckendorf. Représentation des nombres naturels par une somme de nombres de Fibonacci ou de nombres de Lucas. Bull. Soc. R. Sci. Liège **41**, 179–182 (1972) (french).

Simon Breneis and Aicke Hinrichs

6 Fibonacci lattices have minimal dispersion on the two-dimensional torus

Abstract: We study the size of the largest rectangle containing no point of a given point set in the two-dimensional torus, the dispersion of the point set. A known lower bound for the dispersion of any point set of cardinality $n \geq 2$ in this setting is $2/n$. We show that if n is a Fibonacci number then the Fibonacci lattice has dispersion exactly $2/n$ meeting the lower bound. Moreover, we completely characterize integration lattices achieving the lower bound and provide insight into the structure of other optimal sets. We also treat related results in the nonperiodic setting.

Keywords: Dispersion, integration lattice, Fibonacci lattice

MSC 2010: 52C05

6.1 Introduction and main result

We identify the two-dimensional torus with $[0,1]^2$. Any two points $x, y \in [0,1]^2$ define a rectangle $B(x,y)$ in the two-dimensional torus. If $x = (x_1, x_2)$, $y = (y_1, y_2)$ satisfy $x_1 \leq y_1$ and $x_2 \leq y_2$, this is the ordinary rectangle $B(x,y) = [x_1, y_1] \times [x_2, y_2]$. If $x_1 > y_1$ and $x_2 \leq y_2$, then $B(x,y) = ([0, y_1] \cup [x_1, 1]) \times [x_2, y_2]$ is wrapped around in the direction of the first coordinate axis. Analogously, for $x_1 \leq y_1$ and $x_2 > y_2$, it is wrapped around the direction of the second coordinate axis, and for $x_1 > y_1$ and $x_2 > y_2$ around both axis; see Figure 6.1.

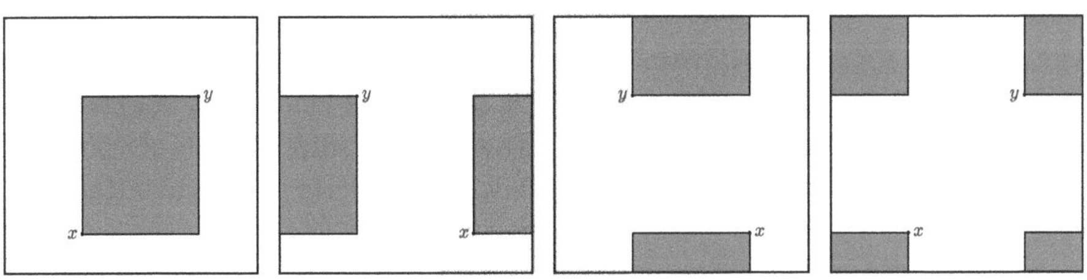

Figure 6.1: Periodic rectangles.

Acknowledgement: Both authors are supported by the Austrian Science Fund (FWF) Project F5513-N26, which is a part of the Special Research Program "Quasi-Monte Carlo Methods: Theory and Applications." AH would like to thank the Isaac Newton Institute for Mathematical Sciences for support and hospitality during the programme "Approximation, sampling and compression in data science" when work on this paper was undertaken. This work was supported by EPSRC Grant Number EP/R014604/1. AH was also supported by a grant from the Simons Foundation.

Simon Breneis, Aicke Hinrichs, Institut für Analysis, Johannes Kepler Universität Linz, Altenbergerstrasse 69, 4040 Linz, Austria, e-mails: simon.breneis@jku.at, aicke.hinrichs@jku.at

https://doi.org/10.1515/9783110652581-006

For a given finite point set $\mathcal{P} \subset [0,1]^2$, the dispersion $\mathrm{disp}(\mathcal{P})$ of $\mathcal{P}$ is the area of the largest rectangle $B(x,y)$ containing no point of $\mathcal{P}$ in the interior. The following lower bound follows as a special case from the result of M. Ullrich in [14] for the d-dimensional torus in the case $d = 2$.

Theorem 6.1. *For any $n \in \mathbb{N}$ with $n \geq 2$ and any point set $\mathcal{P}_n \subset [0,1]^2$ with $\#\mathcal{P}_n = n$, we have*

$$\mathrm{disp}(\mathcal{P}_n) \geq \frac{2}{n}.$$

The main purpose of this note is the investigation whether, and if so, for which sets, this bound is sharp.

It is well understood that the Fibonacci lattice has exceptional uniform distribution properties. We shortly discuss some results in this direction for the discrepancy and the dispersion as measures of uniform distribution. Let $(F_m)_{m\in\mathbb{N}}$ be the sequence of Fibonacci numbers starting with $F_1 = F_2 = 1$ and defined via the recursive relation $F_{m+2} = F_m + F_{m+1}$ for $m \geq 2$. The Fibonacci lattice $\mathcal{F}_m$ is defined as

$$\mathcal{F}_m := \left\{ \left(\frac{k}{F_m}, \left\{ \frac{kF_{m-2}}{F_m} \right\} \right) : k \in \{0,1,\dots,F_m - 1\} \right\}.$$

Here, $\{\alpha\}$ denotes the fractional part of α.

The Fibonacci lattice is an example of an integration lattice. A general integration lattice in dimension $d = 2$ has the form

$$\left\{ \left(\frac{k}{n}, \left\{ \frac{kq}{n} \right\} \right) : k \in \{0,1,\dots,n-1\} \right\}.$$

Here, n and $1 \leq q < n$ are positive integers. The number q is called the generator of the integration lattice. It is sometimes required to be coprime to n. We do not make this additional requirement here. Observe that an integration lattice consists of n points in $[0,1)^2$. For the theory of integration lattices and applications to numerical integration, we refer to [7].

It is well known that the Fibonacci lattice has order optimal L_∞- and L_2-discrepancy. For the L_∞-discrepancy, we refer to the monograph [5] of H. Niederreiter. For the classical L_2-discrepancy, this was first proved by V. Sós and S. K. Zaremba in [9]. For the periodic L_2-discrepancy, it is even conjectured that the Fibonacci lattice is globally optimal among all point sets with the same number of points; see [3]. This is proved in [3] for $n = F_m \leq 13$. Among integration lattices, the Fibonacci lattice has minimal periodic L_2-discrepancy at least if $n = F_m \leq 832040$. This can be shown by a not particularly sophisticated exhaustive search through all integration lattices using a suitable simplification of the Warnock formula for the periodic L_2-discrepancy of integration lattices.

For the dispersion, it was proved by V. Temlyakov in [13] that the Fibonacci lattice is order optimal, that is, that there exists a constant c such that

$$\operatorname{disp}(\mathcal{F}_m) \leq \frac{c}{F_m}.$$

The main purpose of this note is to show that the bound in Theorem 6.1 is actually sharp for the Fibonacci lattices. In particular, we show the following theorem.

Theorem 6.2. *Let $m \geq 3$ be an integer. The Fibonacci lattice $\mathcal{F}_m$ satisfies*

$$\operatorname{disp}(\mathcal{F}_m) = \frac{2}{F_m}.$$

It may be conjectured that, up to torus symmetries, the Fibonacci lattices are the only point sets meeting the lower bound in Theorem 6.1. This is not true. The second purpose of this note is to discuss the structure of general optimal sets. At least for integration lattices, we get a complete characterization.

Theorem 6.3. *Let $\mathcal{P}_n$ be an integration lattice with $n \geq 2$ points. Then $\operatorname{disp}(\mathcal{P}_n) = 2/n$ holds if and only if $n = F_m$ is a Fibonacci number and $\mathcal{P}_n$ is torus symmetric to the Fibonacci lattice $\mathcal{F}_m$ or $n = 2F_m$ is twice a Fibonacci number and $\mathcal{P}_n$ is torus symmetric to the lattice with generator $q = 2F_{m-2}$.*

Clearly, Theorem 6.3 immediately implies Theorem 6.2, so we will prove only Theorem 6.3.

For the convenience of the reader, we provide a short proof of Theorem 6.1 in Section 6.2. In particular, this proof also shows that any point set $\mathcal{P}_n$ with n points satisfying $\operatorname{disp}(\mathcal{P}_n) = 2/n$ has to be of a certain structure. This structure is then further employed in Section 6.4 to give examples of point sets with optimal dispersion that are not integration lattices.

Section 6.3 contains the proof of our main result Theorem 6.3 which, as already mentioned, implies Theorem 6.2. In Section 6.4, we comment on sets with dispersion $2/n$ that are not integration lattices. In Section 6.5, we compute the nonperiodic dispersion of the Fibonacci lattice, which turns out to be only slightly smaller than $2/F_m$. We finish with a final section containing a discussion of related results in particular also in higher dimensions.

6.2 Proof of Theorem 6.1

We now give a simple proof of Theorem 6.1. This proof is basically the same as the proof for general dimension in [14].

Fix a point set $\mathcal{P}_n$ with n points. For $x \in [0,1)$, let $n(x)$ be the number of points in the (periodic) rectangle $B(x) = [x, x + 2/n) \times [0,1)$. Then each point in $\mathcal{P}_n$ is in $B(x)$ for a set of x of measure exactly $2/n$. Hence $\int_0^1 n(x)\,dx = 2$.

Assume first that $n(x) < 2$ for some x. Then, for this x, either $n(x) = 0$ or $n(x) = 1$. Then, for some $\varepsilon > 0$, also the rectangle $B = [x - \varepsilon, x + 2/n) \times [0, 1)$ contains at most one point of $\mathcal{P}_n$. Splitting the box along the second coordinate of this point, if it exists, we obtain a periodic rectangle of size $2/n + \varepsilon$ containing no points of $\mathcal{P}_n$ in its interior showing that $\operatorname{disp}(\mathcal{P}_n) > 2/n$.

If $n(x) > 2$ for some x, then $\int_0^1 n(x)\,dx = 2$ implies that $n(x) < 2$ for some x, again $\operatorname{disp}(\mathcal{P}_n) > 2/n$ follows.

The only case not considered is the case that $n(x) = 2$ for every $x \in [0, 1)$. Let x be the first coordinate of a point in the point set. Then we obtain that there exists exactly one point in the point set with x-coordinate in $(x, x + 2/n)$ and that there is exactly one point in the point set with x-coordinate equal to $x + 2/n$. In particular, splitting the rectangle $(x, x + 2/n) \times [0, 1]$ along the second coordinate of the (only) point in this rectangle gives an empty rectangle of size $2/n$ and $\operatorname{disp}(\mathcal{P}_n) \geq 2/n$ follows.

Moreover, if n is odd, this implies that the x-coordinates of points of $\mathcal{P}_n$ form the set $\{\xi + k/n : k = 0, 1, \ldots, n-1\}$ for some $\xi \in [0, 1/n)$. If n is even, the situation is a little different. Then $n(x) = 2/n$ for every $x \in [0, 1)$ only implies that $\mathcal{P}_n$ is the union of two sets $\{\xi_i + 2k/n : k = 0, 1, \ldots, n/2 - 1\}$ for some $\xi_1, \xi_2 \in [0, 2/n)$. Similar reasoning can be applied to the second coordinate instead of the first coordinate.

Altogether, we proved Theorem 6.1 together with structural properties of point sets meeting the bound. In particular, if n is odd, any point set $\mathcal{P}_n$ with n points satisfying $\operatorname{disp}(\mathcal{P}_n) = 2/n$ is, up to torus symmetries, a lattice point set of the type

$$\left\{\left(\frac{k}{n}, \left\{\frac{\pi(k)}{n}\right\}\right) : k \in \{0, 1, \ldots, n-1\}\right\}$$

for some permutation π of the set $\{0, 1, \ldots, n-1\}$.

6.3 Proof of Theorems 6.2 and 6.3

Throughout this section, we fix an integration lattice

$$\mathcal{P}_n = \left\{\left(\frac{k}{n}, \left\{\frac{kq}{n}\right\}\right) : k \in \{0, 1, \ldots, n-1\}\right\},$$

containing n points with generator $q \in \{1, 2, \ldots, n-1\}$. We will prove Theorem 6.3, Theorem 6.2 is a direct consequence. Our proof of Theorem 6.3 relies on a careful examination of the length of the intervals obtained by splitting the torus with the points $(\{\frac{kq}{n}\})$. To simplify the notation, we will scale the one-dimensional torus by a factor of n and consider the sequence $(n\{\frac{kq}{n}\})$. To this end, let $y : \{0, 1, \ldots, n-1\} \to \{0, 1, \ldots, n-1\}$ be the function defined as

$$y(k) := n\left\{\frac{kq}{n}\right\}.$$

Furthermore, let

$$Y_\ell := (y(k))_{k=0}^{\ell-1}$$

denote the sequence of the first ℓ function values of y.

We now want to consider the distances between consecutive elements of the sequence Y_ℓ.

Definition 6.4. Let $(x_k)_{k=0}^{\ell-1}$ be a sequence of ℓ elements of the one-dimensional torus scaled by n. Let $(y_k)_{k=0}^{\ell-1}$ be the nondecreasing rearrangement of the sequence $(x_k)_{k=0}^{\ell-1}$. For $a, b \in [0, n]$ with $a \neq b$, let $d(a, b)$ denote the oriented scaled torus distance of the points a and b, that is, $d(a, b) = b - a$ for $b < a$ and $d(a, b) = n + b - a$ if $b < a$. We also set $d(a, a) = n$. We say that $c \in (0, n]$ is a distance of the sequence (x_k) if there exists an $i \in \{1, \ldots, \ell - 1\}$ such that

$$d(y_{i-1}, y_i) = c$$

or if

$$d(y_{\ell-1}, y_0) = c.$$

The following lemma is a direct consequence of the three-distance or three-gap theorem conjectured by H. Steinhaus and proved in the late 1950s by V. Sós [8], J. Suránji [11], and S. Świerczkowski [12].

Lemma 6.5. *For any $\ell \in \{1, \ldots, n\}$, the sequence Y_ℓ has at most three different distances. If Y_ℓ has three different distances $d_1 > d_2 > d_3$, then $d_1 = d_2 + d_3$.*

We will now investigate how often those three distances occur. The following definition will be helpful in simplifying the notation. We also refer to Figure 6.2 for an instructive example.

Definition 6.6. Let the sequence Y_ℓ have the distances $d_1 > d_2 > d_3$ a_1, a_2, a_3 times, respectively. Then we say Y_ℓ induces the splitting

$$n = a_1 d_1 + a_2 d_2 + a_3 d_3.$$

Notice that the equality holds if we interpret it algebraically. If there are only one or two distances, the notation is used accordingly. In that case, we also use the notation above and allow $d_3 = d_2$ if $a_3 = 0$ and $d_2 = d_1$ if $a_3 = a_2 = 0$.

As it turns out, if we increase the number ℓ of points considered, we always end up splitting the largest distance.

Lemma 6.7. *By going from Y_ℓ to $Y_{\ell+1}$, the largest distance from Y_ℓ is split.*

<table>
<tr><th>Step l</th><th colspan="13">Splitting</th><th>Distances</th><th>Notation</th></tr>
<tr><td>1</td><td colspan="13">13</td><td>13</td><td>1x13</td></tr>
<tr><td>2</td><td colspan="5">5</td><td colspan="8">8</td><td>8,5</td><td>1x8+1x5</td></tr>
<tr><td>3</td><td colspan="5">5</td><td colspan="5">5</td><td colspan="3">3</td><td>5,3</td><td>2x5+1x3</td></tr>
<tr><td>4</td><td colspan="2">2</td><td colspan="3">3</td><td colspan="5">5</td><td colspan="3">3</td><td>5,3,2</td><td>1x5+2x3+1x2</td></tr>
<tr><td>5</td><td colspan="2">2</td><td colspan="3">3</td><td colspan="2">2</td><td colspan="3">3</td><td colspan="3">3</td><td>3,2</td><td>3x3+2x2</td></tr>
<tr><td>6</td><td colspan="2">2</td><td colspan="3">3</td><td colspan="2">2</td><td colspan="3">3</td><td colspan="2">2</td><td>1</td><td>3,2,1</td><td>2x3+3x2+1x1</td></tr>
<tr><td>7</td><td colspan="2">2</td><td colspan="2">2</td><td>1</td><td colspan="2">2</td><td colspan="3">3</td><td colspan="2">2</td><td>1</td><td>3,2,1</td><td>1x3+4x2+1x1</td></tr>
<tr><td>8</td><td colspan="2">2</td><td colspan="2">2</td><td>1</td><td colspan="2">2</td><td colspan="2">2</td><td>1</td><td colspan="2">2</td><td>1</td><td>2,1</td><td>5x2+3x1</td></tr>
<tr><td>9</td><td>1</td><td>1</td><td colspan="2">2</td><td>1</td><td colspan="2">2</td><td colspan="2">2</td><td>1</td><td colspan="2">2</td><td>1</td><td>2,1</td><td>4x2+5x1</td></tr>
<tr><td>10</td><td>1</td><td>1</td><td colspan="2">2</td><td>1</td><td>1</td><td>1</td><td colspan="2">2</td><td>1</td><td colspan="2">2</td><td>1</td><td>2,1</td><td>3x2+7x1</td></tr>
<tr><td>11</td><td>1</td><td>1</td><td colspan="2">2</td><td>1</td><td>1</td><td>1</td><td colspan="2">2</td><td>1</td><td>1</td><td>1</td><td>1</td><td>2,1</td><td>2x2+9x1</td></tr>
<tr><td>12</td><td>1</td><td>1</td><td>1</td><td>1</td><td>1</td><td>1</td><td>1</td><td colspan="2">2</td><td>1</td><td>1</td><td>1</td><td>1</td><td>2,1</td><td>1x2+11x1</td></tr>
<tr><td>13</td><td>1</td><td>1</td><td>1</td><td>1</td><td>1</td><td>1</td><td>1</td><td>1</td><td>1</td><td>1</td><td>1</td><td>1</td><td>1</td><td>1</td><td>13x1</td></tr>
</table>

Figure 6.2: Consecutive Splittings for $n = 13$ and $q = 5$.

Proof. If Y_ℓ only has one distance, this is trivial. Suppose now that Y_ℓ induces the splitting

$$n = a_1 d_1 + a_2 d_2 + a_3 d_3,$$

where $a_3 \geq 0$, that is, we also consider the case that Y_ℓ only has two different distances. Clearly, Y_n splits the torus into equidistant intervals, that is, Y_n has only one distance. Suppose now that by going from Y_ℓ to $Y_{\ell+1}$, we split either d_2 or d_3. Without loss of generality, we will assume we split d_2. By going from Y_ℓ to $Y_{\ell+1}$, we introduced the point $y(\ell)$. This point has a left neighbor, that is, there exists an $a < \ell$ such that $B(y(a), y(\ell))$ contains no other point of the sequence $Y_{\ell+1}$. In the same way, there exists a right neighbor $y(b)$. Since we split the distance d_2, we know that $d(y(a), y(b)) = d_2$. It is easy to see that for any k such that $\ell + k < n$ we have that $y(\ell + k)$ is in the (periodic) interval $(y(a + k), y(b + k))$. Thus, any point introduced after $y(\ell)$ can only split a distance which is at most d_2. This means that the distance d_1 is never split again. However, since Y_n should induce a splitting with only one distance, this is clearly a contradiction. Thus, we always split d_1. □

From now on, we will make additional assumptions on n and q. On the one hand, we assume without loss of generality that $2q \leq n$. This is possible, since the integration lattices induced by (n, q) and $(n, n-q)$ are torus symmetric. On the other hand, we assume that the integration lattice has optimal dispersion $2/n$, or, since we consider the scaled torus, dispersion $2n$. Since we are only interested in finding all optimal integration lattices, this is no real restriction.

The following lemma will give us an explicit formula for the splitting of Y_ℓ that will then directly imply Theorem 6.3. Here, we also use Fibonacci numbers F_k with $k \leq 0$, which satisfy the same recursion as for $k > 0$.

Lemma 6.8. *Let m, k, j be positive integers satisfying $3 \leq k \leq m$, $F_m \leq n < F_{m+1}$, and $1 \leq j \leq F_{k-2}$.*

If k is odd, then $Y_{F_k - j}$ induces the splitting

$$n = j(F_{k-3}q - F_{k-5}n) + (F_{k-1} - j)(F_{k-4}n - F_{k-2}q) + (F_{k-2} - j)(F_{k-1}q - F_{k-3}n). \quad (6.1)$$

Moreover, the fraction $\frac{n}{q}$ *satisfies*

$$\frac{F_k}{F_{k-2}} \le \frac{n}{q} \le \frac{F_{k-1}}{F_{k-3}}. \tag{6.2}$$

If k is even, then $Y_{F_k - j}$ *induces the splitting*

$$n = j(F_{k-5}n - F_{k-3}q) + (F_{k-1} - j)(F_{k-2}q - F_{k-4}n) + (F_{k-2} - j)(F_{k-3}n - F_{k-1}q). \tag{6.3}$$

Moreover, the fraction $\frac{n}{q}$ *satisfies*

$$\frac{F_{k-1}}{F_{k-3}} \le \frac{n}{q} \le \frac{F_k}{F_{k-2}}. \tag{6.4}$$

Moreover, the inequalities (6.2) *and* (6.4) *for* $3 \le k \le m$ *are not only necessary but also sufficient for* $\mathcal{P}_n$ *to have minimal dispersion* $2/n$.

Proof. We will prove this lemma by induction on k.

Let $k = 3$. The only j we need to consider is $j = 1$. We need to examine the splitting $Y_{F_3 - 1} = Y_1$. Y_1 trivially splits the scaled torus into

$$\begin{aligned} n &= 1(F_0 q - F_{-2}n) + (F_2 - 1)(F_{-1}n - F_1 q) + (F_1 - 1)(F_2 q - F_0 n) \\ &= 1n. \end{aligned}$$

Furthermore, since $2q \le n$, the distances $F_0 q - F_{-2}n \ge F_{-1}n - F_1 q \ge F_2 q - F_0 n$ are ordered. Also (6.2), which reads as

$$\frac{2}{1} = \frac{F_3}{F_1} \le \frac{n}{q} \le \frac{F_2}{F_0} = +\infty,$$

holds since $2q \le n$.

Let $k = 4$. Again, the only j we need to consider is $j = 1$. We need to examine the splitting $Y_{F_4 - 1} = Y_2$. But Y_2 trivially splits the scaled torus into

$$\begin{aligned} n &= 1(F_{-1}n - F_1 q) + (F_3 - 1)(F_2 q - F_0 n) + (F_2 - 1)(F_1 n - F_3 q) \\ &= 1(n - q) + 1q \end{aligned}$$

as claimed. We assumed that the integration lattice has optimal dispersion, that is, there is no empty box of size greater than $2n$. The splitting Y_2 gives us a box of size $3(n - q)$. Thus, we get

$$3(n - q) \le 2n \quad \Longleftrightarrow \quad \frac{n}{q} \le 3.$$

Together with the bound from the case $k = 3$, this implies

$$\frac{2}{1} = \frac{F_3}{F_1} \le \frac{n}{q} \le \frac{F_4}{F_2} = \frac{3}{1},$$

which is (6.2). Moreover, it again follows that the distances of the splitting $F_{-1}n - F_1 q \geq F_2 p - F_0 n \geq F_1 n - F_3 q$ are ordered.

We will now assume the lemma has been proven for $k, k+1$ with k odd and we will prove it for $k+2$ and $k+3$. Of course, we then have to assume $m \geq k+2$ and $m \geq k+3$, respectively.

We start with the proof for $k+2$. We need to consider the splitting $Y_{F_{k+2}-j}$ for $j \in \{1, \dots, F_k\}$. This splitting still exists, since $m \geq k+2$. We know that $Y_{F_{k+1}-1}$ gave us the splitting

$$n = 1(F_{k-4}n - F_{k-2}q) + (F_k - 1)(F_{k-1}q - F_{k-3}n) + (F_{k-1} - 1)(F_{k-2}n - F_k q).$$

Since, by Lemma 6.7, we always split the largest distance and the distances in the splitting above are ordered, $Y_{F_{k+1}}$ gives us

$$n = F_k(F_{k-1}q - F_{k-3}n) + F_{k-1}(F_{k-2}n - F_k q).$$

Now we need to split the largest distance $F_{k+2} - j - F_{k+1}$ more times. Thus, $Y_{F_{k+2}-j}$ gives us

$$n = j(F_{k-1}q - F_{k-3}n) + (F_{k+1} - j)(F_{k-2}n - F_k q) + (F_k - j)(F_{k+1}q - F_{k-1}n).$$

This shows (6.3). It remains to check that the distances are ordered. To this end, we use the minimality with respect to the dispersion. The splitting gives us an empty box of size $(F_{k+2} - j + 1)(F_{k-1}q - F_{k-3}n)$. From the assumption on the dispersion, we conclude that $(F_{k+2} - j + 1)(F_{k-1}q - F_{k-3}n) \leq 2n$. Of course, if this condition is satisfied for $j = 1$, it is satisfied for any $j \in \{1, \dots, F_k\}$. Thus, we have

$$F_{k+2}(F_{k-1}q - F_{k-3}n) \leq 2n \quad \Longleftrightarrow \quad \frac{F_{k+2}}{F_k} = \frac{F_{k+2}F_{k-1}}{F_{k+2}F_{k-3} + 2} \leq \frac{n}{q}.$$

Together with the bounds (6.4) with $k+1$ instead of k, we get the new bounds

$$\frac{F_{k+2}}{F_k} \leq \frac{n}{q} \leq \frac{F_{k+1}}{F_{k-1}}.$$

These are the bounds (6.2) with $k+2$ instead of k, which now also imply that the distances of the splitting were ordered. The proof for $k+3$ is completely analogous. □

Proof of Theorem 6.3. Assume that the integration lattice $\mathcal{P}_n$ satisfies $\operatorname{disp}(\mathcal{P}_n) = 2/n$. Let q be the generator of $\mathcal{P}_n$ and assume that $2q \leq n$, passing to a torus equivalent integration lattice if necessary. Let the positive integer m be such that $n \in \{F_m, \dots, F_{m+1} - 1\}$. Since $n \geq 2$, we have $m \geq 3$. If m is odd, Lemma 6.8 gives us for $k = m$ that $\frac{n}{q}$ must satisfy the inequalities

$$\frac{F_m}{F_{m-2}} \leq \frac{n}{q} \leq \frac{F_{m-1}}{F_{m-3}}.$$

Since fractions of Fibonacci numbers are optimal rational approximations of the golden ration $\varphi = \frac{1+\sqrt{5}}{2}$ (or in that case $\varphi^2 = \frac{3+\sqrt{5}}{2}$), the next rational approximation better than $\frac{F_m}{F_{m-2}}$ and $\frac{F_{m-1}}{F_{m-3}}$ would be $\frac{F_{m+1}}{F_{m-1}}$. But this is not possible, since $n < F_{m+1}$. Thus, if $\frac{n}{q}$ satisfies the above inequality, it has to be equal to one of the two sides.

If $\frac{n}{q} = \frac{F_m}{F_{m-2}}$, then $n = F_m$ and $q = F_{m-2}$ because of the restrictions on n (since $F_{m+1} - 1 < 2F_m$) and we have that $\mathcal{P}_n = \mathcal{F}_m$ is a Fibonacci lattice.

If $\frac{n}{q} = \frac{F_{m-1}}{F_{m-3}}$, then $n = 2F_{m-1}$ because of the restrictions on n. This implies $q = 2F_{m-3}$.

We conclude that the only possible optimal integration lattices are the lattices described in the theorem. Moreover, since the inequalities (6.2) and (6.4) for $3 \le k \le m$ are sufficient for $\mathcal{P}_n$ to have minimal dispersion $2/n$, these lattices indeed have optimal dispersion $2/n$. □

Remark 6.9. The proof of optimality of the Fibonacci lattice without characterizing all optimal integration lattices can be significantly simplified. In fact, it is then easier to directly show the formula

$$\operatorname{disp}(\mathcal{F}_m) = \frac{1}{F_m^2} \max_{3 \le k \le m} F_k F_{m-k+3},$$

which also follows from the above argument. The maximum is attained for $k = 3$ and $k = m$. This was independently observed by M. Ullrich.

6.4 Optimal sets that are not integration lattices

In this section, we give examples of point sets $\mathcal{P}_n$ satisfying $\operatorname{disp}(\mathcal{P}_n) = 2/n$ that are not integration lattices. Of course, the restrictions given in Section 6.2 have to be satisfied. These examples are obtained from the Fibonacci lattices $\mathcal{F}_m$ for even F_m by shifting every other point by a fixed small vector; see Figure 6.3. This leads to the distorted Fibonacci lattices

$$\begin{aligned}\mathcal{F}_{m,\xi,\eta} := &\left\{\left(\frac{k}{F_m}, \left\{\frac{kF_{m-2}}{F_m}\right\}\right) : k \in \{0, 2, \dots, F_m - 2\}\right\} \\ &\cup \left\{\left(\frac{k}{F_m} + \frac{\xi}{F_m}, \left\{\frac{kF_{m-2}}{F_m}\right\} + \frac{\eta}{F_m}\right) : k \in \{1, 3, \dots, F_m - 1\}\right\}\end{aligned}$$

with $0 \le \xi, \eta < 1$. It turns out that for small enough ξ and η, such a distortion does not alter the dispersion of the Fibonacci lattice.

For simplicity, we just analyze the case $\eta = 0$ more closely, that is, half of the Fibonacci lattice is shifted in the direction of the first coordinate. Then the argument from the previous section or the more direct argument mentioned in Remark 6.9 lead to the conclusion that, as long as

$$F_3 F_m \ge \max_{4 \le k \le m-1} (F_k + \xi) F_{m-k+3},$$

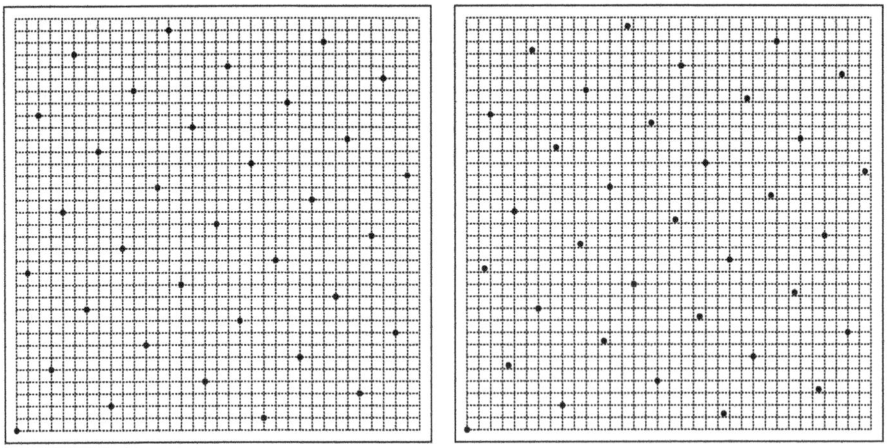

Figure 6.3: Fibonacci lattice, usual and distorted.

the dispersion does not grow. The maximum is attained (at least asymptotically) for $k = 4$ and $k = 5$. So we get the condition

$$F_3 F_m \geq (F_4 + \xi) F_{m-1},$$

which is asymptotically equivalent to

$$\xi \leq \lim_{n\to\infty} \frac{F_3 F_m}{F_{m-1}} - F_4 = 2\varphi - 3 = 0.236068\ldots.$$

The distorted Fibonacci lattices above are neither integration lattices nor lattice point sets. We could not decide if there are lattice point sets with large cardinality that both have optimal dispersion and are not torus equivalent to an integration lattice.

6.5 The nonperiodic case

In this section, we study the dispersion of the Fibonacci lattice in the non-periodic case. Basically, this means that we restrict the allowed rectangles in the definition of the dispersion to rectangles $B(x, y)$ where $x \leq y$ coordinatewise. Let $\operatorname{disp}^*(\mathcal{P}_n)$ denote the corresponding dispersion of a point set $\mathcal{P}_n \subset [0,1]^2$.

The best known lower bound (for $n \geq 16$)

$$\operatorname{disp}^*(\mathcal{P}_n) \geq \frac{5}{4(n+5)} \tag{6.5}$$

was proved in [2]. As far as we know, until now, the best known upper bound in dimension 2 for large n is $\operatorname{disp}(\mathcal{P}_n) \leq 4/n$ if $n = 2^m$ for some positive integer m and a $(0, m, 2)$-net $\mathcal{P}_n$ in base 2, in particular for the Hammersley point set; see [2, Theorem 2] or [1] for higher dimensional versions. For general n, this implies that there exist point sets $\mathcal{P}_n$ of cardinality n with $\mathrm{disp}(\mathcal{P}_n) \le 8/n$. Already the periodic dispersion of the Fibonacci lattices allows an improvement of these upper bounds. We compute the nonperiodic dispersion of the Fibonacci lattice to further improve these bounds.

Theorem 6.10. *Let $m \ge 6$ be an integer. The Fibonacci lattice $\mathcal{F}_m$ without the point $(0, 0)$ satisfies*

$$\mathrm{disp}^*(\mathcal{F}_m \setminus \{(0,0)\}) = \frac{2(F_m - 1)}{F_m^2}.$$

We only sketch the proof here. The proof for the periodic case in particular shows that the maximal periodic boxes containing no points of $\mathcal{F}_m$ in the interior have side length F_j/F_m and F_{m+3-j}/F_m for some $j = 3, 4, \dots, m$ leading to the formula

$$\mathrm{disp}(\mathcal{F}_m) = \frac{1}{F_m^2} \max\{F_j F_{m+3-j} \ : \ j = 3, 4, \dots, m\}.$$

The maximum is attained for $j = 3$ and $j = m$. But the corresponding rectangles with side length $2/F_m$ and $F_m/F_m = 1$ are true periodic rectangles wrapping around one direction. However, it is easy to see that there are still nonperiodic rectangles with side length $2/F_m$ and $(F_m - 1)/F_m$. Those rectangles have an area of $2(F_m - 1)/F_m^2$.

On the other hand, for each $j = 4, 5 \dots, m-1$, there are nonperiodic rectangles with side length F_j/F_m and F_{m+3-j}/F_m that do not contain any point of $\mathcal{F}_m$ in the interior. In the nonperiodic setting, the point $(0, 0)$ can be safely omitted. We arrive at

$$\begin{aligned} &\mathrm{disp}^*(\mathcal{F}_m \setminus \{(0,0)\}) \\ &= \frac{1}{F_m^2} \max\{2(F_m - 1), \max\{F_j F_{m+3-j} \ : \ j = 4, 5, \dots, m-1\}\} \end{aligned}$$

for $m \ge 5$. It is not too hard to check that, for $m \ge 6$, this maximum is attained for $j = 5$. Clearly,

$$\frac{F_5 F_{m-2}}{F_m^2} \le \frac{2(F_m - 1)}{F_m^2}$$

for $m \ge 6$. This leads to the claim of the theorem.

6.6 Further results, final remarks, and open problems

To put our results in the two-dimensional case into perspective, we now also consider the general d-dimensional case. Let $\mathrm{disp}(n, d)$ and $\mathrm{disp}^*(n, d)$ be the minimal dispersion of all point sets $\mathcal{P}_n$ in $[0, 1]^d$ of cardinality n in the periodic and nonperiodic setting, respectively. The modifications in the definitions should be obvious.

The known lower and upper bounds imply that

$$0 < a(d) := \liminf_{n\to\infty} n \operatorname{disp}(n,d) \le \limsup_{n\to\infty} n \operatorname{disp}(n,d) =: b(d) < \infty$$

and

$$0 < a^*(d) := \liminf_{n\to\infty} n \operatorname{disp}^*(n,d) \le \limsup_{n\to\infty} n \operatorname{disp}^*(n,d) =: b^*(d) < \infty.$$

The inequalities

$$a^*(d) \le a(d) \quad \text{and} \quad b^*(d) \le b(d) \tag{6.6}$$

are trivial. It is natural to study these quantities and to determine if $a(d) = b(d)$ and/or $a^*(d) = b^*(d)$, that is, if the limits

$$\lim_{n\to\infty} n \operatorname{disp}(n,d) \quad \text{and/or} \quad \lim_{n\to\infty} n \operatorname{disp}^*(n,d)$$

exist. For $d = 1$, equidistant points are optimal. This implies $a(1) = b(1) = a^*(1) = b^*(1) = 1$.

Already the case $d = 2$ is much more difficult. In the periodic case, Theorems 6.1 and 6.2 show that

$$a(2) = 2 \quad \text{and} \quad b(2) \le \frac{3+\sqrt{5}}{2} = 2.6180339\ldots. \tag{6.7}$$

Here, $b(2)$ is estimated via monotonicity of $\operatorname{disp}(n,d)$ in n together with $\operatorname{disp}(n,d) = 2/n$ if n is a Fibonacci number or twice a Fibonacci number. The question whether $b(2) = a(2)$ remains open. In the nonperiodic case, the lower bound (6.5), Theorem 6.10 and inequalities (6.6) and (6.7) show that

$$\frac{5}{4} \le a^*(2) \le 2 \quad \text{and} \quad b^*(2) \le \frac{3+\sqrt{5}}{2} = 2.6180339\ldots.$$

The exact determination of $a^*(2)$ and $b^*(2)$ remains open.

For general d, we only know that

$$d \le a(d) \le 2^{7d} \quad \text{and} \quad b(d) \le 2^{7d+1}$$

as well as

$$\frac{\log_2 d}{4} \le a^*(d) \le 2^{7d} \quad \text{and} \quad b^*(d) \le 2^{7d+1}.$$

The upper bounds follow from a construction using digital nets due to G. Larcher; see [1]. The lower bound for $a(d)$ follows from the result of [14], the lower bound for $a^*(d)$ from the main result of [1]. Further upper bounds not directly applicable to this problem or yielding worse bounds can be found in the papers [4, 6, 10, 13, 15].

The lower bound from [14] for general dimension d is

$$\operatorname{disp}(\mathcal{P}_n) \geq \frac{d}{n},$$

which is equal to d/n for $d \geq n$. We now discuss the case of integration lattices $\mathcal{P}_n \subset [0,1]^d$ with optimal periodic dispersion. It turns out that, in contrast to the already considered case $d = 2$, such integration lattices can only exist for small n. We restrict the discussion here to the case $d = 3$.

Theorem 6.11. *There are no integration lattices in 3 dimensions which have more than 4 points and satisfy*

$$\operatorname{disp}(\mathcal{P}_n) = \frac{3}{n}.$$

The crucial tool is the following additional information on the splittings induced by a two-dimensional integration lattice. Here, we freely use the notation and language introduced in Section 6.3.

Lemma 6.12. *Let $\mathcal{P}_n \subset [0,1]^2$ be an integration lattice with n points and generator q. Assume that the induced splitting has more than one different distance (it has distances $d_1 > d_2$ and maybe distance $d_3 < d_2$). Then there is an empty interval of distance d_1 and an empty interval of distance at least d_2 which are next to each other.*

Proof. Assume we have an interval of distance d_1. Let the interval on the left and right have distance d_3. This distance d_3 must have come from splitting a distance d_1 or a distance d_2. A distance d_2 has not been split, as there is a distance d_1 remaining and we always split the largest distance. Thus, both distances of length d_3 next to the d_1-distance come from splitting a d_1-distance. A d_1 distance is always split into a d_2 and a d_3 distance. Furthermore, the d_2 is always to the left of the d_3 or the d_2 is always to the right of the d_3. In any way, if there were distances d_1 which were split to both sides of our d_1 distance, then there are now a d_3 and a d_2 distance next to our d_1. □

Proof of Theorem 6.11. Observe that the projection of an integration lattice in dimension 3 onto any of the coordinate planes produces an integration lattice in dimension 2. Let q be the generator of one of those projected lattices. Splitting the 3-dimensional torus along the third coordinate of a point shows that a lower bound for the dispersion of the 3-dimensional integration lattice is given by the maximal size of a periodic rectangle for the 2-dimensional projection containing at most one point in the interior.

Now, assume that the first k points o induce the splitting

$$n = a_1 d_1 + a_2 d_2 + a_3 d_3$$

for the 2-dimensional projected lattice. Then we observe the following: if $a_1 > a_2 + a_3$, an application of the pigeonhole principle shows that there are two empty intervals of size d_1 next to each other. In any other case, Lemma 6.12 tells us that there is an interval of size d_1 next to an interval of size d_2. Thus, in the first case, there is a two-dimensional box of size $(k+1) * (d_1 + d_1)$ and in the second case there is a two-dimensional box of size $(k+1) * (d_1 + d_2)$ which contains only one point. The sizes of those boxes are lower bounds for the 3-dimensional dispersion.

Assume now that the 3-dimensional integration lattice has dispersion $3/n$. We will show a contradiction if n is sufficiently large. Without loss of generality, assume $q \leq n$. We consider now the projected 2-dimensional lattice with generator q. The first two points induce the splitting

$$n = 1(n - q) + 1q$$

and give rise to a relevant box of size

$$(2 + 1) \cdot (n - q + q) = 3n \leq 3n.$$

The first 3 points induce the splitting

$$n = 1(n - 2q) + 2q.$$

The proof will be completed by a giant case distinction:

1. $n - 2q < q \Longleftrightarrow n < 3q$. We have the largest box of size

   $$4(q + q) \leq 3n \Longleftrightarrow \frac{8q}{3} \leq n.$$

   Together, we obtain $8q/3 \leq n < 3q$. The next splitting is

   $$n = 1q + 2(n - 2q) + 1(3q - n).$$

   We again need to distinguish two cases:
   (a) $n - 2q < 3q - n \Longleftrightarrow n < 5q/2$. This is a contradiction.
   (b) $n - 2q \geq 3q - n \Longleftrightarrow n \geq 5q/2$. We have the largest box

   $$5(q + n - 2q) \leq 3n \Longleftrightarrow n \leq \frac{5q}{2}.$$

   This is a contradiction.
2. $n - 2q \geq q \Longleftrightarrow n \geq 3q$. We have the largest box

   $$4(n - 2q + q) \leq 3n \Longleftrightarrow n \leq 4q.$$

   Together, we obtain $3q \leq n \leq 4q$. The next splitting is

   $$n = 1(n - 3q) + 3q.$$

   We again need to distinguish two cases.

(a) $n - 3q \geq q \Longleftrightarrow n \geq 4q$. Together with $n \leq 4q$ we have $n = 4q$. This is a contradiction for $n > 4$.

(b) $n - 3q < q \Longleftrightarrow n < 4q$. We have the largest box

$$5(q + q) \leq 3n \Leftrightarrow \frac{10q}{3} \leq n.$$

Together, we obtain $10q/3 \leq n < 4q$. The next splitting is

$$n = 2q + 2(n - 3q) + 1(4q - n).$$

We again need to distinguish two cases.

i. $n - 3q \leq 4q - n \Longleftrightarrow n \leq 7q/2$. We have the largest box

$$6(q + 4q - n) \leq 3n \Longleftrightarrow 10q/3 \leq n.$$

Together, we obtain $10q/3 \leq n \leq 7q/2$. The next splitting is

$$n = q + 2(4q - n) + 3(n - 3q).$$

Luckily, we do not need a case distinction here, as we already know the ordering of the distances. We have the largest box

$$7(q + 4q - n) \leq 3n \Longleftrightarrow 7q/2 \leq n.$$

Thus, $n = 7q/2$. This is a contradiction for all n, except for $n = 7$ and $q = 2$. This case can be excluded separately by exhaustively trying all possibilities.

ii. $n - 3q > 4q - n \Longleftrightarrow n > 7q/2$. We have the largest box

$$6(q + n - 3q) \leq 3n \Longleftrightarrow n \leq 4q.$$

In total, $7q/2 < n < 4q$. The next splitting is

$$n = q + 3(n - 3q) + 2(4q - n).$$

Again, we do not need a case distinction. We have the largest box

$$7(q + n - 3q) \leq 3n \Longleftrightarrow n \leq 7q/2.$$

This is a contradiction.

Every branch of the case distinction failed for $n > 4$. This proves the theorem. □

Bibliography

[1] C. Aistleitner, A. Hinrichs, D. Rudolf. On the size of the largest empty box amidst a point set. Discrete Appl. Math. **230**, 146–150 (2017).

[2] A. Dumitrescu, M. Jiang. On the largest empty axis-parallel box amidst n points. Algorithmica **66**(2), 225–248, June (2013).

[3] A. Hinrichs, J. Oettershagen. Optimal point sets for quasi-Monte Carlo integration of bivariate periodic functions with bounded mixed derivatives. In: Monte Carlo and Quasi-Monte Carlo Methods. Springer Proc. Math. Stat., vol. 163, pp. 385–405. Springer, Cham (2016).

[4] D. Krieg. On the dispersion of sparse grids. J. Complex. **45**, 115–119 (2018).

[5] H. Niederreiter. Random Number Generation and Quasi-Monte Carlo Methods. CBMS-NSF Regional Conference Series in Applied Mathematics, vol. 63. Society for Industrial and Applied Mathematics (SIAM), Philadelphia, PA (1992).

[6] D. Rudolf. An upper bound of the minimal dispersion via delta covers. In: Contemporary Computational Mathematics – A Celebration of the 80th Birthday of Ian Sloan. Vol. 1, 2, pp. 1099–1108. Springer, Cham (2018).

[7] I. H. Sloan, S. Joe. Lattice Methods for Multiple Integration. Oxford Science Publications. The Clarendon Press, Oxford University Press, New York (1994).

[8] V. T. Sós. On the distribution mod 1 of the sequence $n\alpha$. Ann. Univ. Sci. Budapest, Eötvös Sect. Math. **1**, 127–134 (1958).

[9] V. T. Sós, S. K. Zaremba. The mean-square discrepancies of some two-dimensional lattices. Studia Sci. Math. Hung. **14**(1–3), 255–271 (1982). 1979.

[10] J. Sosnovec. A note on minimal dispersion of point sets in the unit cube. Eur. J. Comb. **69**, 255–259 (2018).

[11] J. Suránji. Über die Anordnung der Vielfachen einer reellen Zahl mod 1. Ann. Univ. Sci. Budapest, Eötvös Sect. Math. **1**, 107–111 (1958).

[12] S. Świerczkowski. On successive settings of an arc on the circumference of a circle. Fundam. Math. **46**, 187–189 (1959).

[13] V. N. Temlyakov. Smooth fixed volume discrepancy, dispersion, and related problems. J. Approx. Theory **237**, 113–134 (2019).

[14] M. Ullrich. A lower bound for the dispersion on the torus. Math. Comput. Simul. **143**, 186–190 (2018).

[15] M. Ullrich, J. Vybíral. An upper bound on the minimal dispersion. J. Complex. **45**, 120–126 (2018).

Gerhard Larcher and Wolfgang Stockinger

7 On pair correlation of sequences

Abstract: We give a survey on the concept of Poissonian pair correlation (PPC) of sequences in the unit interval, on existing and recent results and we state a list of open problems. Moreover, we present and discuss a quite recent multi-dimensional version of PPC.

Keywords: Pair correlation of sequences, uniform distribution

MSC 2010: 11K36

7.1 The concept of Poissonian pair correlation for sequences in [0, 1)

Let $x_1, x_2, \dots,$ be a sequence of real numbers in the unit interval $[0,1)$. In the following, for some $x \in [0,1)$, we denote by $\|x\|$ the distance to the nearest integer, that is, to be precise $\|x\| := \min(x, 1-x)$. Further, in the sequel, $\{\cdot\}$ will denote the fractional part of a real number.

Definition 7.1. We say that $(x_n)_{n\geq 1} \in [0,1)$ has Poissonian pair correlation (PPC), if for all real $s > 0$, we have

$$\lim_{N\to\infty} \frac{1}{N} \#\left\{1 \leq k \neq l \leq N \;\middle|\; \|x_k - x_l\| < \frac{s}{N}\right\} = 2s.$$

To put this in intuitive words, PPC means to study small distances between sequence elements, that is, the concept of PPC deals with a "local" distribution property of a sequence in the unit interval.

It is natural to expect that the pair correlation function, R_N, defined as

$$R_N(s) := \frac{1}{N} \#\left\{1 \leq k \neq l \leq N \;\middle|\; \|x_k - x_l\| < \frac{s}{N}\right\}$$

tends to $2s$. We give the following heuristic explanation for this limit behavior:

Consider a fixed N, and fix a sequence element x_n for some $1 \leq n \leq N$. Then the region around x_n with length $\frac{2s}{N}$ (see Figure 7.1) is expected to contain $2s\frac{N-1}{N}$ of the remaining $(N-1)$ points x_i, for $i = 1, \dots, N$ and $i \neq n$.

Acknowledgement: The first author is supported by the Austrian Science Fund (FWF): Project F5507-N26, which is part of the Special Research Program "Quasi-Monte Carlo Methods: Theory and Applications." The second author is supported by a special Upper Austrian grant.

Gerhard Larcher, Johannes Kepler University, Linz, Austria, e-mail: gerhard.larcher@jku.at
Wolfgang Stockinger, University of Oxford, United Kingdom

https://doi.org/10.1515/9783110652581-007

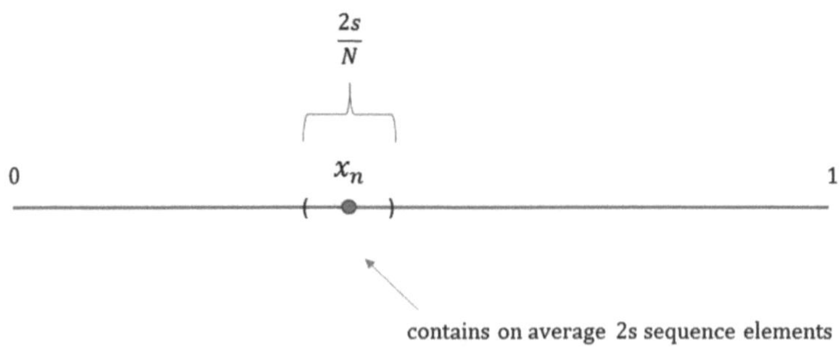

Figure 7.1

Consequently, this means, on average there are $2s\frac{N-1}{N}$ different indices $i = 1, \dots, N$ with $i \neq n$, such that

$$\|x_i - x_n\| < \frac{s}{N}.$$

Since n can attain values between 1 and N, we expect that there are $2s(N-1)$ pairs with

$$\|x_k - x_l\| < \frac{s}{N}, \quad \text{for } 1 \leq k \neq l \leq N.$$

Hence, we expect the quantity $R_N(s)$ to be approximately $2s\frac{N-1}{N}$ and, therefore,

$$\lim_{N\to\infty} R_N(s) = 2s.$$

Indeed, it can be shown that, in a certain sense, almost every sequence $x_1, x_2, \dots$ in $[0,1)$ has PPC. To be precise, if we consider a sequence $(X_n)_{n\geq 1}$ of i. i. d. random variables drawn from the uniform distribution on $[0,1)$, then

$$\lim_{N\to\infty} \frac{1}{N}\#\left\{1 \leq k \neq l \leq N \,\middle|\, \|X_k - X_l\| < \frac{s}{N}\right\} = 2s,$$

almost surely.

We want to briefly mention that the original motivation for the investigation of the PPC property comes from quantum physics. Roughly speaking, the concept is related to the distribution properties of the discrete energy spectrum $\lambda_1, \lambda_2, \dots$ of a Hamiltonian operator of a quantum system. The famous Berry–Tabor conjecture in quantum physics now states, that this discrete energy spectrum (ignoring degenerate cases) has PPC. For more details on the connection to quantum physics, we refer the reader to [1] and the references cited therein.

Note that for several quantum systems the discrete energy spectrum $\lambda_1, \lambda_2, \dots$ has the following special form:

$$(\lambda_n)_{n\geq 1} = (\{a_n\alpha\})_{n\geq 1},$$

where α is a real constant, and $(a_n)_{n\geq 1}$ is a given sequence of positive integers. Therefore, in the 1990s Rudnick, Sarnak, and Zaharescu started to investigate the PPC property of sequences of the form $(\{a_n\alpha\})_{n\geq 1}$ in $[0,1)$ from a purely mathematical point of view.

Whenever, in the following, we consider such sequences, we restrict the setting to strictly increasing sequences of positive integers. The most basic example of such a sequence is the classical Kronecker sequence $(\{n\alpha\})_{n\geq 1}$. This sequence does not have the PPC property for any choice of α. In most of the seminal papers on PPC, this fact was argued by taking the famous three-gap theorem into account (see, e. g., [25–27]).

The three-gap theorem states the following: For every choice of α and for every N, the gaps between neighboring points of the set

$$\{1\alpha\}, \{2\alpha\}, \dots, \{N\alpha\}$$

can have at most three different lengths. A sequence with such a gap structure does not exhibit a random behavior and, therefore, it is reasonable to expect that it cannot have PPC. Nonetheless, it is not immediately clear that this argument is indeed valid.

To argue that, we want to emphasize that the elements of a sequence satisfying such a weak gap structure could be ordered in a way, such that "many different" distances between (not necessarily neighboring) elements can occur (see Figure 7.2). However, for the Kronecker sequence a very simple argument can be given to deduce the fact that it does not have PPC for any choice of α.

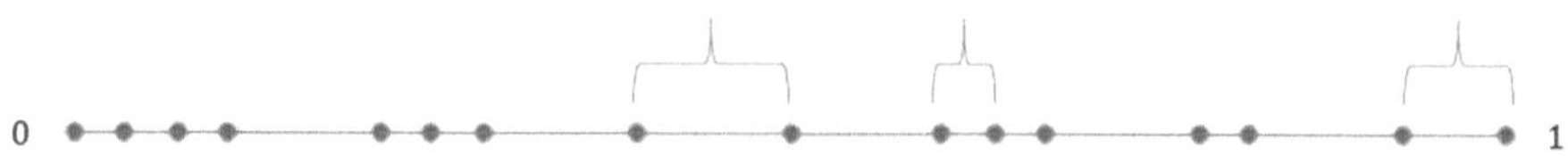

Figure 7.2

Let $\alpha \sim \frac{p_n}{q_n}$ where $\frac{p_n}{q_n}$ is a best approximation fraction to α with $(p_n, q_n) = 1$. It is well known from basic Diophantine approximation theory that $\alpha = \frac{p_n}{q_n} + \theta_n$, with either $0 \leq \theta_n < \frac{1}{2q_n^2}$, or $-\frac{1}{2q_n^2} < \theta < 0$.

Let us assume the first case. Then the set of points

$$\{1\alpha\}, \{2\alpha\}, \dots, \{N\alpha\}$$

equals the set of points

$$\frac{0}{N} + \varphi_0, \frac{1}{N} + \varphi_1, \dots, \frac{N-1}{N} + \varphi_{N-1},$$

with $0 \leq \varphi_i < \frac{1}{2N}$, for $i = 0, \dots, N-1$ (see points in Figure 7.3).

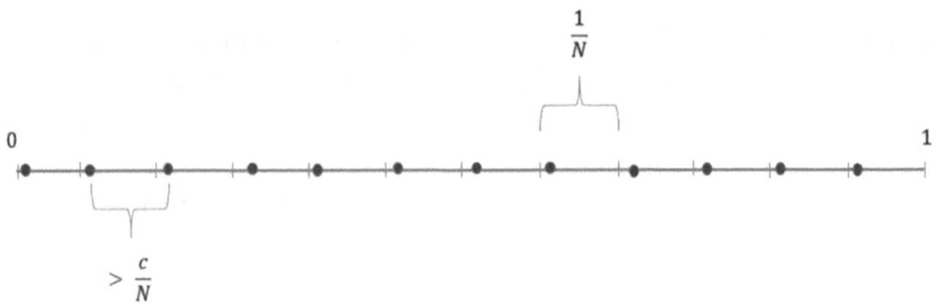

Figure 7.3

Hence, two arbitrary elements of this point set have a distance of at least $\frac{1}{2N}$. Thus, for the choice $s = \frac{1}{4}$, we get that $R_N(s) = 0$, and consequently the pair correlation function cannot tend to $2s = \frac{1}{2}$ for N to infinity.

In fact, a deeper investigation reveals that the three-gap theorem is indeed a valid argument to deduce the result on the PPC structure of the Kronecker sequence. It is an immediate consequence of the following theorem, which was proven in [20], in combination with the three-gap theorem.

Theorem 7.1. *Let $(x_n)_{n\geq 1}$ be a "weak finite-gap-sequence," that is, there exists an integer L and indices $N_1 < N_2 < N_3 < \cdots$ such that for all i the set $x_1, x_2, \ldots, x_{N_i}$ has at most L different gap lengths between neighboring elements. Then $(x_n)_{n\geq 1}$ does not have PPC.*

Let us now come to "positive results" and to the study of the metrical pair correlation theory of sequences of the form $(\{a_n\alpha\})_{n\geq 1}$. In [27], Rudnick and Sarnak showed the following.

Theorem 7.2. *The sequence $(\{n^d\alpha\})_{n\geq 1}$ with an integer $d \geq 2$ has PPC for almost all α.*

The case $d = 2$ is of particular interest, as in this setting the spacings of sequence elements are related to the distances between the energy levels of the so-called "boxed oscillator," that is, the study of the PPC property of $(\{n^2\alpha\})_{n\geq 1}$ is of special importance in quantum physics. The PPC property and also the gap distribution of the sequence $(\{n^2\alpha\})_{n\geq 1}$ was further investigated by several authors (see [15, 22, 26]) and they could also derive the metrical result for $d = 2$. Heath–Brown could even show slightly more.

Theorem 7.3. *The sequence $(\{n^2\alpha\})_{n\geq 1}$ has PPC for almost all real numbers α. Moreover, there is a dense set of constructible values of α for which the PPC property holds, that is, there is an informal algorithm, which, for any closed interval I of positive length, provides a convergent sequence of rational numbers belonging to I, whose limit α satisfies that $(\{n^2\alpha\})_{n\geq 1}$ has PPC.*

These results are only metrical statements and, until now, no single explicit α is known such that $(\{n^2\alpha\})_{n\geq 1}$ (or $(\{n^d\alpha\})_{n\geq 1}$ for any integer $d \geq 2$) has PPC. Nonetheless, we know that it is **not true**, that $(\{n^2\alpha\})_{n\geq 1}$ has PPC for **all** irrational α. Consider the following example of an α that is in a certain sense well approximable: If α is an irrational number such that $|\alpha - \frac{a}{q}| < \frac{1}{4q^3}$ for infinitely many integers a and q, then $(\{n^2\alpha\})_{n\geq 1}$ does not have PPC (see [15]).

On the other hand, it is conjectured that for an α which is not too well approximable, we have PPC for $(\{n^2\alpha\})_{n\geq 1}$. To be precise: Let α be such that for every $\varepsilon > 0$ there is a $c(\varepsilon) > 0$ with $|\alpha - \frac{a}{q}| > c(\varepsilon)\frac{1}{q^{2+\varepsilon}}$ for all $a, q \in \mathbb{Z}$, then $(\{n^2\alpha\})_{n\geq 1}$ has PPC (see, e. g., [15]). This property for an irrational α is often referred to as Diophantine. It is well known that almost all irrationals are Diophantine, for example, every real irrational algebraic number has this property. The above discussion illustrates that the pair correlation theory of sequences $(\{n^d\alpha\})_{n\geq 1}$ is strongly related to the Diophantine properties of α.

The case of lacunary sequences $(a_n)_{n\geq 1}$ was considered, for example, by Rudnick and Zaharescu (see [25]) or by Berkes, Philipp, and Tichy (see [8]). We recall that a sequence $(a_n)_{n\geq 1}$ is a lacunary sequence if there exists a $c > 1$ such that $\frac{a_{n+1}}{a_n} > c$ for all $n \geq N(c)$. Again, they obtained the following metrical result.

Theorem 7.4. *Let $(a_n)_{n\geq 1}$ be a lacunary sequence of positive integers. Then $(\{a_n\alpha\})_{n\geq 1}$ has PPC for almost all α.*

We may again ask for explicit examples of lacunary sequences $(a_n)_{n\geq 1}$ and $\alpha \in \mathbb{R}$ such that $(\{a_n\alpha\})_{n\geq 1}$ has PPC.

One of the most basic examples of a lacunary sequence of integers certainly is the sequence $(2^n)_{n\geq 1}$. In a first step, we may restrict the possible candidates α for which $(\{2^n\alpha\})_{n\geq 1}$ could have PPC. To do so, we consider the following result which has been shown independently by Grepstad and Larcher [14], Aistleitner, Lachmann and Pausinger [4], and Steinerberger [30].

Theorem 7.5. *If the sequence $(x_n)_{n\geq 1}$ in $[0,1)$ has PPC, then $(x_n)_{n\geq 1}$ is uniformly distributed in $[0,1)$.*

Remark. The paper of Grepstad and Larcher also contains a quantitative version of this result. Roughly speaking: If $R_N(s)$ tends to $2s$ "fast in some uniform sense," then the discrepancy D_N of the sequence $(x_n)_{n\geq 1}$ cannot tend to zero "too slowly."

Having this result in mind, we can restrict the set of possible choices for α, such that $(\{2^n\alpha\})_{n\geq 1}$ has PPC. The above theorem implies that for such an α the sequence $(\{2^n\alpha\})_{n\geq 1}$ has to be uniformly distributed, and hence α has to be normal in base 2.

The most well-known example of a real α which is normal in base 2 is the Champernowne number, that is, the number α which has in base 2 the digit representation

$$\alpha = 0.\,01\,10\,11\,100\,101\,110\,111\,1000\ldots$$

However, it was shown by Pirsic and Stockinger in [24], that for α the Champernowne number, the sequence $(\{2^n\alpha\})_{n\geq 1}$ does not have PPC. Also for further concrete examples like Stoneham-numbers or infinite de Bruijn-words, the sequence $(\{2^n\alpha\})_{n\geq 1}$ does not have PPC (see [20]).

Indeed, until now we do not know any concrete example of $(a_n)_{n\geq 1}$, a lacunary sequence, and a real α such that $(\{a_n\alpha\})_{n\geq 1}$ does have PPC.

Recently, a much more general metric result on PPC of sequences of the form $(\{a_n\alpha\})_{n\geq 1}$ was given in [7] which shows that there is an intimate connection between the concept of PPC of sequences $(\{a_n\alpha\})_{n\geq 1}$ and the notion of additive energy of the sequence $(a_n)_{n\geq 1}$. The concept of additive energy plays a central role in additive combinatorics and also appears in the study of the metrical discrepancy theory of sequences $(\{a_n\alpha\})_{n\geq 1}$ (see [2, 6]).

For a strictly increasing sequence $a_1 < a_2 < a_3 < \cdots$ of positive integers, we consider the first N elements $a_1, \ldots, a_N$. The additive energy of $a_1, \ldots, a_N$ is given by

$$E(a_1, \ldots, a_N) := \sum_{\substack{1\leq i,j,k,l\leq N \\ a_i - a_j = a_k - a_l}} 1.$$

It is obvious that $N^2 \leq E(a_1, \ldots, a_N) \leq N^3$ always holds. In [7], the following was shown.

Theorem 7.6. *Let $(a_n)_{n\geq 1}$ be a strictly increasing sequence of integers such that there exists $\varepsilon > 0$ with*

$$E(a_1, \ldots, a_N) = \mathcal{O}(N^{3-\varepsilon}),$$

then $(\{a_n\alpha\})_{n\geq 1}$ has PPC for almost all α.

This result recovers all above mentioned metrical results and implies several new results and examples.

Example 7.1. If $(a_n)_{n\geq 1}$ is lacunary, then $E(a_1, \ldots, a_N) = \mathcal{O}(N^2)$, hence $(\{a_n\alpha\})_{n\geq 1}$ has PPC for almost all α.

Example 7.2. If $(a_n)_{n\geq 1}$ are the values of a polynomial $f(n) \in \mathbb{Z}[x]$ of degree $d \geq 2$, then $E(a_1, \ldots, a_N) = \mathcal{O}(N^{2+\varepsilon})$ for all $\varepsilon > 0$, hence $(\{a_n\alpha\})_{n\geq 1}$ has PPC for almost all α.

Example 7.3. Let $(a_n)_{n\geq 1}$ be a convex sequence, that is, $a_n - a_{n-1} < a_{n+1} - a_n$ for all n, then it was shown by Konjagin [18] that $E(a_1, \ldots, a_N) = \mathcal{O}(N^{\frac{5}{2}})$, hence $(\{a_n\alpha\})_{n\geq 1}$ has PPC for almost all α.

Example 7.4. If $a_n = [\beta n^c]$ for some $\beta > 0$ and $c > 1$, then

$$E(a_1, \ldots, a_N) = \mathcal{O}(\max(N^{\frac{5}{2}}, N^{4-c})),$$

hence $(\{a_n\alpha\})_{n\geq 1}$ has PPC for almost all α (see [29]).

The above theorem immediately raises two natural questions.

Question 7.1. Is it possible for an increasing sequence of distinct integers $(a_n)_{n\geq 1}$ which satisfies $E(a_1,\dots,a_N) = \Omega(N^3)$ that the sequence $(\{a_n\alpha\})_{n\geq 1}$ has PPC for almost all α?

Question 7.2. If, for almost all α, $(\{a_n\alpha\})_{n\geq 1}$ does **not** have PPC, does this imply $E(a_1,\dots,a_N) = \Omega(N^3)$?

Both questions were answered by J. Bourgain in an Appendix to [7]. Concerning question 7.1, Bourgain showed:

– If $E(a_1,\dots,a_N) = \Omega(N^3)$, then there exists a set of positive measure such that $(\{a_n\alpha\})_{n\geq 1}$ does not have PPC for every α in this set.

This result was improved by Lachmann and Technau [19]:

– If $E(a_1,\dots,a_N) = \Omega(N^3)$, then there exists a set of full Lebesgue measure such that $(\{a_n\alpha\})_{n\geq 1}$ does not have PPC for every α contained in this set.

Finally, in [21] this result was improved to its final form.

Theorem 7.7. *If $E(a_1,\dots,a_N) = \Omega(N^3)$, then there is **no** α such that $(\{a_n\alpha\})_{n\geq 1}$ has PPC.*

Concerning question 7.2 Bourgain showed that the answer to this question is **"no"**: He gave a construction for a sequence $(a_n)_{n\geq 1}$ with $E(a_1,\dots,a_N) = o(N^3)$, such that $(\{a_n\alpha\})_{n\geq 1}$ does not have PPC for almost all α. Up to now, we have the following situation:

$E(a_1,\dots,a_N) = \Omega(N^3)$ implies that there is **no** α such that the sequence $(\{a_n\alpha\})_{n\geq 1}$ has PPC.

$E(a_1,\dots,a_N) = \mathcal{O}(N^{3-\varepsilon})$ for some $\varepsilon > 0$ implies PPC for almost all α.

The result by Aistleitner, Larcher, and Lewko, was first extended by Bloom, Chow, Gafni, and Walker, albeit under an additional density condition on the integer sequence $(a_n)_{n\geq 1}$ (see [9]).

Theorem 7.8. *Let $a_1,\dots,a_N$ be the first N elements of an increasing sequence of positive integers $(a_n)_{n\geq 1}$ satisfying the following density condition:*

$$\delta(N) = \Omega_\varepsilon\left(\frac{1}{(\log N)^{2+2\varepsilon}}\right),$$

where $\delta(N) := N^{-1}\#(\{a_1,\dots,a_N\}\cap\{1,\dots,N\})$ and suppose that

$$E(a_1,\dots,a_n) = \mathcal{O}_\varepsilon\left(\frac{N^3}{(\log N)^{2+\varepsilon}}\right),$$

for some $\varepsilon > 0$, then, for almost all α, the sequence $(\{a_n\alpha\})_{n\geq 1}$ has PPC.

Recently, Bloom and Walker (see [10]) improved over this result by showing the following theorem.

Theorem 7.9. *There exists an absolute positive constant C such that the following is true. Suppose that*

$$E(a_1,\dots,a_N)=\mathcal{O}\left(\frac{N^3}{(\log N)^C}\right),$$

then for almost all α, $(\{a_n\alpha\})_{n\geq 1}$ *has PPC.*

A consequence of this result is the following theorem.

Theorem 7.10. *Let* $(a_n)_{n\geq 1}$ *be an arbitrary infinite subset of the squares. Then* $(a_n)_{n\geq 1}$ *is metric Poissonian, that is, for almost all* α, $(\{a_n\alpha\})_{n\geq 1}$ *has PPC.*

To see that this result is valid, we note that, if $a_1,\dots,a_N$ denotes a finite set of squares, then $E(a_1,\dots,a_N)=\mathcal{O}(N^3\exp(-c_1\log^{c_2}N))$ for some absolute positive constants c_1 and c_2; see, for example, [28].

The proof of Theorem 7.9 relies on a new bound for GCD sums with $\alpha=1/2$, which improves over the bound by Bondarenko and Seip (see [11]), if the additive energy of $a_1,\dots,a_N$ is sufficiently large. Note that the constant C was not specified in the above mentioned paper, but the authors thereof conjecture that Theorem 7.9 holds for $C>1$ already. This result would be best possible. To see this, consider the following result by Walker [31].

Theorem 7.11. *Let* $(a_n)_{n\geq 1}=(p_n)_{n\geq 1}$ *be the sequence of primes (note that for the primes we have* $E(p_1,\dots,p_N)\asymp\frac{N^3}{\log N}$*). Then* $(\{p_n\alpha\})_{n\geq 1}$ *does not have PPC for almost all* α.

The region between $\mathcal{O}\left(\frac{N^3}{(\log N)^C}\right)$, $C>1$, and $\Omega(N^3)$ is therefore the interesting region and one might speculate about a sharp threshold which allows to fully describe the metrical pair correlation theory in terms of the additive energy. Further constructions and examples of sequences in this "interesting region," with an even smaller additive energy compared to the primes, were given by Lachmann and Technau [19]:

Theorem 7.12. *There exists a strictly increasing sequence of positive integers* $(a_n)_{n\geq 1}$ *with*

$$E(a_1,\dots,a_N)=\mathcal{O}\left(\frac{N^3}{\log N(\log\log N)}\right)$$

such that $(\{a_n\alpha\})_{n\geq 1}$ *does not have PPC for almost all* α.

On the other hand, they gave a positive result of the following form.

Theorem 7.13. *There exists a strictly increasing sequence of positive integers* $(a_n)_{n\geq 1}$ *with*

$$E(a_1,\dots,a_N)=\Omega\left(\frac{N^3}{\log N(\log\log)^{1+\varepsilon}}\right)$$

for all $\varepsilon>0$, *such that* $(\{a_n\alpha\})_{n\geq 1}$ *has PPC for almost all* α.

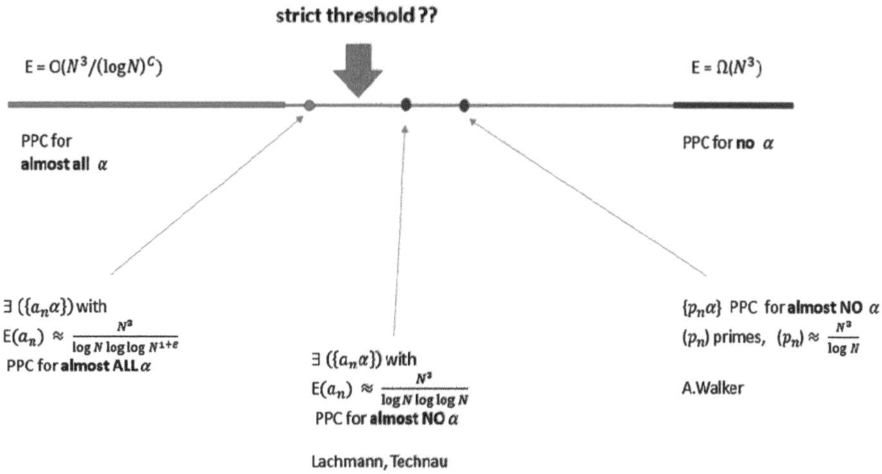

Figure 7.4

Figure 7.4 summarizes the link between the additive energy and PPC properties of $(\{a_n\alpha\})_{n\geq 1}$. The following question therefore is near at hand:

Is there a strict threshold T such that an additive energy of magnitude smaller than T implies PPC of $(\{a_n\alpha\})_{n\geq 1}$ for almost all α and an additive energy of magnitude larger than T implies PPC for $(\{a_n\alpha\})_{n\geq 1}$ for almost no α? The fundamental question concerning such a putative threshold was raised in [10]. The authors of this paper conjectured that there is a sharp Khintchine-type threshold, that is, if $E(a_1,\ldots,a_N) = \Theta(N^3\psi(N))$, for some weakly decreasing function $\psi : \mathbb{Z}_{\geq 1} \to [0,1]$, then, for almost all α, $(\{a_n\alpha\})_{n\geq 1}$ has PPC if and only if

$$\sum_{N\geq 1} \frac{\psi(N)}{N}$$

converges.

The negative answer to this question was given by Aistleitner, Lachmann, and Technau [5]:

Theorem 7.14. *There exists a sequence $(a_n)_{n\geq 1}$ of integers with*

$$E(a_1,\ldots,a_N) = \Omega\left(\frac{N^3}{(\log N)^{\frac{3}{4}+\varepsilon}}\right)$$

such that $(\{a_n\alpha\})_{n\geq 1}$ has PPC for almost all α. Hence a threshold T cannot exist.

To conclude, the additive energy is not enough to fully describe the metrical pair correlation theory. Some further number theoretic properties need to be considered to cope with that problem.

7.2 The concept of Poissonian pair correlation for sequences in $[0,1)^d$

Of course, it makes sense to generalize the concept of PPC to the multi-dimensional setting. One way to generalize the one-dimensional concept to a multidimensional setting was defined and discussed in [16] (for a more general analysis of a multidimensional PPC concept, we refer to the recent work [23]). Here, we present the definition of [16].

Definition 7.2. Let $(x_n)_{n\geq 1}$ be a sequence in the d-dimensional unit-cube $[0,1)^d$. We say that $(x_n)_{n\geq 1}$ has PPC if for all $s > 0$ we have

$$\lim_{N\to\infty} \frac{1}{N}\#\left\{1 \leq k \neq l \leq N \,\middle|\, \|x_k - x_l\|_\infty < \frac{s}{N^{\frac{1}{d}}}\right\} = (2s)^d.$$

For this definition of d-dimensional PPC, it again follows that $(x_n)_{n\geq 1}$ with PPC is uniformly distributed in $[0,1)^d$. Moreover, for many of the above mentioned results in dimension $d = 1$ we have analogous statements in dimension $d \geq 2$. For example, the d-dimensional Kronecker sequence

$$(\{n\alpha_1\}, \{n\alpha_2\}, \ldots, \{n\alpha_d\})_{n\geq 1}$$

never has PPC. The proof of this fact however needs a bit more subtle arguments than in dimension 1.

Naturally, we would also expect that under the same condition on the additive energy as in Theorem 7.9, the sequence

$$(\{a_n\boldsymbol{\alpha}\})_{n\geq 1}$$

has Poissonian pair correlations for almost all instances and, in fact, we have the following even better result, which is a consequence of better bounds on GCD sums for larger exponents than 1/2.

Theorem 7.15. *Let $a_1, \ldots, a_N$ denote the first N elements of $(a_n)_{n\geq 1}$ and suppose that*

$$E(a_1, \ldots, a_N) = \mathcal{O}\left(\frac{N^3}{(\log N)^{1+\varepsilon}}\right), \quad \text{for any } \varepsilon > 0,$$

then for almost all choices of $\boldsymbol{\alpha} = (\alpha_1, \ldots, \alpha_d) \in \mathbb{R}^d$,

$$(\{a_n\boldsymbol{\alpha}\})_{n\geq 1}$$

has PPC.

However, if the additive energy is of maximal order, that is, if we have $E(a_1, \ldots, a_N) = \Omega(N^3)$, then there is no $\boldsymbol{\alpha}$ such that $(\{a_n\boldsymbol{\alpha}\})_{n\geq 1}$ has PPC.

Theorem 7.16. *If $E(a_1,\dots,a_N) = \Omega(N^3)$, then for any choice of $\boldsymbol{\alpha} = (\alpha_1,\dots,\alpha_d) \in \mathbb{R}^d$ the sequence*

$$(\{a_n\boldsymbol{\alpha}\})_{n\ge1},$$

does not have Poissonian pair correlations.

7.3 Open problems

Many questions related to the concept of PPC are still open, and we will state some of them in this section as open problems. Note that Problem 7.4 and Problem 7.5 listed below, were recently solved in [3] and [17], respectively.

Problem 7.1. Is it possible to extend the regions concerning the size of the additive energy of an integer sequence $(a_n)_{n\ge1}$ in Figure 7.4? To be precise: Are there functions $\varphi(n)$ (which increases slower than $(\log N)^C$, for C the constant in Theorem 7.9) and $\psi(n)$ both tending to $+\infty$ for n to infinity, such that:

If $E(a_1,\dots,a_N) = \mathcal{O}\left(\frac{N^3}{\varphi(N)}\right)$, then, for almost **all** α, $(\{a_n\alpha\})_{n\ge1}$ has PPC.

If $E(a_1,\dots,a_N) = \Omega\left(\frac{N^3}{\psi(N)}\right)$, then there is **no** α such that $(\{a_n\alpha\})_{n\ge1}$ has PPC.

Problem 7.2. We know that if $E(a_1,\dots,a_N) = \Omega(N^3)$, then $(\{a_n\alpha\})_{n\ge1}$ has PPC for **no** α. We consider the following question to be of high interest: Is there a sequence $(a_n)_{n\ge1}$ with the property that for almost all α, $(\{a_n\alpha\})_{n\ge1}$ does not have PPC, but there exists a set of zero measure such that $(\{a_n\alpha\})_{n\ge1}$ has PPC for every α contained in this set?

Indeed, we believe, that this is not possible, that is, we have the following.

Conjecture 7.1. *If, for almost all α, $(\{a_n\alpha\})_{n\ge1}$ does not have PPC, then it has PPC for* ***no*** *α.*

For example (by the result of A. Walker), this would imply: $(\{p_n\alpha\})_{n\ge1}$ has PPC for **no** α.

Problem 7.3. Although the metrical theory of sequences of the form $(\{a_n\boldsymbol{\alpha}\})_{n\ge1}$ seems to be well established, we do not know any explicit construction of $\boldsymbol{\alpha}$ (not even in the one-dimensional case) such that $(\{a_n\boldsymbol{\alpha}\})_{n\ge1}$ has Poissonian pair correlations. It is in general very hard to construct sequences on the torus having the PPC property. The only known explicit examples—to the best of our knowledge—of sequences with this property are $\{\sqrt{n}\}_{n\ge1}$ (see [12]) and certain directions of vectors in an affine Euclidean lattice (see [13]). Hence, of course, it would be of high interest to find more concrete examples of sequences with PPC.

Problem 7.4. This problem concerns a possible extension of Theorem 7.1 mentioned above. We recall that we have shown that a sequence $(x_n)_{n\ge1}$ with a weak finite gap property never has PPC. We wonder whether this result can be improved by showing that it still holds if we have to deal with a sequence having a "slowly-growing-gap" property, that is:

There is a (slowly growing) function L and a sequence of indices $N_1 < N_2 < N_3 < \cdots$ such that $x_1, x_2, \ldots, x_{N_i}$ always has gaps of at most $L(N_i)$ different lengths.

Problem 7.5. Let $(\boldsymbol{x}_n)_{n\geq 1}$ be the d-dimensional Halton-sequence in any bases $q_1, \ldots, q_d$, where $d \geq 2$. Does $(\boldsymbol{x}_n)_{n\geq 1}$ have PPC or not? Of course, we strongly conjecture that it does not have PPC. In dimension $d = 1$, the Halton-sequence is the well-known van der Corput sequence. In this case, the PPC property trivially does not hold. In fact, it should not be too hard to prove this in the multidimensional case, too.

The last problem we want to state concerns a multidimensional version of the metrical PPC result for the primes.

Problem 7.6. Is it true that for almost all instances of $\boldsymbol{\alpha}$ the sequence $(\{p_n\boldsymbol{\alpha}\})_{n\geq 1}$, where $(p_n)_{n\geq 1}$ denotes the primes, does not have PPC?

Bibliography

[1] I. Aichinger, C. Aistleitner, G. Larcher. On quasi-energy-spectra, pair correlations of sequences and additive combinatorics. In: J. Dick, F. Y. Kuo, H. Wozniakowski (eds.) Contemporary Computational Mathematics – A Celebration of the 80th Birthday of Ian Sloan, pp. 1–16. Springer-Verlag (2018).

[2] C. Aistleitner, R. Hofer, G. Larcher. On evil Kronecker sequences and lacunary trigonometric products. Ann. Inst. Fourier (2016).

[3] C. Aistleitner, T. Lachmann, P. Leonetti, P. Minelli. On the number of gaps of sequences with Poissonian Pair Correlations (2019). arXiv:1908.06292.

[4] C. Aistleitner, T. Lachmann, F. Pausinger. Pair correlations and equidistribution. J. Number Theory **182**, 206–220 (2018).

[5] C. Aistleitner, T. Lachmann, N. Technau. There is no Khintchine threshold for metric pair correlations (2018). arXiv:1802.02659.

[6] C. Aistleitner, G. Larcher. Additive energy and irregularities of distribution. Unif. Distrib. Theory **12**(1), (2017).

[7] C. Aistleitner, G. Larcher, M. Lewko. Additive energy and the Hausdorff dimension of the exceptional set in metric pair correlation problems. Israel J. Math. **222**(1), 463–485 (2017). With an appendix by Jean Bourgain.

[8] I. Berkes, W. Philipp, R. Tichy. Pair correlations and U-statistics for independent and weakly dependent random variables. Ill. J. Math. **45**(2), 559–580, Summer (2001).

[9] T. Bloom, S. Chow, A. Gafni, A. Walker. Additive energy and the metric Poissonian property. Mathematika **64**(3), 679–700 (2018).

[10] T. F. Bloom, A. Walker. GCD sums and sum-product estimates (2018). arXiv:1806.07849.

[11] A. Bondarenko, K. Seip. GCD sums and complete sets of square-free numbers. Bull. Lond. Math. Soc. **47**(1), 29–41 (2015).

[12] D. El-Baz, J. Marklof, I. Vinogradov. The two-point correlation function of the fractional parts of $\sqrt{n}$ is Poisson. Proc. Am. Math. Soc. **143**(7), 2815–2828 (2015).

[13] D. El-Baz, J. Marklof, I. Vinogradov. The distribution of directions in an affine lattice: two-point correlations and mixed moments. Int. Math. Res. Not. **2015**, 1371–1400 (2015).
[14] S. Grepstad, G. Larcher. On pair correlation and discrepancy. Arch. Math. **109**, 143–149 (2017).
[15] D. R. Heath-Brown. Pair correlation for fractional parts of αn^2. Math. Proc. Camb. Philos. Soc. **148**(3), 385–407 (2010).
[16] A. Hinrichs, L. Kaltenböck, G. Larcher, W. Stockinger, M. Ullrich. On a multi-dimensional Poissonian pair correlation concept and uniform distribution (2018). arXiv:1809.05672.
[17] R. Hofer, L. Kaltenböck. Multi-Dimensional Pair Correlations of Niederreiter and Halton Sequences are not Poissonian (2019). arXiv:1908.11147.
[18] S. V. Konyagin. An estimate of the L1-norm of an exponential sum (in Russian). In: The Theory of Approximations of Functions and Operators, Abstracts of Papers of the International Conference Dedicated to Stechkins 80th Anniversary (2000).
[19] T. Lachmann, N. Technau. On exceptional sets in the metric Poissonian pair correlations problem. Monatshefte Math. **189**, 137–156 (2019).
[20] G. Larcher, W. Stockinger. Some negative results related to Poissonian pair correlation problems (2018). arXiv:1803.052361.
[21] G. Larcher, W. Stockinger. Pair correlation of sequences $(\{a_n\alpha\})_{n\in\mathbb{N}}$ with maximal order of additive energy. Accepted by Math. Proc. of Cambridge Philosophical Society, preprint (2018). arXiv:1802.02901.
[22] J. Marklof, A. Strömbergsson. Equidistribution of Kronecker sequences along closed horocycles. Geom. Funct. Anal. **13**(6), 1239–1280 (2003).
[23] J. Marklof. Pair correlation and equidistribution on manifolds (2019). preprint arxiv.
[24] Í. Pirsic, W. Stockinger. The Champernowne constant is not Poissonian. Funct. Approx. Comment. Math. **60**(2), 253–262 (2019).
[25] Z. Rudnick, A. Zaharescu. A metric result on the pair correlation of fractional parts of sequences. Acta Arith. **89**(3), 283–293 (1999).
[26] Z. Rudnick, P. Sarnak, A. Zaharescu. The distribution of spacings between the fractional parts of $n^2\alpha$. Invent. Math. **145**(1), 37–57 (2001).
[27] Z. Rudnick, P. Sarnak. The pair correlation function of fractional parts of polynomials. Commun. Math. Phys. **194**(1), 61–70 (1998).
[28] T. Sanders. On the Bogolyubov-Ruzsa lemma. Anal. PDE **5**(3), 627–655 (2012).
[29] I. I. Sapiro-Pyateckii. On a variant of the Waring-Goldbach problem. Mat. Sb. **30**(72), 105–120 (1952).
[30] S. Steinerberger. Localized quantitative criteria for equidistribution. Acta Arith. **180**, 183–199 (2017).
[31] A. Walker. The primes are not metric Poissonian. Mathematika **64**, 230–236 (2018).

Aleksandar Nikolov

8 Some of Jiří Matoušek's contributions to combinatorial discrepancy theory

Abstract: Jiří Matoušek, who passed away in 2015, made important advancements in combinatorial discrepancy theory, especially for geometric set systems. This survey covers some of his work in this area, including recent work of Matoušek that has not been surveyed before. Throughout, our goal is to emphasize how Matoušek used concepts from computational and discrete geometry in his discrepancy work, and to present his results in the context of recent developments in the field.

Keywords: Combinatorial discrepancy theory, computational geometry

MSC 2010: 05D05, 11K38, 52C99

8.1 Combinatorial discrepancy

How well can a probability measure μ on $\mathbb{R}^d$ be approximated by the uniform distribution on n points? This question, which arises in many settings, ranging from number theory to numerical analysis and computer science, lies at the heart of discrepancy theory. In order to formalize it, we need to define what it means for one probability measure to approximate another. The approach taken in discrepancy theory (and also in algorithmic randomness, computational pseudorandomness, and other fields dealing with similar questions) is to say that the measures are similar if they "look the same" to a family of *test sets*. More formally, suppose that μ and ν are two probability measures defined on the same measurable space (X, Σ), and let $\mathcal{S}$ be a family of measurable subsets (the test sets). Then we can say that ν *ε-approximates* μ with respect to $\mathcal{S}$ if

$$\sup_{S \in \mathcal{S}} |\mu(S) - \nu(S)| \le \varepsilon.$$

In measure-theoretic discrepancy theory, typically μ is the Lebesgue measure λ^d on $[0,1]^d$ or some other Borel probability measure on $\mathbb{R}^d$, and ν is the discrete measure ν_P defined by $\nu_P(S) = \frac{|S \cap P|}{|P|}$ for some finite set $P \subset \mathbb{R}^d$. Quantitatively, we are interested in the smallest ε for which μ is ε-approximated by some ν_P for P of size n. Following the standard notation, we define the discrepancy of $P \subseteq X$ with respect to $\mathcal{S}$ and the measure μ by

$$D(\mathcal{S}, P; \mu) = \sup_{S \in \mathcal{S}} \left| \mu(S) - \frac{|S \cap P|}{|P|} \right|,$$

Aleksandar Nikolov, University of Toronto, Department of Computer Science, Canada, e-mail: anikolov@cs.toronto.edu

https://doi.org/10.1515/9783110652581-008

and the discrepancy of $\mathcal{S}$ with respect to μ by

$$D(\mathcal{S}, n; \mu) = \inf\{D(P, \mathcal{S}; \mu) : P \subseteq X, |P| = n\}.$$

There is a large body of work studying the growth rate of the function $D(\mathcal{S}, n; \lambda^d)$ with n for different families of sets $\mathcal{S}$, for example, axis-aligned boxes, half-spaces, and convex sets. In this survey, however, we focus on a combinatorial analogue of this rich theory, in which, instead of asking how well discrete measures can approximate a continuous one, we ask how well discrete measures with small support can approximate a measure with larger support. In the simplest variant of this question, we can take X to be a finite set of size N, $\mathcal{S}$ to be a family of subsets of X, and take $\mu = \nu_X$. If the approximating set P has cardinality $|P| = \frac{N}{2}$, we have

$$D(\mathcal{S}, P; \nu_X) = \max_{S \in \mathcal{S}} \left| \frac{|S|}{N} - \frac{2|S \cap P|}{N} \right| = \frac{1}{N} \max_{S \in \mathcal{S}} \big||S \setminus P| - |S \cap P|\big|.$$

Then $D(\mathcal{S}, \frac{N}{2}; \nu_X)$ is achieved by the set P of size $\frac{N}{2}$ that minimizes the quantity $\max_{S \in \mathcal{S}} \big||S \setminus P| - |S \cap P|\big|$. Dropping the size requirement for P gives the *combinatorial discrepancy* of $(X, \mathcal{S})$, defined by

$$\operatorname{disc}(\mathcal{S}) = \min_{P \subseteq X} \max_{S \in \mathcal{S}} \big||S \setminus P| - |S \cap P|\big|.$$

It is common to encode membership in P by a *coloring* χ, in which the color of any $p \in P$ is $\chi(p) = +1$, and the color of any $p \in X \setminus P$ is $\chi(p) = -1$. The discrepancy of $\chi : X \to \{-1, +1\}$ with respect to X is then defined by

$$\operatorname{disc}(\mathcal{S}, \chi) = \max_{S \in \mathcal{S}} |\chi(S)|,$$

where $\chi(S) = \sum_{p \in S} \chi(p)$.[1] With this notation, the discrepancy of $\mathcal{S}$ becomes $\operatorname{disc}(\mathcal{S}) = \min\{\operatorname{disc}(\mathcal{S}, \chi) : \chi : X \to \{-1, +1\}\}$.

At first, it may seem like something may be lost in these simplifications. Nevertheless, it turns out that combinatorial discrepancy is a powerful tool for designing low discrepancy sets of points, and provides an upper bound on $D(\mathcal{S}, n; \mu)$ under very mild assumptions on μ. Suppose that $\mathcal{S}$ is a family of Borel subsets of $\mathbb{R}^d$, and μ is a Borel measure, such that $D(\mathcal{S}, n; \mu) = o(n)$. The restriction of $\mathcal{S}$ to a finite set $P \subset \mathbb{R}^d$ is defined by $\mathcal{S}|_P = \{S \cap P : S \in \mathcal{S}\}$. The combinatorial discrepancy function

$$\operatorname{disc}(\mathcal{S}, n) = \max\{\operatorname{disc}(\mathcal{S}|_P) : P \subset \mathbb{R}^d, |P| = n\}$$

bounds $nD(\mathcal{S}, n; \mu)$ from above up to a universal constant, as long as $\operatorname{disc}(\mathcal{S}, n) = o(n)$.[2] This connection goes back at least to Beck's work on the Tusnády problem [8].

1 In the rest of the survey, we use $\chi(S) = \sum_{p \in S} \chi(p)$ for arbitrary functions $\chi : X \to \mathbb{R}$.

2 See [34, 1] for precise statements.

The connection between the combinatorial and measure-theoretic notions of discrepancy motivates the study of combinatorial discrepancy for natural families of geometric shapes $\mathcal{S}$. Then bounds on $\text{disc}(\mathcal{S}, n)$ can be used to bound $D(\mathcal{S}, n; \mu)$ uniformly for *any* sufficiently nice measure μ. This research direction, initiated by Beck, was advanced spectacularly in the work of Jiří Matoušek. The present survey will give a glimpse into some of the beautiful work of Matoušek in combinatorial discrepancy theory, both for geometric sets and in more abstract combinatorial settings. We will see the interplay between computational and discrete geometry and combinatorial discrepancy in his work. We will also cover his most recent work in discrepancy, which has not been surveyed before. For many results, we will sketch simpler proofs that are made possible by recent developments in discrepancy theory.

Matoušek was one of the great expositors of mathematics, and the interested reader is encouraged to refer to his book, Geometric Discrepancy [34], for a thorough introduction to this subject. The author of this survey is inspired by Matoušek's beautifully lucid style, even if he cannot hope to truly match it.

8.2 The discrepancy of half-spaces and VC-dimension

In order to gain intuition into combinatorial discrepancy, let us explore some basic bounds on the discrepancy of an arbitrary family $\mathcal{S}$ of m subsets of a finite set X of size n. (From now on, we will call the pair $(\mathcal{S}, X)$ a set system, and when X is clear from the context, we will refer to $\mathcal{S}$ itself as the set system.) An elementary application of the probabilistic method (see, e. g., Lecture 4 of [49]) shows that $\text{disc}(\mathcal{S}) = O(\sqrt{n \log(m)})$, and this discrepancy is achieved, with high probability, by independently assigning each $p \in X$ a uniformly random color in $\{-1, +1\}$. When m is sufficiently large, this bound cannot be improved. For example, in one extreme case, we can take $\mathcal{S}$ to be the powerset of X. Then, whatever coloring $\chi : X \to \{-1, +1\}$ we choose, the sets $\{p \in X : \chi(p) = -1\}$ and $\{p \in X : \chi(p) = +1\}$ are both in $\mathcal{S}$, and the larger of them has discrepancy at least $n/2$. Surprisingly, when m is linear in n or smaller, one can do better than a uniformly random coloring. Spencer [48] showed that $\text{disc}(\mathcal{S}) = O(\sqrt{n \log(m/n)})$ when $m > n$, and $\text{disc}(\mathcal{S}) = O(\sqrt{m})$ when $m \le n$. Later in this survey we will come back to the methods used in Spencer's proof.

The discussion above concerns entirely unstructured collections of sets, for which specialized methods offer only logarithmic factor improvements over simple probabilistic constructions. By contrast, the set systems that arise in applications of discrepancy typically have more structure. An example, considered in the work of Matoušek, Welzl, and Wernisch [30], is the construction of an ε-approximation for geometric families of sets. An *ε-approximation* of a probability measure μ with respect to $\mathcal{S}$ is simply a point set P such that $D(\mathcal{S}, P; \mu) \le \varepsilon$. Let us use $\mathcal{H}_d$ to denote the family of half-spaces in $\mathbb{R}^d$, that is, all sets of the form $\{p \in \mathbb{R}^d : \langle a, p\rangle \le b\}$ where $a \in \mathbb{R}^d$, $b \in \mathbb{R}$, and $\langle\cdot,\cdot\rangle$ is the standard inner product. Constructing small ε-approximations of discrete distributions μ with respect to $\mathcal{H}_d$ (and other geometric families) is of interest in computational geometry, since often a computational problem can be solved on the ε-approximation rather than on a full specification of μ, resulting in much better running time at the cost of some loss in accuracy. As discussed in the previous section, the existence of small ε-approximations is implied by upper bounds on $\mathrm{disc}(\mathcal{H}_d, n)$, and this motivated Matoušek, Welzl, and Wernisch to study this quantity. They identified combinatorial properties that distinguish $\mathcal{H}_d$, and other geometric families of "low complexity," from unstructured set systems, and result in nontrivially small combinatorial discrepancy. Subsequent work of Matoušek further sharpened their results. We review this line of work below.

8.2.1 The primal shatter function

Let P be a set of n points in $\mathbb{R}^d$, and recall that $\mathcal{H}_d|_P$ is the collection of subsets of P induced by half-spaces, that is, $\mathcal{H}_d|_P = \{P \cap H : H \in \mathcal{H}_d\}$. One simple but key property of half-spaces is that, for any fixed d, $\mathcal{H}_d|_P$ has size polynomial in n. This is because any d-dimensional half-space H can be shifted and rotated so that it rests on d points of P, without changing $H \cap P$. So, any set in $\mathcal{H}_d|_P$ is determined by a subset of d points in P and a choice of whether they lie in or out of the half-space, giving a bound of $|\mathcal{H}_d|_P| \le 2^d\binom{n}{d} = O(n^d)$.[3] Many geometric families of sets share this property, and Matoušek, Welzl, and Wernisch showed that it is sufficient for establishing nontrivial discrepancy upper bounds. To state their results, let us introduce the standard definition of the *primal shatter function* $\pi_{\mathcal{S}} : \mathbb{N} \to \mathbb{N}$ of a family $\mathcal{S}$ of subsets of $\mathbb{R}^d$, given by $\pi_{\mathcal{S}}(n) = \max\{|\mathcal{S}|_P| : P \subset \mathbb{R}^d, |P| = n\}$. We just established that $\pi_{\mathcal{H}_d}(n) = O(n^d)$; a similar argument shows, for example, that $\pi_{\mathcal{B}_d}(n) = O(n^{d+1})$, where $\mathcal{B}_d$ is the family of all Euclidean balls in $\mathbb{R}^d$, and that $\pi_{\mathcal{R}_d}(n) = O(n^{2d})$, where $\mathcal{R}_d$ is the family of axis-aligned boxes in $\mathbb{R}^d$. More generally, recall that the *Vapnik–Chervonenkis (VC)-dimension* $\mathrm{VCdim}(\mathcal{S})$ of $\mathcal{S}$ is defined to be the largest n such that there exists a $P \subset \mathbb{R}^d$ of size n for which $\mathcal{S}|_P$ is the powerset of P. Then the Sauer–Shelah lemma shows that $\pi_{\mathcal{S}}(n) = O(n^{\mathrm{VCdim}(\mathcal{S})})$ [53, 46, 47]. Nevertheless, direct bounds on the shatter function are often tighter and easier to establish than bounds via the VC-dimension.

We are now ready to state one of the main results in [30].

Theorem 8.1 (Matoušek, Welzl, and Wernisch [30]). *Suppose that a family $\mathcal{S}$ of subsets of $\mathbb{R}^d$ satisfies $\pi_{\mathcal{S}}(n) = O(n^\kappa)$ for $\kappa > 1$. Then* $\mathrm{disc}(\mathcal{S}, n) = O(n^{\frac{1}{2}-\frac{1}{2\kappa}} \log(n)^{\frac{1}{2}+\frac{1}{2\kappa}})$.

3 Here and in the rest of the survey, the asymptotic notation treats d as a fixed constant.

Roughly speaking, Theorem 8.1 shows that the discrepancy of half-spaces is polynomially smaller than the discrepancy achievable by a simple random coloring. The key property of families of sets with polynomially bounded shatter function that enables this nontrivial upper bound is a form of combinatorial compactness. Recall that, by a simple volume argument, the number of d-dimensional Euclidean balls of radius δ that we can fit inside a unit Euclidean ball is no more than $(1/\delta)^d$. A deep consequence of the polynomial bound on the shatter function is that, for any $P \subset \mathbb{R}^d$, the sets in $\mathcal{H}_d|_P$ satisfy an analogous packing bound. In the context of the following lemma, we say that a collection $\mathcal{P}$ of subsets of a finite set P is *δ-separated* if for any two $P', P'' \in \mathcal{P}$ we have $|P' \triangle P''| \geq \delta|P|$, where $P' \triangle P'' = (P' \setminus P'') \cup (P'' \setminus P')$ is the symmetric difference.

Lemma 8.2 (Haussler [25]). *Suppose that a family $\mathcal{S}$ of subsets of $\mathbb{R}^d$ satisfies $\pi_{\mathcal{S}}(n) = O(n^\kappa)$. Then for any set $P \subset \mathbb{R}^d$, and any $\delta \in (0,1]$, if $\mathcal{P} \subseteq \mathcal{S}|_P$ is δ-separated, then $|\mathcal{P}| = O(\delta^{-\kappa})$.*

In fact, [30] uses a simpler to prove version of the lemma with an additional $\log(1/\delta)^\kappa$ factor in the bound on $|\mathcal{P}|$, which is due to Dudley [20]. Because of this, the bound stated in their paper is larger than the one we stated in Theorem 8.1 by a $\sqrt{\log n}$ factor. Plugging in Lemma 8.2 into their argument immediately gives the bound we stated.

A typical way to use Lemma 8.2 is to construct a small net of representative sets for $\mathcal{S}|_P$, as captured in the following lemma. In the statement of the lemma, $\mathcal{S}'$ are the representative sets, which have the property that any set in $\mathcal{S}|_P$ is close to a set in $\mathcal{S}'$, up to a small "correction." The set of corrections is denoted $\mathcal{T}$ in the lemma.

Lemma 8.3. *Suppose $\mathcal{S}, \kappa, P$, and δ are as in Lemma 8.2. Then there exist two set systems $\mathcal{S}'$ and $\mathcal{T}$ on P such that*
- $\mathcal{S}' \subseteq \mathcal{S}|_P$,
- $|\mathcal{S}'| = O(\delta^{-\kappa})$ *and* $|\mathcal{T}| \leq 2|\mathcal{S}|_P|$,
- $|T| < \delta|P|$ *for all* $T \in \mathcal{T}$,
- *any* $S \in \mathcal{S}|_P$ *can be written as* $S = (S' \setminus T) \cup T'$ *for some* $S' \in \mathcal{S}'$ *and* $T, T' \in \mathcal{T}$ *such that* $T \subseteq S'$ *and* $T' \cap S' = \emptyset$.

Proof. Take $\mathcal{S}'$ be an inclusion-maximal δ-separated subset of $\mathcal{S}|_P$. By Lemma 8.2, $|\mathcal{S}'| = O(\delta^{-\kappa})$. Then any $S \in \mathcal{S}|_P$ satisfies $|S \triangle S'| < \delta|P|$ for some $S' \in \mathcal{S}'$, or otherwise, we could add S to $\mathcal{S}'$, contradicting the latter's maximality. We fix an arbitrary such S' for any $S \in \mathcal{S}|_P$ and add $T = S' \setminus S$ and $T' = S \setminus S'$ to $\mathcal{T}$. Both $|T|$ and $|T'|$ are bounded by $|S \triangle S'| < \delta|P|$, and all the desired properties of $\mathcal{S}'$ and $\mathcal{T}$ hold. □

Lemma 8.3 shows that, under the assumption that $\mathcal{S}$ has polynomially bounded shatter function, the sets in the set system $\mathcal{S}|_P$ induced by $\mathcal{S}$ can be decomposed into small sets $\mathcal{T}$, and a small number of large sets $\mathcal{S}'$. To show that $\mathcal{S}|_P$ has small discrepancy, it suffices to do so for $\mathcal{S}' \cup \mathcal{T}$, as, for any coloring $\chi : P \to \{-1, +1\}$, the representation $S = (S' \setminus T) \cup T'$ implies $\chi(S) = \chi(S') - \chi(T) + \chi(T')$. It is easy to see that a set system consisting of small sets has small discrepancy, for example, by assigning every point an independent random color. Intuitively, the same is true for set systems with a small number of large sets: requiring that a set has small discrepancy poses a constraint on the allowable colorings, and if there are not too many sets, then there are not enough constraints to disallow all colorings. The influential partial coloring lemma of Beck, formulated next, gives a precise form of these informal observations.

Lemma 8.4 (Beck [9]). *Let $\mathcal{S}'$ and $\mathcal{T}$ be set systems on a set P of size n such that $|T| \le s$ for all $T \in \mathcal{T}$, and*

$$\prod_{S' \in \mathcal{S}'} (|S'| + 1) \le 2^{(n-1)/5}.$$

Then there exists a partial coloring[4] $\chi : P \to \{-1, 0, +1\}$ *such that* $\chi(S') = 0$ *for all* $S' \in \mathcal{S}'$, $|\chi(T)| = O(\sqrt{s \log(|\mathcal{T}|)})$ *for all* $T \in \mathcal{T}$, *and, for at least* $\frac{n}{10}$ *points* p *in* P, *we have* $\chi(p) \ne 0$.

Proof sketch. If we pick $\chi' : P \to \{-1, +1\}$ to be a uniformly random coloring of P, a standard application of Hoeffding's inequality shows that

$$\Pr\Big(\max_{T \in \mathcal{T}} |\chi'(T)| > \sqrt{2s \ln(4|\mathcal{T}|)}\Big) < \frac{1}{2}.$$

Equivalently, the set $\mathcal{C}$ of colorings χ' of P such that $\max_{T \in \mathcal{T}} |\chi'(T)| \le \sqrt{2s \ln(4|\mathcal{T}|)}$ has size at least 2^{n-1}.

On the other hand, the number of different values that the vector $(\chi'(S'))_{S' \in \mathcal{S}'}$ can take for different colorings $\chi' : P \to \{-1, +1\}$ of P is at most $\prod_{S' \in \mathcal{S}'} (|S'| + 1)$, which is bounded by $2^{(n-1)/5}$ by assumption. Then the pigeonhole principle implies that there must be a set $\mathcal{C}' \subseteq \mathcal{C}$ of colorings of P of size at least $2^{4(n-1)/5}$ so that any two $\chi', \chi'' \in \mathcal{C}'$ satisfy $\chi'(S') = \chi''(S')$ for all $S' \in \mathcal{S}'$. Let us fix one such $\chi' \in \mathcal{C}'$. Since the size of $\mathcal{C}'$ is much larger than the number of colorings of P which agree with χ' on $> \frac{9n}{10}$ of the points in P, there must be at least one $\chi'' \in \mathcal{C}'$ for which $\chi'(p) \ne \chi''(p)$ for at least $\frac{n}{10}$ points $p \in P$. We can then take $\chi = \frac{1}{2}(\chi' - \chi'')$ as our partial coloring. □

We now have everything in place to sketch the proof of Theorem 8.1. Once again, our goal is to show that for an arbitrary $P \subset \mathbb{R}^d$ of size n, the discrepancy is bounded as $\mathrm{disc}(\mathcal{S}|_P) = O(n^{\frac{1}{2} - \frac{1}{2\kappa}} \log(n)^{\frac{1}{2} + \frac{1}{2\kappa}})$. We define $\mathcal{S}'$ and $\mathcal{T}$ as in Lemma 8.3 with δ on the order of $O(\frac{\log(n)^{1/\kappa}}{n^{1/\kappa}})$. Then $\mathcal{S}'$ and $\mathcal{T}$ satisfy the assumptions of Lemma 8.4 with $s = O(n^{1-\frac{1}{\kappa}} \log(n)^{\frac{1}{\kappa}})$ and $\mathcal{T} \le 2|\mathcal{S}|_P| = O(n^{\kappa})$. The lemma gives us a partial coloring $\chi : P \to \{-1, 0, +1\}$ such that at least $\frac{n}{10}$ points receive a nonzero color, and the discrepancy of any set $S \in \mathcal{S}|_P$ is bounded by

$$|\chi(S)| \le |\chi(S')| + |\chi(S_1'')| + |\chi(S_2'')| = O(n^{\frac{1}{2} - \frac{1}{2\kappa}} \log(n)^{\frac{1}{2} + \frac{1}{2\kappa}}),$$

---

4 We call a function $\chi : P \to \{-1, 0, +1\}$ a *partial* coloring, since we can think of the elements receiving color 0 as uncolored.

where $S = (S' \setminus T) \cup T'$ is the representation of S from Lemma 8.3. We can now inductively find a full coloring $\chi' : P_0 \to \{-1,+1\}$ of the set P_0 of points $p \in P$ for which $\chi(p) = 0$, and define a full coloring of P by $\chi + \chi'$. The induction to complete the partial coloring to a full one increases the discrepancy by only a constant factor, since the size of the set of points we are working with decreases geometrically with every application of Lemma 8.4.

Subsequently, Matoušek strengthened the discrepancy upper bound in Theorem 8.1 to $\mathrm{disc}(\mathcal{S}, n) = O(n^{\frac{1}{2}-\frac{1}{2\kappa}})$ [32]. The key to this improvement is a refined version of the partial coloring lemma, which is due to Spencer [48], used in conjunction with a more careful decomposition lemma. Spencer introduced his partial coloring lemma in the proof of his upper bound on the discrepancy of arbitrary set systems with m sets and n elements, which we mentioned above. We give a more recent, stronger version of the lemma, due to Lovett and Meka, and proved using different methods.

Lemma 8.5 (Lovett and Meka [29]). *Let $\mathcal{S}$ be a set system on a set P of size n, let $\chi_0 : P \to [-1,+1]$ be a* fractional coloring, *and let $\lambda : \mathcal{S} \to \mathbb{R}_{\geq 0}$ be such that*

$$\sum_{S \in \mathcal{S}} e^{-\frac{\lambda(S)^2}{16}} \leq \frac{n}{16}.$$

Then there exists a fractional coloring $\chi : P \to [-1,+1]$ such that $|\chi(S) - \chi_0(S)| \leq \lambda(S)\sqrt{|S|}$ for any $S \in \mathcal{S}$, and, for at least $\frac{n}{10}$ points p in P, we have $\chi(p) \in \{-1,+1\}$.

To gain some intuition, let us compare this lemma to Lemma 8.4. Let us take $\mathcal{S} = \mathcal{S}' \cup \mathcal{T}$, where $\mathcal{S}'$ and $\mathcal{T}$ are as in Lemma 8.4. It makes sense to assume that any set in $\mathcal{S}'$ has size at least s, as we can move the set to $\mathcal{T}$ otherwise. Then the condition on $\mathcal{S}'$ implies that $|\mathcal{S}'| \leq \frac{n-1}{5\log_2(s)}$, which is less than $\frac{n-1}{16}$ for all large enough s. Setting $\lambda(S') = 0$ for $S' \in \mathcal{S}'$, and $\lambda(T) = 4\sqrt{\log(16|\mathcal{T}|)}$ for $T \in \mathcal{T}$, we get that

$$\sum_{S \in \mathcal{S}} e^{-\frac{\lambda(S)^2}{16}} \leq \frac{n}{16},$$

and the condition in Lemma 8.5 is satisfied. Ignoring the distinction between a partial and a fractional coloring, which is usually immaterial, Lemma 8.4 can be thought of as a very special case of Lemma 8.5, in which the value of λ on any set must be either 0 or logarithmic in the total number of sets. Defining λ in a more flexible way leads to tighter discrepancy bounds in many cases.

In order to utilize (Spencer's variant of) Lemma 8.4, Matoušek refined the decomposition in Lemma 8.3, by applying it recursively. Applying Lemma 8.3 once with $\delta = \frac{2}{n}$, where $n = |P|$, we get that any set S in $\mathcal{S}|_P$ is "close" to a set S' in the subcollection $\mathcal{S}' \subseteq \mathcal{S}|_P$, in the sense that we can write $S = (S' \setminus T) \cup T'$ for small sets $T, T' \in \mathcal{T}$. Let us rename $\mathcal{S}|_P$ to $\mathcal{S}_0$, $\mathcal{S}'$ to $\mathcal{S}_1$, and the set of "corrections" $\mathcal{T}$ to $\mathcal{T}_1$. Restating Lemma 8.3 in this notation, we know that

- $\mathcal{S}_1 \subseteq \mathcal{S}_0$,
- $|\mathcal{S}_1| = O((n/2)^\kappa)$ and $|\mathcal{T}_1| \leq 2|\mathcal{S}_0|$,

- $|T| < 2$ for all $T \in \mathcal{T}_1$,
- any $S \in \mathcal{S}_0$ can be written as $S = (S' \setminus T) \cup T'$ for some $S' \in \mathcal{S}_1$ and $T, T' \in \mathcal{T}_1$ such that $T \subseteq S'$ and $T' \cap S' = \emptyset$.

The first property implies that $\mathcal{S}_1$ again satisfies the assumptions of Lemma 8.3, and we can apply it again with $\delta = \frac{4}{n}$, giving us a subcollection $\mathcal{S}_2 = \mathcal{S}_1'$ of size $O((n/4)^\kappa)$ and a set of corrections $\mathcal{T}_2$, where each set $T \in \mathcal{T}_2$ is of size $|T| < 4$, and $|\mathcal{T}_2| \le 2|\mathcal{S}_1|$. Continuing in this manner, we get collections of sets $\mathcal{T}_1, \dots, \mathcal{T}_\ell$ such that $|\mathcal{T}_i| = O((n/2^{i-1})^\kappa)$ and $|T| < 2^i$ for every $T \in \mathcal{T}_i$. We stop when $\mathcal{T}_\ell$ has size 1, which happens for some $\ell = O(\log(n))$. We can write any $S \in \mathcal{S}|_P$ as

$$S = (\dots(((T_\ell \setminus T_{\ell-1}) \cup T'_{\ell-1}) \setminus T_{\ell-2}) \cup \dots \setminus T_1) \cup T'_1, \tag{8.1}$$

where $T_i, T'_i \in \mathcal{T}_i$, all unions are between disjoint sets, and all set differences remove a set from a superset of it. This recursive decomposition is reminiscent of the technique of chaining used in the theory of stochastic processes; see, for example, [51]. It is also of independent interest, and was used, for example, by the author of this survey in work on algorithms for private data analysis [41].

We can now apply Lemma 8.5 to $\mathcal{T}_1 \cup \dots \cup \mathcal{T}_\ell$, and a fractional coloring χ_0 set to be 0 on all points in P. We define λ so that $\lambda(T) = 2^{-i/2} n^{\frac{1}{2}-\frac{1}{2\kappa}} \phi(i)$ for any set $T \in \mathcal{T}_i$ and a function ϕ which we will choose shortly. Let us ignore, for simplicity, the constant in the asymptotic notation and assume that $|\mathcal{T}_i| \le \frac{n^\kappa}{2^{\kappa(i-1)}}$. Then we have

$$\sum_{i=1}^{\ell} \sum_{T \in \mathcal{T}_i} e^{-\frac{\lambda(T)^2}{16}} \le 2^{1+\kappa} \sum_{i=1}^{\ell} \frac{n^\kappa}{2^{\kappa i}} \exp\left(-\frac{n^{1-\frac{1}{\kappa}} \phi(i)^2}{2^{i+4}}\right).$$

Let us focus on the term on the right-hand side corresponding to the value of i closest to $i_0 = \log_2(n^{1-\frac{1}{\kappa}})$. We have $\frac{n^\kappa}{2^{\kappa i_0}} = n$, and $\frac{n^{1-\frac{1}{\kappa}} \phi(i_0)^2}{2^{i_0+4}} = \frac{\phi(i_0)^2}{16}$. Setting $\phi(i_0)$ to be a large enough constant ensures that this term is much smaller than $\frac{n}{16}$. We want to make sure that this is the dominating term, which happens for a large class of nicely behaved functions ϕ. For example, we can choose $\phi(i) = c2^{-|i-i_0|/64}$ for a sufficiently large $c > 0$. Now the entire sum is bounded by $\frac{n}{16}$, and the conditions of Lemma 8.5 are satisfied. The lemma guarantees the existence of a fractional coloring χ such that, for at least $\frac{n}{10}$ points p, we have $\chi(p) \in \{-1, +1\}$, and for any $T \in \mathcal{T}_i$, $|\chi(T)| \le cn^{\frac{1}{2}-\frac{1}{2\kappa}} 2^{-|i-i_0|/64}$. Taking an arbitrary $S \in \mathcal{S}|_P$ and representing it as in (8.1), we have

$$|\chi(S)| \le \sum_{i=1}^{\ell} |\chi(T_i)| + |\chi(T'_i)| \le 2c \sum_{i=1}^{\ell} n^{\frac{1}{2}-\frac{1}{2\kappa}} 2^{-|i-i_0|/64} = O(n^{\frac{1}{2}-\frac{1}{2\kappa}}).$$

It is now apparent why we didn't just set ϕ to be a constant function: such a choice would not be enough to get the sum above to converge. The final step is to repeat this construction with the points $p \in P$ for which $-1 < \chi(p) < +1$. We can treat their color

$\chi(p)$ as $\chi_0(p)$ in a new application of Lemma 8.5, and continue in this fashion until all colors become -1 or $+1$. This inductive process only increases the discrepancy by a constant, as the number of points with colors between -1 and $+1$ drops exponentially fast. The result is summarized in the following theorem.

Theorem 8.6 (Matoušek [32]). *Suppose that a family $\mathcal{S}$ of subsets of $\mathbb{R}^d$ satisfies $\pi_{\mathcal{S}}(n) = O(n^\kappa)$ for $\kappa > 1$. Then* $\operatorname{disc}(\mathcal{S}, n) = O(n^{\frac{1}{2}-\frac{1}{2\kappa}})$.

Theorem 8.6 is tight up to constants, even in the special case when $\mathcal{S} = \mathcal{H}_d$. This was shown by Alexander [2], and a more direct and elementary proof of the lower bound was given by Chazelle, Matoušek, and Sharir [19].

8.2.2 The dual shatter function

In the previous section, we explored the consequences for combinatorial discrepancy of the observation that there are at most $O(n^d)$ sets induced by half-spaces on a set of n points in $\mathbb{R}^d$. Another sense in which half-spaces and hyperplanes are well behaved is that a bounded number of hyperplanes divide $\mathbb{R}^d$ into relatively few regions. More precisely, any arrangement of m lines in $\mathbb{R}^2$ divides the plane into at most $O(m^2)$ cells, which can be easily verified with Euler's formula. In general, an arrangement of m hyperplanes in $\mathbb{R}^d$ divides $\mathbb{R}^d$ into $O(m^d)$ cells. These observations can be captured abstractly by the *dual shatter function* $\pi^*_{\mathcal{S}} : \mathbb{N} \to \mathbb{N}$ of a collection of subsets $\mathcal{S}$ of $\mathbb{R}^d$. Given a subcollection $\mathcal{S}' \subseteq \mathcal{S}$, let us call a set of points $P \subset \mathbb{R}^d$ *distinguishable* by $\mathcal{S}'$ if, for any two distinct points p and q in P, there is some $S \in \mathcal{S}'$ such that $|S \cap \{p, q\}| = 1$. Then we define

$$\pi^*_{\mathcal{S}}(m) = \max\{|P| : P \subset \mathbb{R}^d, \mathcal{S}' \subseteq \mathcal{S}, |\mathcal{S}'| = m, P \text{ is distinguishable by } \mathcal{S}'\}.$$

The fact about the number of cells in an arrangement of hyperplanes above can be expressed by the bound $\pi^*_{\mathcal{H}_d}(m) = O(m^d)$. There is a sense in which the dual shatter function is, in fact, the dual of the shatter function. In particular, let us define the *dual set system* $\mathcal{S}^*$ of $\mathcal{S}$ as the collection of sets S_p, indexed by points $p \in \mathbb{R}^d$, and defined by $S_p = \{S \in \mathcal{S} : p \in S\}$. Then $\pi^*_{\mathcal{S}} = \pi_{\mathcal{S}^*}$, since a point set P being distinguishable by $\mathcal{S}' \subseteq \mathcal{S}$ is equivalent to the sets $S_p \cap \mathcal{S}'$ being distinct for all $p \in P$ and, therefore, the largest set of points distinguishable by $\mathcal{S}'$ equals the cardinality of $\mathcal{S}^*|_{\mathcal{S}'}$. Moreover, the classical duality between points and hyperplanes gives an isomorphism between $\mathcal{H}_d$ and $\mathcal{H}^*_d$, which implies that $\pi^*_{\mathcal{H}_d} = \pi_{\mathcal{H}^*_d} = \pi_{\mathcal{H}_d}$. These observations explain the bound $\pi^*_{\mathcal{H}_d}(m) = O(m^d)$ as being just the primal shatter function bound in disguise. Of course, this sort of self-duality is rare, and often the primal and dual shatter function can have different orders of growth. For example, the set of Euclidean balls $\mathcal{B}_d$ in $\mathbb{R}^d$ satisfies $\pi_{\mathcal{B}_d}(n) = O(n^{d+1})$ and $\pi^*_{\mathcal{B}_d}(m) = O(m^d)$. In his book on geometric discrepancy, Matoušek describes how bounds on the dual shatter function for semi-algebraic sets can be derived from basic results in real algebraic geometry [34].

The fact that the dual shatter function can sometimes have a slower order of growth than the primal shatter function motivates the question whether $\mathrm{disc}(\mathcal{S}, n)$ can be bounded in terms of $\pi^*_{\mathcal{S}}$. Such a bound was also proved by Matoušek, Welzl, and Wernisch, and is the second result from their paper that we discuss.

Theorem 8.7 (Matoušek, Welzl, and Wernisch [30]). *Suppose that a family $\mathcal{S}$ of subsets of $\mathbb{R}^d$ satisfies $\pi^*_{\mathcal{S}}(m) = O(m^\kappa)$ for $\kappa > 1$. Then* $\mathrm{disc}(\mathcal{S}, n) = O(n^{\frac{1}{2}-\frac{1}{2\kappa}}\sqrt{\log(n)})$.

The proof of Theorem 8.7 relies on a deep theorem of Welzl establishing the existence of paths which are not crossed too many times by sets in a collection $\mathcal{S}$ with polynomially bounded dual shatter function. In the following, we say that a pair of points $\{p, q\}$ in $\mathbb{R}^d$ is *crossed* by a set $S \in \mathcal{S}$ if $|S \cap \{p, q\}| = 1$.

Theorem 8.8 (Welzl [54]). *Suppose that a family $\mathcal{S}$ of subsets of $\mathbb{R}^d$ satisfies $\pi^*_{\mathcal{S}}(m) = O(m^\kappa)$ for some constant $\kappa > 1$. Then, for any set P of n points in $\mathbb{R}^d$, there exists an ordering $p_1, \dots, p_n$ of P so that any set $S \in \mathcal{S}$ crosses at most $O(n^{1-\frac{1}{\kappa}})$ pairs of consecutive points $\{p_i, p_{i+1}\}$.*

In fact, Welzl states the weaker bound $O(n^{1-\frac{1}{\kappa}} \log n)$ on the number of crossed pairs, but once again, this is due only to using a weaker packing bound than that in Lemma 8.2. Because of this, the discrepancy bound stated in [30] is also weaker than the one in Theorem 8.7, but plugging in Theorem 8.8 in their argument immediately yields the bound we give.

In order to give some intuition behind Theorem 8.8, let us fix some set of points P, and consider a game between two players: the Max player proposes a (multi-)set $\mathcal{S}'$ of sets from $\mathcal{S}|_P$, and the Min player answers with a pair of points $p, q \in P$. The Min player pays to the Max player an amount equal to the fraction of sets in $\mathcal{S}'$ that cross $\{p, q\}$. The Max player wants to maximize the payment he receives, while the Min player wants to minimize the payment she makes. An application of Lemma 8.2 to the dual set system $(\mathcal{S}|_P)^*$ shows that for any choice of $\mathcal{S}'$, Min can find a pair $\{p, q\}$ crossed by at most $O(\frac{|\mathcal{S}'|}{n^{1/\kappa}})$ sets in $\mathcal{S}'$. Then the von Neumann's minimax theorem in game theory implies that one can switch the order of the players, without changing the value of the game: Min can go first, and play a (multi)set E of pairs of points from P, so that Max cannot find any set $S \in \mathcal{S}|_P$ which crosses more than $O(\frac{|E|}{n^{1/\kappa}})$ pairs in E. This is not quite enough to prove Theorem 8.8, because there is no guarantee that E forms a path, or that it spans all the points in P. These issues can be fixed by adapting a constructive proof of the minimax theorem via the multiplicative weights update method [21, 3].

With Theorem 8.8 in hand, Theorem 8.7 follows by an ingenious application of the probabilistic method. We split the ordering from Theorem 8.8 into disjoint pairs $\{p_1, p_2\}, \{p_3, p_4\}, \dots, \{p_{2k-1}, p_{2k}\}$, possibly leaving a singleton point p_n if n is odd. Then we choose a random coloring $\chi : P \to \{-1, +1\}$ so that, for any pair $\{p_{2i-1}, p_{2i}\}$, $\chi(p_{2i-1})$ is uniform in $\{-1, +1\}$ and $\chi(p_{2i-1}) = -\chi(p_{2i})$, while the colors for points not in the same pair are independent. The only pairs of points that contribute a nonzero value to the discrepancy $\chi(S)$ of a set $S \in \mathcal{S}|_P$ are those crossed by S. Moreover, each of the $O(n^{1-\frac{1}{\kappa}})$ crossed pairs contributes either -1 or $+1$ to $\chi(S)$, independently and with equal probability. A standard application of Hoeffding's inequality then shows that, for any $S \in \mathcal{S}|_P$,

$$\Pr(|\chi(S)| \geq ctn^{\frac{1}{2}-\frac{1}{2\kappa}}) \leq 2e^{-t^2/4},$$

for a large enough constant $c > 0$. Setting $t = 2\sqrt{\log(4m)}$ for $m = |\mathcal{S}|_P|$ and applying the union bound, we get that, with probability at least $\frac{1}{2}$, $\operatorname{disc}(\mathcal{S}|_P, \chi) = O(n^{\frac{1}{2}-\frac{1}{2\kappa}}\sqrt{\log m})$. The proof is complete by the fact that $\pi_{\mathcal{S}}$ is polynomially bounded if and only if $\pi^*_{\mathcal{S}}$ is, so $m \leq \pi_{\mathcal{S}}(n) = O(n^{\kappa'})$ for some constant κ'.

We remark that Theorem 8.8 was originally proved by Welzl for the purpose of constructing a range counting data structure[5] for geometric sets. In the next section, we will see more examples from Matoušek's work of using ideas from range searching data structures to prove discrepancy upper bounds.

Theorem 8.7 and the bound $\pi_{\mathcal{B}_d}(m) = O(m^d)$ imply that $\operatorname{disc}(\mathcal{B}_d, n) = O(n^{\frac{1}{2}-\frac{1}{2d}}\sqrt{\log n})$, which is tighter than the bound $O(n^{\frac{1}{2}-\frac{1}{2d+2}})$ implied by Theorem 8.6. It is natural to wonder whether this bound can be further improved to $\operatorname{disc}(\mathcal{B}_d, n) = O(n^{\frac{1}{2}-\frac{1}{2d}})$, and this remains a fascinating open question. More generally, one may ask if the conclusion of Theorem 8.7 holds without the $\sqrt{\log n}$ factor, in analogy with Theorem 8.6. Perhaps surprisingly, Matoušek showed this is not true even for $\kappa = 2$.

Theorem 8.9 (Matoušek [33]). *For any natural number n and for $\kappa = 2$ or $\kappa = 3$, there exists a set system $\mathcal{S}$ on n elements[6] with $\pi^*_{\mathcal{S}}(m) = O(m^\kappa)$ such that* $\operatorname{disc}(\mathcal{S}) = \Omega(n^{\frac{1}{2}-\frac{1}{2\kappa}}\sqrt{\log n})$.

We describe the construction used by Matoušek to prove Theorem 8.9 in the case $\kappa = 2$. Let $\mathbb{F}$ be a finite field of order q. Matoušek starts with the set system $\mathcal{T}$ containing all sets $T_f \subseteq \mathbb{F} \times \mathbb{F}$ defined by a degree 2 polynomial f as $T_f = \{(x, f(x)) : x \in \mathbb{F}\}$. $\mathcal{T}$ is a set system over a universe of size $n = q^2$, in which every set has size $q = \sqrt{n}$ and $|\mathcal{T}| = q^2(q-1) = n^{3/2} - n$. Moreover, for any two distinct $T_f, T_g \in \mathcal{T}$, $|T_f \cap T_g| \leq 2$, since quadratic polynomials that agree on more than 2 points must be identical. A counting argument reveals that $\pi^*_{\mathcal{T}}(m) \leq m^2$. Indeed, fix any m distinct sets $T_1, \ldots, T_m$ from $\mathcal{T}$, and, for any $(x, y) \in \mathbb{F} \times \mathbb{F}$, define the vector $u(x, y) \in \{0, 1\}^m$ by setting $u(x, y)_i = 1$ if and only if $(x, y) \in T_i$. This vector indicates which sets T_i the ordered pair (x, y) belongs to, and the maximum number of elements (x, y) distinguishable by $\{T_1, \ldots, T_m\}$

5 A range counting data structure for $\mathcal{S}$ is a way to store a set of points $P \subset \mathbb{R}^d$ in a computer's memory, so that the count $|S \cap P|$ can be computed quickly for every $S \in \mathcal{S}$. The goal is to be able to compute $|S \cap P|$ in fewer than $O(|P|)$ steps.

6 So far, we have only discussed shatter functions for subsets of $\mathbb{R}^d$, but it should be clear that the definitions make sense in the abstract setting of arbitrary set systems, too.

is equal to the number of distinct values taken by $u(x,y)$ over $x,y \in \mathbb{F}$. The number of distinct vectors $u(x,y)$ with only a single 1 entry is obviously bounded by m. Suppose that $(x_1,y_1),\dots,(x_k,y_k)$ are the pairs for which the vector $u(x_\ell,y_\ell)$ has at least two entries equal to 1, and let us count triples (ℓ,i,j) such that $u(x_\ell,y_\ell)_i = u(x_\ell,y_\ell)_j = 1$ in two ways in order to bound k. On one hand, the number of such triples is at least k. On the other hand, for each choice of i and j, there are at most two triples (ℓ,i,j) and (ℓ',i,j) of this type, because $|T_i \cap T_j| \leq 2$, and any triple (ℓ,i,j) corresponds to a pair (x_ℓ,y_ℓ) that belongs to both T_i and T_j. Therefore, k is bounded by the total number of triples, which is, in turn, bounded by $2\binom{m}{2}$. Putting things together, we have that $|\{u(x,y) : x,y \in \mathbb{F}\}| \leq m + 2\binom{m}{2} = m^2$. In fact, one can check that this argument still holds if we replace any set in $\mathcal{T}$ with a subset of it. Matoušek constructs $\mathcal{S}$ by independently taking a uniformly random subset of each $T \in \mathcal{T}$. He shows that, with high probability, the discrepancy of $\mathcal{S}$ is bounded from below as in Theorem 8.9. The fact that $\mathcal{T}$ has a superlinear (in n) number of sets of size $\sqrt{n}$ is crucial for this probabilistic argument.

As is apparent from the construction, the set system $\mathcal{S}$ does not have a natural geometric structure: every set in $\mathcal{S}$ is a random subsample of the graph of a polynomial function over a finite field. Proving Theorem 8.9 for any natural collection of geometric sets in $\mathbb{R}^d$ (e. g., the set of Euclidean balls $\mathcal{B}_d$) is an interesting challenge.

8.3 Sets with product structure

So far, we saw examples of geometric set systems that have combinatorial discrepancy polynomially smaller than that of an arbitrary set system. It turns out that sometimes the improvement can be even exponential, and the discrepancy depends only logarithmically on the number of points. The simplest example is $\mathcal{R}_d$, the family of axis-aligned boxes in $\mathbb{R}^d$, that is, sets of the form $[a_1,b_1] \times \cdots \times [a_d,b_d]$ for $a,b \in \mathbb{R}^d$. The problem of determining $\operatorname{disc}(\mathcal{R}_d,n)$ was posed by the Hungarian probabilist, Tusnády, in the early 1980s, and has a rich history (see [38] and [44] for references to the relevant literature). The family of axis-aligned boxes can be generalized in at least two directions: we can consider polytopes that are the intersection of half-spaces whose normals do not necessarily form an orthogonal basis; we can also consider sets which are products of lower-dimensional sets other than intervals. Using insights from range searching data structures, Matoušek proved discrepancy upper bounds for axis-aligned boxes, as well as for both of these generalizations.

Before we describe Matoušek's results related to sets with product structure, let us briefly mention one important motivation for studying $\operatorname{disc}(\mathcal{R}_d,n)$. Recall that, by the general connection between combinatorial and measure-theoretic discrepancy, $\frac{1}{n}\operatorname{disc}(\mathcal{R}_d,n)$ bounds $D(\mathcal{R}_d,n;\mu)$ from above for any Borel measure μ. Because of the Koksma–Hlawka inequality (see, e. g., [43]), estimates on $D(\mathcal{R}_d,n;\mu)$ are of interest in numerical integration. It is a classical fact that for the Lebesgue measure λ^d on $[0,1]^d$, $D(\mathcal{R}_d, n; \lambda^d) = O(\frac{\log(n)^{d-1}}{n})$ [23, 24]. The best known upper bound on $\mathrm{disc}(\mathcal{R}_d, n)$ [44] extends this result and implies that $D(\mathcal{R}_d, n; \mu) = O(\frac{\log(n)^{d-\frac{1}{2}}}{n})$ for any Borel measure μ. In the combinatorial setting, we know a nearly tight lower bound of $\Omega(\frac{\log(n)^{d-1}}{n})$ on $\frac{1}{n}\mathrm{disc}(\mathcal{R}_d, n)$, due to work by Matoušek, Talwar, and the author [40, 38]. By contrast, the largest known lower bound for the Lebesgue measure discrepancy is $\Omega(\frac{\log(n)^{\frac{d-1}{2}+\eta_d}}{n})$ for a small constant $\eta_d > 0$ that goes to 0 with the dimension [14]. Closing the gap between the upper and lower bounds on $D(\mathcal{R}_d, n; \lambda^d)$ is often called the great open problem in discrepancy theory. The combinatorial lower bounds suggest that the correct order of growth may be the one given by the current upper bound.

We start with Matoušek's upper bound on the discrepancy of axis-aligned boxes, which was the best known result until very recently.

Theorem 8.10 (Matoušek [35]). *For any positive integer d,*

$$\mathrm{disc}(\mathcal{R}_d, n) = O(\log(n)^{d+\frac{1}{2}}\sqrt{\log\log(n)}).$$

Matoušek's proof was based on Beck's original partial coloring lemma (Lemma 8.4). We will sketch his proof, but will replace Lemma 8.4 by Lemma 8.5. This slight change simplifies the calculations, and also improves the upper bound to $O(\log(n)^{d+\frac{1}{2}})$. In the special case $d = 2$, this latter bound was also proved by Srinivasan [50] using a slightly different approach. The first published proof of the $O(\log(n)^{d+\frac{1}{2}})$ upper bound for all d is due to Larsen [26], and uses different methods, to which we will return in the next section.

The key tool in the proof of Theorem 8.10 is a decomposition inspired by the range tree data structure in computational geometry [13]. This is the second example we see of ideas from data structures used by Matoušek in the context of discrepancy theory. For simplicity, let us describe the construction in two dimensions. Suppose that $P \subset \mathbb{R}^2$ is a n-point set; without loss of generality, we may assume that n is a power of 2, and that all points in P have distinct x- and y-coordinates. For any subset Q of P we define a collection $\mathcal{C}^x(Q)$ of sets by ordering Q as $p_1, \dots, p_k$, in increasing order of the x-coordinate, and defining $C^x_{\ell,i}(Q) = \{p_{(i-1)2^\ell+1}, \dots, p_{i2^\ell}\}$ and

$$\mathcal{C}^x(Q) = \{C^x_{\ell,i}(Q) : 0 \le \ell \le \lfloor \log_2(k) \rfloor, 1 \le i \le \lfloor k/2^\ell \rfloor\}.$$

We define $C^y_{\ell,i}(Q)$ and $\mathcal{C}^y(Q)$ analogously, but ordering Q in increasing order of the y-coordinate. When $Q = P$, we write simply $\mathcal{C}^x$ for $\mathcal{C}^x(P)$. $\mathcal{C}^x(Q)$ and $\mathcal{C}^y(Q)$ are called, respectively, *canonical intervals* in the x- and the y-coordinate. We then form the collection $\mathcal{C}$ of *canonical boxes* by taking pairwise intersections, that is, $\mathcal{C} = \{C \cap C' : C \in \mathcal{C}^x, C' \in \mathcal{C}^y(C)\}$. We can think of $\mathcal{C}^x$ as a complete binary tree T: each set in $\mathcal{C}^x$ is a node in T, with singleton sets as the leaves, and any set $C^x_{\ell,i}$ of size 2^ℓ for $\ell > 0$ is connected to the two sets of size $2^{\ell-1}$ contained in it. To represent all of $\mathcal{C}$, we associate a similar binary tree with each node of T, where the binary tree T_C associated with $C \in \mathcal{C}^x$ has the sets $\{C \cap C' : C' \in \mathcal{C}^y(C)\}$ as its nodes. These are the range trees mentioned earlier.

The key fact about $\mathcal{C}$, and range trees, is that any box in $\mathcal{R}_2|_P$ can be written as the union of $O(\log(n)^2)$ disjoint canonical boxes from $\mathcal{C}$. To do so, it is enough to take at most two canonical intervals in $\mathcal{C}^x$ from each level of the binary tree T, and intersect each such interval C with at most two canonical intervals from every level of T_C. In data structures, this decomposition allows efficiently listing all the points in any set in $\mathcal{R}_2|_P$ by navigating through the range tree. In the context of discrepancy, we use it as a way to get an analogue of Lemma 8.3 for boxes, formulated in the lemma below. We use the notation $\mathcal{C}^y_{\geq t}(Q)$ for the subset of $\mathcal{C}^y(Q)$ containing only sets with at least t points, and also $\mathcal{C}_{\geq t} = \{C \cap C' : C \in \mathcal{C}^x, C' \in \mathcal{C}^y_{\geq t}(C)\}$.

Lemma 8.11. *For any positive integer t, any box $R \in \mathcal{R}_2|_P$ can be written as the union of $O(\log(n)^2)$ sets in $\mathcal{C}_{\geq t}$, and a set T_R of size $O(t\log(n))$, where all the sets in the union are disjoint.*

To see why the lemma is true, consider the decomposition of any $R \in \mathcal{R}_2|_P$ into canonical boxes described above. We remove all boxes which do not belong to $\mathcal{C}_{\geq t}$ from this decomposition, and define T_R to be their union. For each level of the binary tree representing $\mathcal{C}^x$, this construction adds $O(t)$ elements to T_R, which gives the upper bound on $|T_R|$.

We can now complete the proof of Theorem 8.10. For a choice of t to be determined soon, let $\mathcal{T} = \{T_R : R \in \mathcal{R}_2|_P\}$, with T_R defined as in Lemma 8.11, and define $\mathcal{S} = \mathcal{C}_{\geq t} \cup \mathcal{T}$. We apply Lemma 8.5 to $\mathcal{S}$ with $\lambda(S) = 0$ for all $S \in \mathcal{C}_{\geq t}$ and $\lambda(T_R) = c\sqrt{\log_2(2n)}$ for a sufficiently large constant $c > 0$. We claim that $|\mathcal{C}_{\geq t}| = O(\frac{n\log n}{t})$. Indeed, for each fixed $C^x_{\ell,i} \in \mathcal{C}^x$,

$$|\{C^x_{\ell,i} \cap C' : C' \in \mathcal{C}^y_{\geq t}(C^x_{\ell,i})\}| = O\left(\frac{|C^x_{\ell,i}|}{t}\right) = O\left(\frac{2^\ell}{t}\right),$$

so, summing over all ℓ and i, we get that, for a constant $c' > 0$,

$$|\mathcal{C}_{\geq t}| \leq c' \sum_{\ell=0}^{\log_2(n)} \frac{n}{2^\ell} \cdot \frac{2^\ell}{t} = \frac{c'n(1+\log_2(n))}{t}.$$

Therefore, we can set $t \geq 16c'(1+\log_2(n))$, and, since $|\mathcal{T}| \leq |\mathcal{R}_2|_P| \leq n^4$, for a large enough constant c the set system $\mathcal{S}$ and the function λ satisfy the assumptions of Lemma 8.5. The lemma then implies a fractional coloring $\chi : P \to [-1,+1]$ so that $|\chi(S)| \leq \lambda(S)\sqrt{|S|}$ for any $S \in \mathcal{S}$, and for at least $\frac{n}{10}$ points p in P, we have $\chi(p) \in \{-1,+1\}$. Since, by Lemma 8.11, any $R \in \mathcal{R}_2|_P$ can be written as the disjoint union of sets in $\mathcal{C}_{\geq t}$, which have discrepancy 0 under χ, and a single set in $\mathcal{T}$, which has discrepancy $O(\log(n)^{3/2})$, we get that $\mathrm{disc}(\mathcal{R}_2|_P, \chi) = O(\log(n)^{3/2})$. Inductively completing the fractional coloring to a true coloring increases the discrepancy by at most a $O(\log(n))$ factor, giving the final bound of $O(\log(n)^{5/2})$. The proof for the higher dimensional case is analogous, and only the notation becomes slightly more cumbersome.

The decomposition in Lemma 8.11 (and its higher dimensional variant) was recently used by Bansal and Garg, together with a more advanced method of using fractional colorings, in order to give the improved bound $\mathrm{disc}(\mathcal{R}_d, n) = O(\log(n)^d)$ [7]. Subsequently, the author improved the upper bound to $O(\log(n)^{d-\frac{1}{2}})$ using different methods [44]. The best known lower bound of $\Omega(\log(n)^{d-1})$ is due to the author and Matoušek, and will be discussed in the next section.

As already mentioned, Theorem 8.10 can be extended in several directions. Suppose that A is a finite set of vectors in $\mathbb{R}^d$, and let $\mathrm{POL}(A)$ be the set of polytopes of the form $\bigcap_{i=1}^{\ell} H_i$, where, for each i, H_i is a half-space $\{p \in \mathbb{R}^d : \langle a, p\rangle \le t\}$ for some $t \in \mathbb{R}$ and $a \in A$. In other words, $\mathrm{POL}(A)$ contains all polytopes whose facets have normal vectors parallel to vectors in A. Then $\mathcal{R}_d$ equals $\mathrm{POL}(A)$ for $A = \{\pm e_1, \dots, \pm e_d\}$, where e_i is the ith standard basis vector of $\mathbb{R}^d$. Since $\mathcal{R}_d|_P$, as a set system over P, does not change after applying the same invertible linear transformation to P and to the sets in $\mathcal{R}_d$, we get that

$$\mathrm{disc}(\mathcal{R}_d, n) = \mathrm{disc}(\mathrm{POL}(A \cup -A), n)$$

when A is a basis of $\mathcal{R}_d$. When A is arbitrary, Matoušek showed a decomposition of any polytope in $\mathrm{POL}(A)$ into a finite number of sets from $\mathrm{POL}(B_1) \cup \dots \cup \mathrm{POL}(B_k)$, where each B_i is a basis of $\mathbb{R}^d$. Together with the techniques he used to bound $\mathrm{disc}(\mathcal{R}_d, n)$, this decomposition allowed him to prove the following theorem.

Theorem 8.12 (Matoušek [35]). *For any positive integer d, and any set of vectors A in $\mathbb{R}^d$, $\mathrm{disc}(\mathrm{POL}(A), n) = O(\log(n)^{d+\frac{1}{2}}\sqrt{\log\log(n)})$, where the constant in the asymptotic notation may depend on d and A.*

Once again, Matoušek used Lemma 8.4 rather than Lemma 8.5 in his proof, and using the latter, as we did for boxes above, immediately improves the bound in Theorem 8.12 by a $O(\sqrt{\log\log(n)})$ factor. Similar to the case of boxes, Bansal and Garg first improved this bound to $O(\log(n)^d)$ and then the author improved it further to $O(\log(n)^{d-\frac{1}{2}})$ [7, 44].

Another natural generalization of the family of axis-aligned boxes is to take families $\mathcal{S}_1, \dots, \mathcal{S}_k$ of "nice" sets, where $\mathcal{S}_i$ is a collection of subsets of $\mathbb{R}^{d_i}$ for some positive integers d_i such that $d_1 + \dots + d_k = d$, and consider the discrepancy of

$$\mathcal{S} = \{S_1 \times \dots \times S_k : S_1 \in \mathcal{S}_1, \dots, S_k \in \mathcal{S}_k\}. \tag{8.2}$$

Boxes are the special case when all the d_i are equal to 1, and each $\mathcal{S}_i$ equals the family of all intervals of the real line. Motivated by work of Beck and Chen [10, 11], Matoušek studied families $\mathcal{S}_i$ of *Tarski cells*, that is, subsets of $\mathbb{R}^{d_i}$ determined by a Boolean formula over polynomial inequalities with real coefficients. Each $\mathcal{S}_i$ is determined by such a Boolean formula, and a setting of the coefficients of the polynomials determines a set $S \in \mathcal{S}_i$. For example, we can express all intersections of two discs in $\mathbb{R}^2$ as the Tarski cell

$$\{x \in \mathbb{R}^2 : (x_1 - a_1)^2 + (x_2 - a_2)^2 \le r^2\} \cap \{x \in \mathbb{R}^2 : (x_1 - b_1)^2 + (x_2 - b_2)^2 \le s^2\},$$

parametrized by $a, b \in \mathbb{R}^2$ and $r, s \in \mathbb{R}$. It is worth noting that a collection $\mathcal{S}$ of Tarski cells in $\mathbb{R}^d$ has dual shatter function $\pi^*(m) = O(m^d)$ (see [34]), and so, by Theorem 8.7, it has discrepancy $\text{disc}(\mathcal{S}, n) = O(n^{\frac{1}{2}-\frac{1}{2d}}\sqrt{\log(n)})$. Matoušek proved the following theorem on the discrepancy of products of Tarski cells.

Theorem 8.13 (Matoušek [36]). *Given positive integers $d_1, \ldots, d_k$ such that $d_1 + \cdots + d_k = d$, and collections of sets $\mathcal{S}_1, \ldots, \mathcal{S}_k$, where $\mathcal{S}_i$ is a collection of Tarski cells in $\mathbb{R}^{d_i}$, the collection $\mathcal{S}$ defined as in* (8.2) *has discrepancy* $\text{disc}(\mathcal{S}, n) = O(n^{\frac{1}{2}-\frac{1}{2d'}+o(1)})$ *for $d' = \max\{d'_1, \ldots d'_k\}$ and $d'_i = \max\{d_i, 2d_i - 3\}$.*

Further, Matoušek conjectured that the d'_i factors can be replaced with d_i, which would give a nearly tight result. Roughly, the theorem states that the discrepancy of the set system of products of Tarski cells is not much bigger than what we can expect for the worst-case discrepancy of the individual families $\mathcal{S}_i$ of Tarski cells.

The proof of the theorem uses a range tree similar to the one used for Theorem 8.10, but instead of the binary trees used in Lemma 8.11, we need to use a more complicated tree construction that uses the structure of the families $\mathcal{S}_i$. The key ingredient are partitions of finite pointsets into parts of a given size, so that most parts are either contained in or disjoint from any given $S \in \mathcal{S}_i$. Recursively taking such partitions allows building a tree for each $\mathcal{S}_i$ analogous to the ones representing $\mathcal{C}^x$ and $\mathcal{C}^y$; then these trees are combined into a *multilevel partition tree*. While the high-level ideas are similar to those in the proof of Theorem 8.10, the actual constructions rely on deep results for range searching data structures, in which Matoušek was an expert. We refer the reader to his survey [31] of this interesting area of computational geometry.

8.4 General discrepancy results

So far, we focused on Matoušek's results for set systems with geometric structure. It is not surprising that most of his work on combinatorial discrepancy focused on geometric set systems, as he was an expert on computational and discrete geometry. Nevertheless, he also proved important theorems on the combinatorial discrepancy of abstract set systems, and some of these theorems also have implications for concrete geometric set systems. We survey some of this work in this final technical section.

One difficulty when studying discrepancy of abstract set systems $(\mathcal{S}, X)$ is that $\text{disc}(\mathcal{S})$ is not robust to otherwise trivial changes to $\mathcal{S}$. For example, if we add a copy

p' of every element $p \in X$ to every set that p belongs to, then the resulting set system $\mathcal{S}'$ trivially has discrepancy 0, regardless of what the discrepancy of $\mathcal{S}$ was to begin with. This means that even set systems with discrepancy 0 can hide arbitrarily complicated substructures within them. This lack of robustness of discrepancy also has computational consequences: under the widely believed conjecture in complexity theory that $\mathsf{P} \neq \mathsf{NP}$, no algorithm that takes a polynomial number of steps can distinguish set systems $\mathcal{S}$ of $O(|\mathcal{S}|)$ sets with discrepancy 0, from those with discrepancy $\Omega(\sqrt{|X|})$ [18]. These issues motivate studying a more robust notion of discrepancy instead, and one that has been particularly fruitful is *hereditary discrepancy*, introduced by Lovász, Spencer, and Vesztergombi [28], and defined by

$$\operatorname{herdisc}(\mathcal{S}) = \max_{Y \subseteq X} \operatorname{disc}(\mathcal{S}|_Y).$$

Note that the combinatorial discrepancy function $\operatorname{disc}(\mathcal{S}, n)$ of a collection $\mathcal{S}$ of subsets of $\mathbb{R}^d$ can also be equivalently defined in terms of hereditary discrepancy as

$$\operatorname{disc}(\mathcal{S}, n) = \max\{\operatorname{herdisc}(\mathcal{S}|_P) : P \subset \mathbb{R}^d, |P| = n\},$$

so, in this sense, we are not losing anything by focusing on this more robust notion. On the other hand, it is usually much easier to prove lower bounds and general structural theorems about hereditary discrepancy.

One powerful tool in proving lower bounds on hereditary discrepancy is the *determinant lower bound*, also due to Lovász, Spencer, and Vesztergombi. Let us define the function detlb on an $m \times n$ matrix A by

$$\operatorname{detlb}(A) = \max_k \max |\det(B)|^{1/k},$$

where the second maximum ranges over $k \times k$ submatrices B of A. Let us also define the *incidence matrix* of a set system $(\mathcal{S}, X)$ to be the matrix A whose rows are indexed by $\mathcal{S}$, and columns by X, and has entries

$$a_{S,p} = \begin{cases} 1 & p \in S \\ 0 & p \notin S \end{cases}.$$

We are going to write $\operatorname{detlb}(\mathcal{S}) = \operatorname{detlb}(A)$ for the incidence matrix A of $\mathcal{S}$. With these definitions out of the way, we can state the lower bound.

Theorem 8.14 (Lovász, Spencer, Vesztergombi [28]). *For any finite set system $\mathcal{S}$ whose incidence matrix is A,*

$$\operatorname{herdisc}(\mathcal{S}) \geq \frac{1}{2} \operatorname{detlb}(\mathcal{S}). \tag{8.3}$$

A starting point for the proof of Theorem 8.14 is the observation that for any set system $\mathcal{S}$ on a universe X of size n, which for convenience we identify with $\{1, \dots, n\}$, we can write

$$\operatorname{disc}(\mathcal{S}) = \min_{x \in \{-1,+1\}^n} \|Ax\|_\infty.$$

In fact, Lovász, Spencer, and Vesztergombi define the right-hand side above as the discrepancy of an *arbitrary* matrix A; then they define the hereditary discrepancy $\operatorname{herdisc}(A)$ as the maximum discrepancy over submatrices of A. Theorem 8.14 also holds for arbitrary matrices, and states that $\operatorname{herdisc}(A) \geq \frac{1}{2}\operatorname{detlb}(A)$. The core of the proof is the fact that, for any $m \times n$ matrix A, and $t = \operatorname{herdisc}(A)$, the hypercube $[-1, 1]^n$ can be covered by 2^n copies of the polytope $K = \{x \in \mathbb{R}^n : \|Ax\|_\infty \leq 2t\}$, one centered at each point in $\{-1, +1\}^n$. The theorem then follows by a volume argument.

It was conjectured in [28] that inequality (8.3) can be reversed up to a constant factor, that is, that $\operatorname{herdisc}(\mathcal{S}) \leq c \operatorname{detlb}(\mathcal{S})$ for some absolute constant c. One piece of evidence for this conjecture is a theorem of Ghouila–Houri, which states that, for matrices A with entries in $\{-1, 0, +1\}$, $\operatorname{herdisc}(A) = 1$ if and only if $\operatorname{detlb}(A) = 1$ [22]. Nevertheless, the conjecture was refuted by a beautiful example of Hoffman, which we sketch next. Take X to be the edges of a complete k-ary tree of depth k, take $\mathcal{S}_1$ to contain the sets of edges in any root to leaf path, and take $\mathcal{S}_2$ to contain, for any vertex in the tree, the set of edges connecting the vertex to its children. $\mathcal{S}_1$ and $\mathcal{S}_2$ both have hereditary discrepancy 1, and, by Ghouila-Houri's theorem, $\operatorname{detlb}(\mathcal{S}_1) = \operatorname{detlb}(\mathcal{S}_2) = 1$. Then it can be shown that $\operatorname{detlb}(\mathcal{S}_1 \cup \mathcal{S}_2)$ must also be bounded by a constant. (We will state a general theorem to this effect shortly.) It is, however, not hard to see that any coloring of X leaves at least one set in $\mathcal{S}_1 \cup \mathcal{S}_2$ monochromatic: if every set in $\mathcal{S}_2$ is bichromatic, then one can find a root to leaf path consisting entirely of edges colored +1. This means that, if we write $n = |X| = k^k$, then $\operatorname{disc}(\mathcal{S}_1 \cup \mathcal{S}_2) = \Omega(\frac{\log(n)}{\log\log(n)})$. Other, more complicated examples show that in fact the gap between $\operatorname{herdisc}(\mathcal{S})$ and $\operatorname{detlb}(\mathcal{S})$ can be as large as $\Omega(\log n)$ [45, 42]. Resolving a longstanding open question, Matoušek showed that these examples are essentially the worst possible.

Theorem 8.15 (Matoušek [37]). *There exists a constant $c > 0$ such that, for any finite set system $(\mathcal{S}, X)$,*

$$\operatorname{herdisc}(\mathcal{S}) \leq c \log_2(2|\mathcal{S}||X|)^{3/2} \operatorname{detlb}(\mathcal{S}).$$

The proof of Theorem 8.15 relies on a useful relaxation of discrepancy, called *vector discrepancy* and defined for a set system $\mathcal{S}$ over a universe X of size n by

$$\operatorname{vdisc}(\mathcal{S}) = \min_{\chi : X \to \mathbb{S}^{n-1}} \max_{S \in \mathcal{S}} \Big\| \sum_{p \in S} \chi(p) \Big\|_2,$$

where $\mathbb{S}^{n-1}$ is the unit Euclidean sphere in $\mathbb{R}^n$. We recover the standard notion of discrepancy if we restrict the values of χ to two antipodal points on $\mathbb{S}^{n-1}$. Because of this, we always have $\mathrm{vdisc}(\mathcal{S}) \le \mathrm{disc}(\mathcal{S})$. Vector discrepancy is in some ways a more tractable quantity than the discrepancy: for example, it can be computed efficiently, and satisfies a duality property that we will use shortly. Unfortunately, it can often be far from the discrepancy. Surprisingly, however, the hereditary versions of the two quantities are never too far from each other. If we define the *hereditary vector discrepancy* of $\mathcal{S}$, analogously to the hereditary discrepancy, by $\mathrm{hervdisc}(\mathcal{S}) = \max_{Y\subseteq X} \mathrm{vdisc}(\mathcal{S}|_Y)$, then we have the following result, proved in the seminal work of Bansal [6].

Theorem 8.16 (Bansal [6]). *There exists a constant $c > 0$ such that, for any finite set system $(\mathcal{S}, X)$,*

$$\mathrm{herdisc}(\mathcal{S}) \le c\log_2(2|\mathcal{S}||X|)\,\mathrm{hervdisc}(\mathcal{S}).$$

Because of Theorem 8.16, the proof of Theorem 8.15 reduces to showing an inequality between vector discrepancy and the determinant lower bound. Indeed, Matoušek proved that there exists a constant $c > 0$ such that for any set system $(\mathcal{S}, X)$,

$$\mathrm{vdisc}(\mathcal{S}) \le c\sqrt{\log_2(2|X|)}\,\mathrm{detlb}(\mathcal{S}). \tag{8.4}$$

To prove (8.4), Matoušek used the fact that vector discrepancy can be formulated as the value of a convex optimization problem, that is, the minimum of a convex function over a convex set. Such problems exhibit a beautiful duality theory: under some technical conditions, any such minimization problem is associated with an equivalent problem of maximizing a concave function over a convex set, and the two problems have the same value (see, e. g., [16]). Applying this theory to vector discrepancy yields an equivalent formulation of $\mathrm{vdisc}(\mathcal{S})$ as the largest value $t > 0$ for which there exist functions $w : \mathcal{S} \to \mathbb{R}_{\ge 0}$ and $z : X \to \mathbb{R}$ such that

$$\sum_{S\in\mathcal{S}} w(S) = \sum_{p\in X} z(p) = 1,$$

and for all $\chi : X \to \mathbb{R}$, we have

$$\sum_{S\in SS} w(S)\Big(\sum_{p\in S}\chi(p)\Big)^2 \ge t^2 \sum_{p\in X} z(p)\chi(p)^2. \tag{8.5}$$

In the special case when w and z are both constant, (8.5) is equivalent to the requirement that the incidence matrix A satisfies $\frac{1}{|S|}\|Ax\|_2^2 \ge t^2\frac{1}{|X|}\|x\|_2^2$, for any $|X|$-dimensional vector x. By the Courant–Fischer theorem, this is also equivalent to requiring that the smallest singular value of A is at least $t\sqrt{\frac{|S|}{|X|}}$. If it were also the case that $|S| = |X|$, that is, A is a square matrix, then we would have that $|\det(A)|^{1/|X|}$, which equals the geometric mean of the singular values of A, is at least the smallest singular value of

A, which is at least $t = \mathrm{vdisc}(\mathcal{S})$. This establishes (8.4), without the $\sqrt{\log_2(2|X|)}$ factor, under the many assumptions we made. It turns out that all these assumptions can be removed with only constant loss in the bounds, except the assumption that w is constant, whose removal seems to require losing the factor of $\sqrt{\log_2(2|X|)}$.

Theorems 8.14 and 8.15 together imply that, up to logarithmic factors, hereditary discrepancy and the determinant lower bound are always equal. This is useful because it allows proving theorems about hereditary discrepancy by proving them about the determinant lower bound instead. For example, a very natural question is how hereditary discrepancy behaves under taking unions of set systems. Hoffman's example above shows that it can grow by a logarithmic factor. The next theorem, also shown by Matoušek, implies that the growth cannot be much faster than that.

Theorem 8.17 (Matoušek [37]). *For any k finite set systems $\mathcal{S}_1, \ldots \mathcal{S}_k$ over a common universe, and their union $\mathcal{S} = \bigcup_{i=1}^k \mathcal{S}_i$, we have*

$$\mathrm{detlb}(\mathcal{S}) \leq \sqrt{ek} \cdot \max_{i=1}^{k} \mathrm{detlb}(\mathcal{S}_i).$$

Therefore, there exists a constant $c > 0$ such that for any such set systems we have

$$\mathrm{herdisc}(\mathcal{S}) \leq c \log(|\mathcal{S}||X|)^{3/2} \sqrt{k} \max_{i=1}^{k} \mathrm{herdisc}(\mathcal{S}_i).$$

This type of theorem appears hard to prove using direct combinatorial arguments, and no such proof is known.

While the determinant lower bound is often more tractable than discrepancy itself, it is often also hard to estimate directly, since it is defined as a maximum over an exponential number of submatrices, and there is no known efficient algorithm for computing it. This motivates searching for a yet more nicely behaved function that approximates hereditary discrepancy. Unlike the determinant lower bound, we will find such a function by considering natural *upper* bounds on hereditary discrepancy. One simple upper bound is given by the maximum size of a set in the set system $\mathcal{S}$: by a standard probabilistic method argument, if all sets in $\mathcal{S}$ have size bounded by s, then $\mathrm{disc}(\mathcal{S}) = O(\sqrt{s \log(|\mathcal{S}|)})$. Moreover, this bound also holds for the hereditary discrepancy, since the maximum size of a set in any restriction of $\mathcal{S}$ is also at most s. A much deeper result of Banaszczyk [4] shows that, if every element of the universe appears in at most t sets, then $\mathrm{disc}(\mathcal{S}) = O(\sqrt{t \log(|\mathcal{S}|)})$.[7] Again, this bound applies also to hereditary discrepancy, because taking a restriction of $\mathcal{S}$ does not increase t. In terms of the incidence matrix A, the largest set size s can be written as the largest squared Euclidean norm of a row of A, which we denote by $\rho(A)^2$. Similarly, the largest

7 Beck and Fiala famously conjectured that the correct bound is $O(\sqrt{t})$ [12]. They showed the bound $2t - 1$, and the current best bound that only depends on t is due to Bukh [17] and equals $2t - f(t)$ for a slowly growing function $f(t)$. Beck and Fiala's conjecture is wide open.

number t of sets in which any element appears is $\rho(A^\top)^2$, where $A^\top$ is the transpose of A. It was the insight of Larsen that these two bounds can be combined into a single one. Let us define the *γ_2-factorization constant* of a matrix A as

$$\gamma_2(A) = \inf\{\rho(B)\rho(C^\top) : BC = A\},$$

where B and C range over matrices of the appropriate dimensions. This function is, in fact, a norm, and has been studied extensively in functional analysis (see, e. g., [52]), and, more recently, in complexity theory [27]. The next theorem connects it to combinatorial discrepancy, as well. In it, and in the remainder of this section, we use $\gamma_2(\mathcal{S})$ to denote $\gamma_2(A)$, where A is the incidence matrix of $\mathcal{S}$.

Theorem 8.18 (Larsen [26]). *There exists a constant $c > 0$ such that for any finite set system $\mathcal{S}$ we have*

$$\operatorname{herdisc}(\mathcal{S}) \le c\sqrt{\log_2(2|\mathcal{S}|)}\,\gamma_2(\mathcal{S}). \tag{8.6}$$

Curiously, Larsen proved Theorem 8.18 as a tool for giving lower bounds on the complexity of range searching data structure, giving us another connection between data structures and discrepancy. The formulation in terms of the γ_2 norm is from [38].

Let us sketch the proof of Theorem 8.18. Once again, it is enough to establish the inequality (8.6) for $\operatorname{disc}(\mathcal{S})$, since it is easy to check that for any restriction $\mathcal{S}|_Y$ of $\mathcal{S}$ to a subset Y of the universe we have $\gamma_2(\mathcal{S}|_Y) \le \gamma_2(\mathcal{S})$. The proof of (8.6) for $\operatorname{disc}(\mathcal{S})$ uses the following powerful theorem of Banaszczyk.

Theorem 8.19 (Banaszczyk [4]). *For any $m \times n$ matrix A such that $\rho(A^\top) \le \frac{1}{5}$, and any convex subset K of $\mathbb{R}^m$ with standard Gaussian measure at least $\frac{1}{2}$, there exists a vector $x \in \{-1, +1\}^n$ such that $Ax \in K$.*

To use Theorem 8.19 in order to prove Theorem 8.18, we take a factorization $BC = A$ of the incidence matrix A such that $\rho(C^\top) = \frac{1}{5}$, and $\rho(B) = 5\gamma_2(A)$, and define $K = \{x : \|Bx\|_\infty \le c\sqrt{\log_2(2|\mathcal{S}|)}\,\gamma_2(A)\}$. For a large enough constant c, standard concentration of measure bounds show that K has Gaussian measure at least $\frac{1}{2}$, and Theorem 8.19 implies that there exists some vector x with coordinates in $\{-1, +1\}$ such that

$$Cx \in K \iff \|BCx\|_\infty \le c\sqrt{\log_2(2|\mathcal{S}|)}\,\gamma_2(A) \iff \|Ax\|_\infty \le c\sqrt{\log_2(2|\mathcal{S}|)}\,\gamma_2(A). \tag{8.7}$$

The last inequality is equivalent to $\operatorname{disc}(\mathcal{S}) \le c\sqrt{\log_2(2|\mathcal{S}|)}\,\gamma_2(A)$, proving the theorem.

A result proved by Matoušek, jointly with the author and with Talwar, shows an inequality in the reverse direction.

Theorem 8.20 (Matoušek, Nikolov, Talwar [38]). *There exists a constant $c > 0$ such that for any finite set system $\mathcal{S}$,*

$$\gamma_2(\mathcal{S}) \le c\log_2(2|\mathcal{S}|)\operatorname{detlb}(\mathcal{S}) \le \frac{c}{2}\log_2(2|\mathcal{S}|)\operatorname{herdisc}(\mathcal{S}). \tag{8.8}$$

Like vector discrepancy, the γ_2 norm of a matrix can be written as the value of a convex optimization problem, and, once again, the proof of Theorem 8.18 relies on the duality theory of such problems. The dual formulation of the γ_2 norm of a matrix A is

$$\gamma_2(A) = \max\{\|PAQ\|_* : P, Q \text{ diagonal}, \operatorname{tr}(P^2) = \operatorname{tr}(Q^2) = 1\}, \tag{8.9}$$

where $\|\cdot\|_*$ is the matrix *trace norm*, equal to the sum of the singular values. If we think about the special case when A is a square $n \times n$ matrix, and P and Q are both equal to $\frac{1}{\sqrt{n}}I$, where I is the $n \times n$ identity matrix, then $\|PAQ\|_*$ is just the arithmetic mean of the singular values of A. Since $|\det(A)|^{1/n}$ is the geometric mean of the singular values, (8.8) appears to go in the wrong direction. Nevertheless, it turns out that, while $|\det(A)|^{1/n}$ may be much smaller than $\frac{1}{n}\|A\|_*$ (e. g., when A is singular), A must contain a *submatrix* whose determinant is comparable to $\frac{1}{n}\|A\|_*$. This fact is related to the *restricted invertibility theorem* of Bourgain and Tzafriri [15], but has a much simpler linear algebraic proof, given in [38].

The γ_2 norm enjoys many pleasant properties, which make it a powerful tool to study combinatorial discrepancy. In contrast to the determinant lower bound, there are known algorithms which compute $\gamma_2(\mathcal{S})$ efficiently [27]. Moreover, it satisfies the following properties:

1. γ_2 is a norm, that is, for any two matrices A and B with equal dimensions, $\gamma_2(A + B) \le \gamma_2(A) + \gamma_2(B)$;
2. for any two set systems $\mathcal{S}_1$ and $\mathcal{S}_2$, $\gamma_2(\mathcal{S}_1 \cup \mathcal{S}_2)^2 \le \gamma_2(\mathcal{S}_1)^2 + \gamma_2(\mathcal{S}_2^2)$;
3. for any matrix A, $\gamma_2(A) = \gamma_2(A^\top)$; this implies that for the dual $\mathcal{S}^*$ of a set system $\mathcal{S}$, $\gamma_2(\mathcal{S}^*) = \gamma_2(\mathcal{S})$;
4. for any two matrices A and B, and their Kronecker product $A \otimes B$, $\gamma_2(A \otimes B) = \gamma_2(A)\gamma_2(B)$; this implies that for two set systems $\mathcal{S}_1$ and $\mathcal{S}_2$, and the product set system $\mathcal{S} = \{S_1 \times S_2 : S_1 \in \mathcal{S}_1, S_2 \in \mathcal{S}_2\}$, we have $\gamma_2(\mathcal{S}) = \gamma_2(\mathcal{S}_1)\gamma_2(\mathcal{S}_2)$.

Via (8.6) and (8.8), these properties all hold, up to logarithmic factors, for hereditary discrepancy, as well. For example, this partially explains why we saw similar discrepancy bounds for set systems with polynomially bounded primal or dual shatter function in Theorems 8.6 and 8.7. Moreover, using these properties and the inequalities (8.6) and (8.8), it is often possible to give surprisingly short proofs of upper and lower bounds on discrepancy, which are difficult to establish with other methods. Perhaps the most striking example is the application of this method to axis-aligned boxes in $\mathbb{R}^d$, for which we can give a shorter proof of the upper bound we discussed in the previous section, as well as a proof of the best known lower bound, captured in the following theorem.

Theorem 8.21 (Matoušek, Nikolov, Talwar [40, 38]). *For any positive integer d,*

$$\operatorname{disc}(\mathcal{R}_d, n) = \Omega(\log(n)^{d-1}).$$

Proof sketch. Let us define $[n] = \{1, \ldots, n\}$, and further define $\mathcal{I}$ to be the set of intervals on $[n]$. The discrepancy of $\mathcal{I}$ is 1 (just consider a coloring with alternating +1 and −1), and so is its hereditary discrepancy, because any restriction of $\mathcal{I}$ is a set system of the same type. Then, by Theorem 8.20, $\gamma_2(\mathcal{I}) = O(\log(n))$. In fact, using the dual formulation (8.9), one can also show that $\gamma_2(\mathcal{I}) = \Omega(\log(n))$, as well. It is easy to verify that $\mathcal{R}_d|_{[n]^d}$ consists of d-wise products of sets in $\mathcal{I}$, and so, applying the fourth property above repeatedly, we get that

$$\gamma_2(\mathcal{R}_d|_{[n]^d}) = \gamma_2(\mathcal{I})^d = \Theta(\log(n)^d). \tag{8.10}$$

Theorem 8.20 and (8.10) now imply that $\mathrm{disc}(\mathcal{R}_d, n) = \Omega(\log(n)^{d-1})$. □

In fact, (8.10) also implies the upper bound $\mathrm{disc}(\mathcal{R}_d, n) = O(\log(n)^{d+\frac{1}{2}})$. This is because, up to shifting and scaling, we can think of any n-point set P in $\mathbb{R}^d$ as a subset of $[n]^d$. Then $\mathcal{R}_d|_P$ can be thought of as a restriction of $\mathcal{R}_d|_{[n]^d}$, and Theorem 8.18 and (8.10) imply the upper bound. This result is due to Larsen [26], with essentially the same proof. As mentioned above, the upper bound was recently improved to $O(\log(n)^{d-\frac{1}{2}})$ by the author [44]. This improvement uses a variant of the γ_2 method, utilizing a recent result of Banaszczyk [5] in place of Theorem 8.19.

8.5 Conclusion

There are a number of combinatorial discrepancy results of Matoušek we did not cover in this survey. Perhaps the most important one among them is the tight upper bound on the discrepancy of arithmetic progressions, proved with Spencer [39]. The author encourages the reader to also read Matoušek's original papers, which are beautifully and clearly written, as well as his book [34]. The aim of this survey is to be a guide to his body of work on combinatorial discrepancy, and to help to put it in the context of more recent developments, and also in the context of related results in computational and discrete geometry.

On a personal note, writing the paper [38] with Matoušek felt like a master class in mathematical writing, and I can only regret that I did not have enough time with him to learn more. His clarity of thought and his ability to see the core idea of a proof behind technical details is something I deeply admire.

Bibliography

[1] C. Aistleitner, D. Bilyk, A. Nikolov. Tusnády's problem, the transference principle, and non-uniform QMC sampling. In: Monte Carlo and Quasi-Monte Carlo Methods, pp. 169–180. Springer International Publishing (2016).

[2] R. Alexander. Geometric methods in the study of irregularities of distribution. Combinatorica **10**(2), 115–136 (1990).

[3] S. Arora, E. Hazan, S. Kale. The multiplicative weights update method: a meta-algorithm and applications. Theory Comput. **8**(1), 121–164 (2012).

[4] W. Banaszczyk. Balancing vectors and Gaussian measures of n-dimensional convex bodies. Random Struct. Algorithms **12**(4), 351–360 (1998).

[5] W. Banaszczyk. On series of signed vectors and their rearrangements. Random Struct. Algorithms **40**(3), 301–316 (2012).

[6] N. Bansal. Constructive algorithms for discrepancy minimization. In: Foundations of Computer Science (FOCS), 2010 51st Annual IEEE Symposium on, pp. 3–10. IEEE (2010).

[7] N. Bansal, S. Garg. Algorithmic discrepancy beyond partial coloring. In: STOC, pp. 914–926. ACM (2017).

[8] J. Beck. Balanced two-colorings of finite sets in the square i. Combinatorica **1**(4), 327–335 (1981).

[9] J. Beck. Roth's estimate of the discrepancy of integer sequences is nearly sharp. Combinatorica **1**(4), 319–325 (1981).

[10] J. Beck, W. W. L. Chen. Note on irregularities of distribution. Mathematika **33**(1), 148–163 (1986).

[11] J. Beck, W. W. L. Chen. Note on irregularities of distribution. II. Proc. Lond. Math. Soc. (3) **61**(2), 251–272 (1990).

[12] J. Beck, T. Fiala. Integer-making theorems. Discrete Appl. Math. **3**(1), 1–8 (1981).

[13] J. L. Bentley, M. I. Shamos. A problem in multivariate statistics: algorithm, data structure, and applications. In: Proceedings of the Fifteenth Allerton Conference on Communication, Control, and Computing (1977).

[14] D. Bilyk, M. Lacey, A. Vagharshakyan. On the small ball inequality in all dimensions. J. Funct. Anal. **254**(9), 2470–2502 (2008).

[15] J. Bourgain, L. Tzafriri. Invertibility of large submatrices with applications to the geometry of Banach spaces and harmonic analysis. Isr. J. Math. **57**(2), 137–224 (1987).

[16] S. Boyd, L. Vandenberghe. Convex Optimization. Cambridge University Press, Cambridge (2004).

[17] B. Bukh. An improvement of the Beck-Fiala theorem. Comb. Probab. Comput. **25**(3), 380–398 (2016).

[18] M. Charikar, A. Newman, A. Nikolov. Tight hardness results for minimizing discrepancy. In: SODA 2011, pp. 1607–1614. SIAM (2011).

[19] B. Chazelle, J. Matoušek, M. Sharir. An elementary approach to lower bounds in geometric discrepancy. Discrete Comput. Geom. **13**(1), 363–381 (1995).

[20] R. M. Dudley. Central limit theorems for empirical measures. Ann. Probab. **6**(6), 899–929 (1979). 1978.

[21] Y. Freund, R. Schapire. Game theory, on-line prediction and boosting. In: COLT, pp. 325–332. ACM (1996).

[22] A. Ghouila-Houri. Caractérisation des matrices totalement unimodulaires. C. R. Acad. Sci. Paris **254**, 1192–1194 (1962).

[23] J. H. Halton. On the efficiency of certain quasi-random sequences of points in evaluating multi-dimensional integrals. Numer. Math. **2**, 84–90 (1960).

[24] J. M. Hammersley. Monte Carlo methods for solving multivariable problems. Ann. N.Y. Acad. Sci. **86**, 844–874 (1960). 1960.

[25] D. Haussler. Sphere packing numbers for subsets of the Boolean n-cube with bounded Vapnik-Chervonenkis dimension. J. Comb. Theory, Ser. A **69**(2), 217–232 (1995).

[26] K. G. Larsen. On range searching in the group model and combinatorial discrepancy. SIAM J. Comput. **43**(2), 673–686 (2014).
[27] N. Linial, S. Mendelson, G. Schechtman, A. Shraibman. Complexity measures of sign matrices. Combinatorica **27**(4), 439–463 (2007).
[28] L. Lovász, J. Spencer, K. Vesztergombi. Discrepancy of set-systems and matrices. Eur. J. Comb. **7**(2), 151–160 (1986).
[29] S. Lovett, R. Meka. Constructive discrepancy minimization by walking on the edges. SIAM J. Comput. **44**(5), 1573–1582 (2015).
[30] J. Matoušek, E. Welzl, L. Wernisch. Discrepancy and approximations for bounded VC-dimension. Combinatorica **13**(4), 455–466 (1993).
[31] J. Matoušek. Geometric range searching. ACM Comput. Surv. **26**(4), 421–461 (1994).
[32] J. Matoušek. Tight upper bounds for the discrepancy of halfspaces. Discrete Comput. Geom. **13**(1), 593–601 (1995).
[33] J. Matoušek. On discrepancy bounds via dual shatter function. Mathematika **44**(1), 42–49 (1997).
[34] J. Matoušek. Geometric Discrepancy (An Illustrated Guide). Springer (1999).
[35] J. Matoušek. On the discrepancy for boxes and polytopes. Monatshefte Math. **127**(4), 325–336 (1999).
[36] J. Matoušek. On the discrepancy for Cartesian products. J. Lond. Math. Soc. (2) **61**(3), 737–747 (2000).
[37] J. Matoušek. The determinant bound for discrepancy is almost tight. Proc. Am. Math. Soc. **141**(2), 451–460 (2013).
[38] J. Matoušek, A. Nikolov, K. Talwar. Factorization norms and hereditary discrepancy. Int. Math. Res. Not. (2018).
[39] J. Matoušek, J. Spencer. Discrepancy in arithmetic progressions. J. Am. Math. Soc. **9**(1), 195–204 (1996).
[40] J. Matouvsek, A. Nikolov. Combinatorial discrepancy for boxes via the gamma_2 norm. In: Symposium on Computational Geometry. LIPIcs, vol. 34, pp. 1–15. Schloss Dagstuhl – Leibniz-Zentrum fuer Informatik (2015).
[41] S. Muthukrishnan, A. Nikolov. Optimal private halfspace counting via discrepancy. In: STOC 2012, pp. 1285–1292. ACM (2012).
[42] A. Newman, O. Neiman, A. Nikolov. Beck's three permutations conjecture: a counterexample and some consequences. In: FOCS 2012, pp. 253–262. IEEE Computer Society (2012).
[43] H. Niederreiter. Random Number Generation and Quasi-Monte Carlo Methods. CBMS-NSF Regional Conference Series in Applied Mathematics, vol. 63. Society for Industrial and Applied Mathematics (SIAM), Philadelphia, PA (1992).
[44] A. Nikolov. Tighter bounds for the discrepancy of boxes and polytopes. Mathematika **63**(3), 1091–1113 (2017).
[45] D. Pálvölgyi. Indecomposable coverings with concave polygons. Discrete Comput. Geom. **44**(3), 577–588 (2010).
[46] N. Sauer. On the density of families of sets. J. Comb. Theory, Ser. A **13**, 145–147 (1972).
[47] S. Shelah. A combinatorial problem; stability and order for models and theories in infinitary languages. Pac. J. Math. **41**, 247–261 (1972).
[48] J. Spencer. Six standard deviations suffice. Trans. Am. Math. Soc. **289**, 679–706 (1985).
[49] J. Spencer. Ten Lectures on the Probabilistic Method, 2nd edn. CBMS-NSF Regional Conference Series in Applied Mathematics, vol. 64. Society for Industrial and Applied Mathematics (SIAM), Philadelphia, PA (1994).
[50] A. Srinivasan. Improving the discrepancy bound for sparse matrices: better approximations for sparse lattice approximation problems. In: Proceedings of the Eighth Annual ACM-SIAM

Symposium on Discrete Algorithms (New Orleans, LA, 1997), pp. 692–701. ACM, New York (1997).
[51] M. Talagrand. The Generic Chaining. Upper and Lower Bounds of Stochastic Processes. Springer Monographs in Mathematics. Springer-Verlag, Berlin (2005).
[52] N. Tomczak-Jaegermann. Banach-Mazur Distances and Finite-Dimensional Operator Ideals. Pitman Monographs and Surveys in Pure and Applied Mathematics, vol. 38. J. Wiley, New York (1989).
[53] V. N. Vapnik, A. Ja. Chervonenkis. The uniform convergence of frequencies of the appearance of events to their probabilities. Teor. Veroâtn. Primen. **16**, 264–279 (1971).
[54] E. Welzl. Partition trees for triangle counting and other range searching problems. In: Proceedings of the Fourth Annual Symposium on Computational Geometry (Urbana, IL, 1988), pp. 23–33. ACM, New York (1988).

Luca Brandolini and Giancarlo Travaglini

9 Fourier analytic techniques for lattice point discrepancy

Abstract: Counting integer points in large convex bodies with smooth boundaries containing isolated flat points is oftentimes an intermediate case between balls (or convex bodies with smooth boundaries having everywhere positive curvature) and cubes (or convex polytopes). In this paper, we provide a detailed description of several discrepancy problems in the particular planar case where the boundary coincides locally with the graph of the function $\mathbb{R} \ni t \mapsto |t|^\gamma$, with $\gamma > 2$. We consider both *integer points* problems and *irregularities of distribution* problems. The above "restriction" to a particular family of convex bodies is compensated by the fact that many proofs are elementary. The paper is entirely self-contained.

Keywords: Convex bodies, flat points, decay of Fourier transforms, discrepancy, integer points, irregularities of distribution

MSC 2010: 11H06, 11K38, 42B05

9.1 Introduction

The word *discrepancy* comes from its Latin counterpart *discrepantia* (disagreement, contrast) and here expresses the deviation of a *discrete volume* of a convex body from its *(continuous) volume*. Much of this paper is devoted to the study of lattice points discrepancy in dimension two: for a given convex body $C \subset \mathbb{R}^2$ (i. e., a compact convex set with nonempty interior) and a large real positive parameter R we compare the number of points with integer coordinates contained in the dilated body

$$RC = \{t \in \mathbb{R}^2 : t/R \in C\}$$

and its area. More precisely, we consider the discrepancy

$$\mathcal{D}(RC) := -R^2|C| + \operatorname{card}(RC \cap \mathbb{Z}^2) = -R^2|C| + \sum_{n \in \mathbb{Z}^2} \chi_{RC}(n)$$

where χ_A denotes the characteristic (indicator) function of the set A.

Luca Brandolini, Dipartimento di Ingegneria Gestionale, dell'Informazione e della Produzione, Università degli Studi di Bergamo, Viale Marconi 5, 24044 Dalmine BG, Italy, e-mail: luca.brandolini@unibg.it
Giancarlo Travaglini, Dipartimento di Matematica e Applicazioni, Università di Milano-Bicocca, Ed. U5, Via Cozzi 55, 20125 Milano, Italy, e-mail: giancarlo.travaglini@unimib.it

https://doi.org/10.1515/9783110652581-009

The problem of estimating $\mathcal{D}(RC)$ for large values of R has a long history and several connections to different branches of mathematics (see, e. g., [4, 11, 17, 20, 25, 29, 30, 40]).

Here, we are interested in the following specific family of convex bodies.

Definition 9.1. Let $\mathbb{R} \ni \gamma > 2$. We denote by C_γ any planar compact convex set, contained in the square $(-1/2, 1/2)^2$, whose boundary ∂C_γ coincides, in a small neighborhood U of the origin, with the graph of the function $\mathbb{R} \ni x \mapsto |x|^\gamma$. We also assume that, outside $\frac{1}{2}U$, ∂C_γ is smooth with curvature $\geqslant c > 0$.

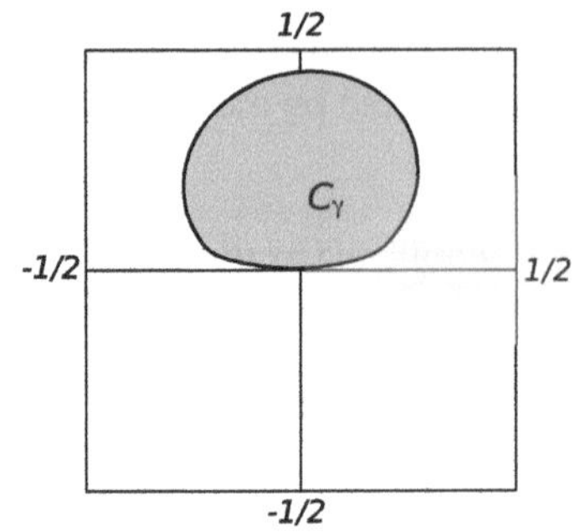

Our interest in the above class of convex bodies comes from the fact that a large part of *Geometric discrepancy* has been developed for rectangles (or parallelepipeds or polytopes) and discs (or balls, or convex bodies having smooth boundary with everywhere positive Gaussian curvature). See the above list of references and also [3, 19, 31, 34, 36]. The above index γ provides a sort of "bridge" between, say, a disc and a square, which respectively can be roughly seen as the cases $\gamma = 2$ and $\gamma = \infty$. Anyway in the last section we shall see a situation where C_γ does not have this *intermediate position*, and a sort of dichotomy appears.

The proofs in this paper are essentially Fourier analytic and several arguments come from [8, 11, 15], and [21]. All the results in this paper are essentially known, except Theorem 9.26.

We set the notation.

We identify the torus $\mathbb{T}^2 = \mathbb{R}^2/\mathbb{Z}^2$ with the unit square $[-1/2, 1/2)^2$. Let $f \in L^1(\mathbb{T}^2)$ and for every $k \in \mathbb{Z}^2$ let

$$\hat{f}(k) = \int_{\mathbb{T}^2} f(t) e^{-2\pi i t \cdot k} \, dt$$

be the Fourier coefficient of $f(t)$, which therefore has Fourier series

$$\sum_{k \in \mathbb{Z}^2} \hat{f}(k) e^{2\pi i t \cdot k}.$$

The points in $\mathbb{Z}^2$ are termed *integer points*. If $g \in L^1(\mathbb{R}^2)$ and $\xi \in \mathbb{R}^2$, then

$$\widehat{g}(\xi) = \int_{\mathbb{R}^2} g(t)e^{-2\pi i t\cdot\xi}\, dt$$

denotes the Fourier transform of $g(t)$.

The connection between the above discrepancy and Fourier analysis is a consequence of the following simple observation. Let C be a convex body in $\mathbb{R}^2$ and, for every $t \in \mathbb{R}^2$, define the *discrepancy function*

$$\begin{aligned}\mathcal{D}_R(t) = \mathcal{D}(RC + t) &= -R^2|C| + \operatorname{card}((RC + t) \cap \mathbb{Z}^2)\\ &= -R^2|C| + \sum_{n\in\mathbb{Z}^2} \chi_{RC}(n - t)\,.\end{aligned}$$

The function $\mathcal{D}_R(t)$ is periodic with Fourier series

$$\sum_{0\neq m\in\mathbb{Z}^2} \widehat{\mathcal{D}}_R(m)e^{2\pi i m\cdot t} = \sum_{0\neq m\in\mathbb{Z}^2} \widehat{\chi}_{RC}(m)e^{2\pi i m\cdot t}\,. \tag{9.1}$$

Indeed,

$$\begin{aligned}\widehat{\mathcal{D}}_R(0) &= \int_{\mathbb{T}^2}\Big(-R^2|C| + \sum_{n\in\mathbb{Z}^2}\chi_{RC}(n - t)\Big)\, dt\\ &= -R^2|C| + \sum_{n\in\mathbb{Z}^2}\int_{\mathbb{T}^2}\chi_{RC}(n - t)\, dt = -R^2|C| + \int_{\mathbb{R}^2}\chi_{RC}(t)\, dt = 0\,,\end{aligned}$$

and for $m \neq 0$,

$$\begin{aligned}\widehat{\mathcal{D}}_R(m) &= \int_{\mathbb{T}^2}\Big(-R^d|C| + \sum_{n\in\mathbb{Z}^2}\chi_{RC}(n - t)\Big)e^{-2\pi i m\cdot t}\, dt\\ &= \int_{RC} e^{-2\pi i R m\cdot t}\, dt = \widehat{\chi}_{RC}(m)\,.\end{aligned}$$

Observe that the two sides of the equality $\widehat{\mathcal{D}}_R(m) = \widehat{\chi}_{RC}(m)$ have a different nature. On the LHS, the terms $\widehat{\mathcal{D}}_R(m)$ are the Fourier coefficients of the periodic function $\mathcal{D}_R(t)$ (defined on $\mathbb{T}^2$), while on the RHS, the terms $\widehat{\chi}_{RC}(m)$ are the restriction (to $\mathbb{Z}^2$) of the Fourier transform $\widehat{\chi}_{RC}(\xi)$ of the function $\chi_{RC}(t)$ (which is defined on $\mathbb{R}^2$).

Throughout the paper, $c, c_1, c_2, \ldots$ denote constants which may change from step to step.

9.2 Integer points in large convex bodies

First, we recall the *circle problem* and the *Hardy–Voronoi identity*. Let R be a positive real number. The circle problem asks for a significant estimate of the sum

$$A(R) = \sum_{0 \le k \le R^2} r(k)$$

of the arithmetic function

$$r(k) = \operatorname{card}\{(m_1, m_2) \in \mathbb{Z}^2 : m_1^2 + m_2^2 = k\},$$

that is the number of ways of writing a nonnegative integer as a sum of two squares. Let $B = B(0,1) = \{t \in \mathbb{R}^2 : |t| \leqslant 1\}$ be the disc of unit radius centered at the origin. More generally, we write $B(\tau, r) := \{t \in \mathbb{R}^2 : |t - \tau| \leqslant r\}$.

More than 200 years ago, C. F. Gauss observed that the average of $r(k)$ reduces to counting the integer points in the dilated disc $RB = \{t \in \mathbb{R}^2 : |t/R| \leqslant 1\}$, for $R > 1$. Then it is easy to observe that $\operatorname{card}(RB \cap \mathbb{Z}^2)$ equals the area $R^2\pi$ of the disc plus an error term smaller, in absolute value, than ($\sqrt{2}$ times) the length of the boundary of the dilated disc. That is,

$$\operatorname{card}(RB \cap \mathbb{Z}^2) = R^2\pi + \mathcal{D}(RB),$$

with $\mathcal{D}(RB) = \mathcal{O}(R)$. The error bound $\mathcal{O}(R)$ has been improved several times during the last century. In 1906, W. Sierpiński proved that $|\mathcal{D}(RB)| \leqslant cR^{2/3}$. The best result so far ($\leqslant cR^{0.627\cdots}$) has been recently obtained by J. Bourgain and N. Watt [6].

In 1916, G. Hardy proved that the exponent 1/2 is not large enough and conjectured that $|\mathcal{D}(RB)| \leqslant cR^{1/2+\varepsilon}$.

Earlier in 1915, G. Hardy proved the following result (previously conjectured by G. Voronoi):

$$R \sum_{k=1}^{+\infty} \frac{r(k)}{\sqrt{k}} J_1(2\pi\sqrt{k}R) = \frac{A(R^+) + A(R^-)}{2} - \pi R^2, \tag{9.2}$$

where $A(R^+)$ and $A(R^-)$ denote the right and left limits at R, respectively, of the discontinuous function $A(x)$, and

$$J_1(x) = \frac{x}{2} \int_{-1}^{1} (1 - t^2)^{1/2} e^{itx}\, dt$$

is a Bessel function, thereby giving an analytic expression for the discrepancy. See [7, 27].

The series in (9.2) is the spherical Fourier series (see (9.1))

$$\sum_{0 \ne m \in \mathbb{Z}^2} \widehat{\chi}_{RB}(m) e^{2\pi i m \cdot t} = \lim_{K \to +\infty} \sum_{0 < |m| \leqslant K} \widehat{\chi}_{RB}(m) e^{2\pi i m \cdot t}$$

of the discrepancy function $\mathbb{T}^2 \ni t \longmapsto \mathcal{D}(RB+t)$, evaluated at the origin. Indeed, for every $0 \neq \xi \in \mathbb{R}^2$, we have

$$\widehat{\chi}_B(\xi) = |\xi|^{-1} J_1(2\pi|\xi|)$$

(see, e. g., [40, p. 216]) and, therefore, after summing on the integers points m on all circles of radius $\sqrt{k}$, we obtain, at $t = 0$,

$$\begin{aligned} \sum_{m\neq 0} \widehat{\chi}_{RB}(m) &= R^2 \sum_{m\neq 0} \widehat{\chi}_B(Rm) = R \sum_{m\neq 0} |m|^{-1} J_1(2\pi R|m|) \\ &= R \sum_{j=1}^{+\infty} \frac{r(k)}{\sqrt{k}} J_1(2\pi R\sqrt{k})\,. \end{aligned} \tag{9.3}$$

The above series is not absolutely convergent and, in spite of its explicit expression, does not seem to help us in funding a sharp bound for the discrepancy, unless we apply a smoothing argument of E. Hlawka which turns the above series into an absolutely convergent one, and provides a new proof of Sierpiński's estimate (see, e. g., [40, p. 162] or the proof of Theorem 9.13 below).

More generally, when C is a convex planar body, the discrepancy function

$$\mathcal{D}_R(t) = -R^2|C| + \operatorname{card}((RC+t) \cap \mathbb{Z}^2)$$

is a periodic piecewise constant function (observe that $\mathcal{D}_R(t)$ may change value only when, moving t, we hit or we leave integer points). The above Hardy–Voronoi identity falls within the framework of pointwise convergence of Fourier series of piecewise smooth functions. A simple nice result in this field says that if the graph of $f(t)$ has the shape in the following figure, about a point t_0, then the spherical means of the above Fourier series converge, at the point t_0, to the number $b\beta/2\pi$

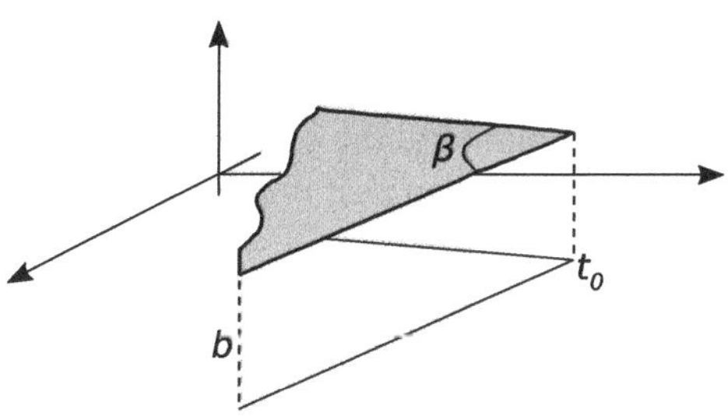

(see, e. g., [7]).

The situation may improve if we introduce an L^2 average (over translations) of the discrepancy function $\mathcal{D}_R(t)$.

9.2.1 Kendall's argument

D. Kendall [28] was the first one to write explicitly the Fourier series of the discrepancy function (and, therefore, to point out the identity (9.3)). Then he used the Parseval identity to prove that for, say, the unit disc B we have

$$\left\{\int_{\mathbb{T}^2} |\mathcal{D}(RB+t)|^2\,dt\right\}^{1/2} \leqslant cR^{1/2}.$$

Indeed it is known (by the asymptotics of Bessel functions or by Theorem 9.3 below) that

$$|\widehat{\chi}_B(\xi)| \leqslant c(1+|\xi|)^{-3/2}.$$

Therefore,

$$\int_{\mathbb{T}^2} |\mathcal{D}(RB+t)|^2\,dt = R^4 \sum_{m\neq 0} |\widehat{\chi}_B(Rm)|^2 \leqslant cR \sum_{m\neq 0} |m|^{-3} = cR. \tag{9.4}$$

Kendall's result for the disc can be extended to the case of an arbitrary planar convex body C as long as we introduce an average over rotations. A. Podkorytov (see [33], see also [40, p. 176], [13]) proved that for every planar convex body C we have

$$\int_0^{2\pi} |\widehat{\chi}_C(\rho\Theta)|^2\,d\theta \leqslant c\,\rho^{-3},$$

where $\Theta = (\cos\theta, \sin\theta)$ and $\rho \geqslant 2$. This and Kendall's argument yield

$$\left\{\int_{SO(2)} \int_{\mathbb{T}^2} |\mathcal{D}(\sigma(RC)+t)|^2\,dt\,d\sigma\right\}^{1/2} \leqslant cR^{1/2} \tag{9.5}$$

for every planar convex body C. Note that, within the family of convex planar bodies having piecewise smooth boundary, the upper bound (9.5) can be inverted (see [41, 15]) if and only if C is not a polygon that is symmetric and can be inscribed in a circle.

Kendall's L^2 result for the disc can be extended to L^p spaces provided $p < 4$ (see [26, 9]).

Theorem 9.2. *Let B be the unit disc. Then*

$$\left\{\int_{\mathbb{T}^2} |\mathcal{D}(RB+t)|^p\,dt\right\}^{1/p} \leqslant c \begin{cases} R^{1/2} & \text{if } 1 \leqslant p < 4, \\ R^{1/2}\log^{1/4}(R) & \text{if } p = 4, \\ R^{2/3(1-1/p)} & \text{if } p > 4. \end{cases} \tag{9.6}$$

The idea for the proof of (9.6) is that in Kendall's argument the series $\sum_{m\neq 0} |m|^{-3}$ converges "more than enough" and we have room for a few positive results when $p > 2$. Actually the upper bounds in Theorem 9.2 are known to be sharp in the range $1 \leqslant p < 4$. The case $p \geqslant 4$ uses Hlawka's smoothing argument and it does not seem to be sharp.

9.2.2 Integer points in large polygons

The study of integer points in polyhedra is another topic with several applications in different parts of mathematics (see, e. g., [2, 5, 37]).

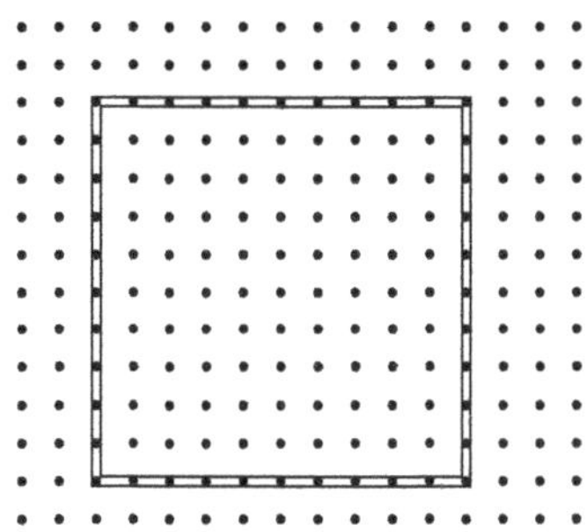

As a first (trivial) example, we consider a square having sides parallel to the axes. Then it is easy to check that the discrepancy is $\approx R$ for infinitely many large values of R. Indeed we see that the two squares of side $\approx R$ in the previous figure have essentially the same area, but one has $\approx R$ integer points more than the other.

A suitable rotation of the square may make the discrepancy for the square very small. H. Davenport (see [19]) has proved that if a square Q has slope (say) $\sqrt{2}$ then

$$\int_{\mathbb{T}^2} |\mathcal{D}(RQ + t)|^2 \, dt \leqslant c \, \log(R) .$$

A logarithmic estimate holds true also after averaging over rotations. In [10], it is proved that the discrepancy associated to a polygon P satisfies, for $R \geqslant 2$,

$$\int_{SO(2)} |\mathcal{D}(R\sigma(P))| \, d\sigma \leqslant c \, \log^2(R) . \tag{9.7}$$

Moreover, this estimate is almost sharp in the following sense. For a triangle $S \subset \mathbb{R}^2$, we have

$$\int_{\mathbb{T}^2} \int_{SO(2)} |\mathcal{D}(R\sigma(S) + t)| \, d\sigma dt \geqslant c \, \log(R) .$$

9.3 Pointwise estimates for $\widehat{\chi}_{C_\gamma}(\xi)$

To study the discrepancy for C_γ, we need careful estimates of the Fourier transform of the function $\chi_{C_\gamma}(t)$. We start with a general result; see [33] and also [16] for a result in higher dimension.

Theorem 9.3. *Let $C \subset \mathbb{R}^2$ be a strictly convex body with piecewise smooth boundary. We write $\Theta = (\cos\theta, \sin\theta)$ and, for $0 \leqslant \theta < 2\pi$ and small $\delta > 0$, let*

$$\lambda(\delta,\theta) = \left\{t \in C : \delta + t\cdot\Theta = \sup_{y\in C}(y\cdot\Theta)\right\}$$

be the chord perpendicular to Θ "at distance δ from the boundary" ∂C of C (see the following figure). Then there exist c_1 and c_2 independent of θ such that, for $\rho > c_1$, we have

$$|\widehat{\chi}_C(\rho\Theta)| \leqslant c_2\rho^{-1}\left(|\lambda(\rho^{-1},\theta)| + |\lambda(\rho^{-1},\theta+\pi)|\right),$$

where $|\lambda|$ denotes the length of the segment λ.

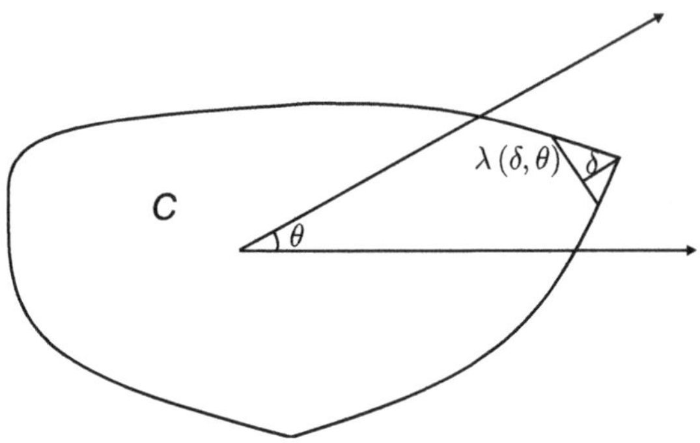

Remark 9.4. A slightly longer proof shows that the stricly convex assumption can be removed. See e.g. [15].

Proof. We may assume $\Theta = (1, 0)$, so that we consider

$$\widehat{\chi}_C(\xi,0) = \int_{-\infty}^{+\infty}\left(\int_{-\infty}^{+\infty}\chi_C(t_1,t_2)\,dt_2\right)e^{-2\pi i\xi t_1}\,dt_1 = \widehat{h}(\xi),$$

where $h(x)$ is the length of the segment given by the intersection of C with the line $t_1 = x$ (we can say that the 2-dimensional Fourier transform is a 1-dimensional Fourier transform of a Radon transform). Observe that the function $h(x)$ is continuous on $\mathbb{R}$ and strictly concave on its support, which we may assume to be the interval $[-1, 1]$. We may assume that $h(x)$ attains its maximum at some $\beta \geqslant 0$ (the other case being similar).

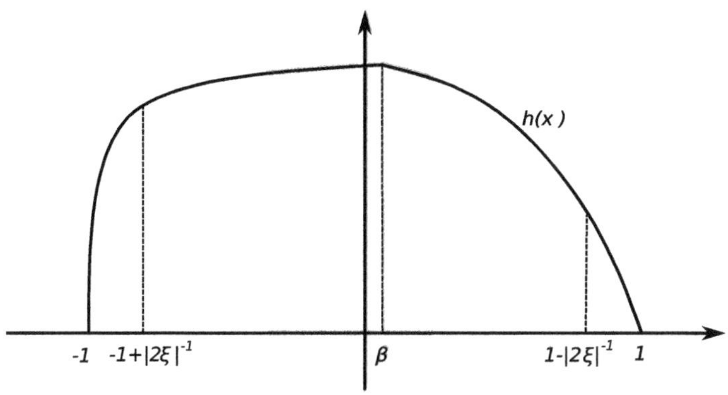

The strict convexity implies the continuity of $h(x)$, so that $h(-1) = h(1) = 0$. We may assume $\xi > 1$. Then integration by parts yields

$$\widehat{h}(\xi) = \int_{-1}^{1} h(x)e^{-2\pi i\xi x}\,dx = \frac{1}{2\pi i\xi}\int_{-1}^{1} h'(x)e^{-2\pi i\xi x}\,dx$$
$$= \frac{-1}{2\pi i\xi}\int_{-1+(2\xi)^{-1}}^{1+(2\xi)^{-1}} h'\left(x - \frac{1}{2\xi}\right)e^{-2\pi i\xi x}\,dx\,.$$

Hence

$$\begin{aligned}2(2\pi i\xi)\widehat{h}(\xi) &= \int_{-1}^{-1+(2\xi)^{-1}} h'(x)e^{-2\pi i\xi x}\,dx\\ &+ \int_{-1+(2\xi)^{-1}}^{1} \left(h'(x) - h'\left(x - \frac{1}{2\xi}\right)\right)e^{-2\pi i\xi x}\,dx\\ &+ \int_{1}^{1+(2\xi)^{-1}} h'\left(x - \frac{1}{2\xi}\right)e^{-2\pi i\xi x}\,dx\\ &= I_1 + I_2 + I_3\,,\end{aligned}$$

say. Since $h(x)$ is increasing on $-1 \leqslant x \leqslant 0$, we have

$$|I_1| \leqslant \int_{-1}^{-1+(2\xi)^{-1}} |h'(x)|\,dx = \int_{-1}^{-1+(2\xi)^{-1}} h'(x)\,dx = h\left(-1 + \frac{1}{2\xi}\right).$$

In the same way, since $h'(x)$ is decreasing, we have

$$\begin{aligned}|I_2| &\leqslant -\int_{-1+(2\xi)^{-1}}^{1}\left(h'(x)-h'\left(x-\frac{1}{2\xi}\right)\right)dx \\ &= h\left(-1+\frac{1}{2\xi}\right)+h\left(1-\frac{1}{2\xi}\right).\end{aligned}$$

In order to estimate I_3, we consider two cases. Let $\beta \in [0,1]$ be the point where $h(x)$ attains its maximum. If $\beta \leqslant 1-(2\xi)^{-1}$, we argue as we did for I_1. If $1-(2\xi)^{-1} \leqslant \beta < 1$, we have

$$\begin{aligned}|I_3| &= \left|\int_{1}^{1+(2\xi)^{-1}} h'\left(x-\frac{1}{2\xi}\right)e^{-2\pi i\xi x}\,dx\right| \\ &\leqslant \left|\int_{1}^{\beta+(2\xi)^{-1}} h'\left(x-\frac{1}{2\xi}\right)e^{-2\pi i\xi x}\,dx\right| + \left|\int_{\beta+(2\xi)^{-1}}^{1+(2\xi)^{-1}} h'\left(x-\frac{1}{2\xi}\right)e^{-2\pi i\xi x}\,dx\right| \\ &\leqslant \int_{1}^{\beta+(2\xi)^{-1}} \left|h'\left(x-\frac{1}{2\xi}\right)\right|dx + \int_{\beta+(2\xi)^{-1}}^{1+(2\xi)^{-1}} \left|h'\left(x-\frac{1}{2\xi}\right)\right|dx \\ &\leqslant \int_{1}^{\beta+(2\xi)^{-1}} h'\left(x-\frac{1}{2\xi}\right)dx - \int_{\beta+(2\xi)^{-1}}^{1+(2\xi)^{-1}} h'\left(x-\frac{1}{2\xi}\right)dx \\ &= 2h(\beta)-h\left(1-\frac{1}{2\xi}\right) \leqslant 4h(0)-h\left(1-\frac{1}{2\xi}\right) \leqslant 3h\left(1-\frac{1}{2\xi}\right),\end{aligned}$$

by the concavity of $h(x)$. This completes the proof. □

Corollary 9.5. *Let C be a planar convex body having smooth boundary with strictly positive curvature. Then, for every $|\xi| \geqslant 1$, we have*

$$|\widehat{\chi}_C(\xi)| \leqslant \kappa\,|\xi|^{-3/2} \tag{9.8}$$

(where κ depends on C).

Proof. We choose a point in ∂C, which we may assume to be the origin. We also assume that C is contained in the right half-plane and that C contains a ball of radius 1. For the sake of simplicity, we may also assume that ∂C is locally (i. e., for $|y| \leqslant c$) the graph of an even function $g(y)$ satisfying $g(0) = g'(0) = 0$ and $|g'(y)| \leqslant c$. Hence we consider only $0 \leqslant y \leqslant c$, so that $2g(y)$ is the inverse of the function $h(x)$ described at the beginning of the proof of Theorem 9.3. Moreover, our assumptions imply that (see again Theorem 9.3 for the notation)

$$h(\delta) = \frac{1}{2}|\lambda(\delta,-\pi)|$$

and $h(\delta)$ is strictly increasing for $0 \leqslant \delta \leqslant 1$. The curvature $K(y)$ at the point $(g(y), y) \in \partial C$ satisfies $c_1 \leqslant K(y) \leqslant c_2$ (where c_1 and c_2 depend on the convex body C). Since

$$g''(y) = \left(1 + [g'(y)]^2\right)^{3/2} K(y),$$

we have

$$g(y) = \int_0^y (y-t) g''(t)\, dt \approx \int_0^y (y-t)\, dt \approx y^2,$$

where $A \approx B$ means that A and B are positive and, for suitable constants c_1, c_2, we have $c_1 A \leqslant B \leqslant c_2 A$.

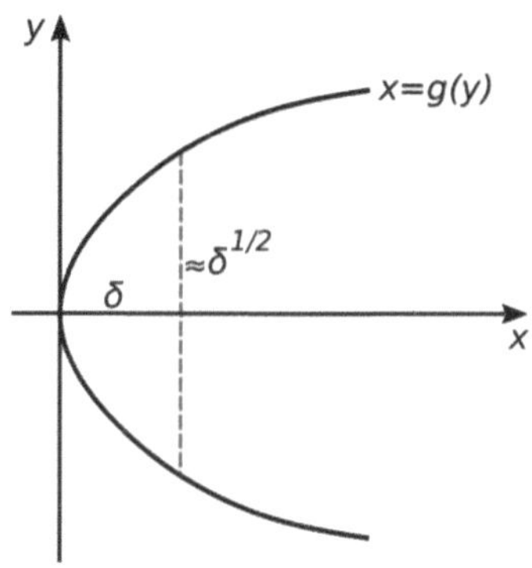

Then Theorem 9.3 yields

$$|\lambda(\delta, -\pi)| = h(\delta) \approx \delta^{1/2}$$

and, therefore, (9.8). □

Remark 9.6. The estimate (9.8) still holds under the less strict and more geometric assumption that C is a convex body that can roll unimpeded inside a disc. See [9]. Observe that no convex polygon or convex body with smooth boundary having a flat point of order > 2 can roll unimpeded inside a disc.

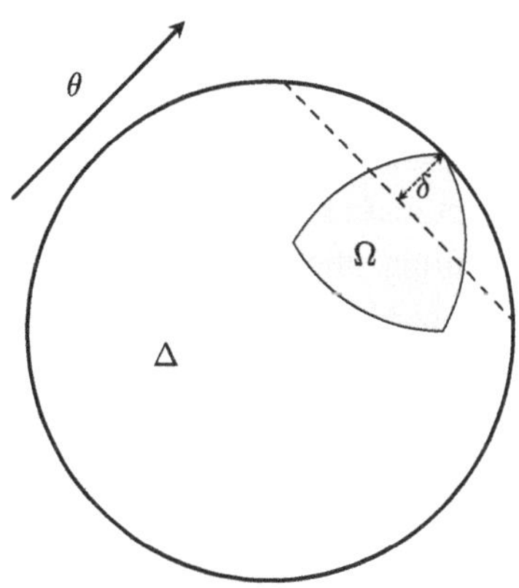

Remark 9.7. Assume that C is a convex planar body with piecewise smooth boundary. Without any assumptions on the curvature, the estimate (9.8) may fail. However, Theorem 9.3 shows that

$$|\widehat{\chi}_C(\xi)| \leqslant c|\xi|^{-1}, \tag{9.9}$$

whenever $|\xi| \geqslant 1$.

We can now state and prove some useful pointwise estimates for the decay of $\widehat{\chi}_{C_\gamma}(\xi)$; see [15].

Theorem 9.8. *Let $\gamma > 2$ and let C_γ be as in the Introduction, let $\psi \in (-\pi/2, \pi/2]$, let either $\theta = \psi - \pi/2$ or $\theta = \psi + \pi/2$, and let $\Theta = (\cos\theta, \sin\theta)$. Then, for $\rho \geqslant 2$ we have (for small $\varepsilon > 0$ and suitable positive constants c, c_1)*

$$|\widehat{\chi}_{C_\gamma}(\rho\Theta)| \leqslant \begin{cases} c\rho^{-1-1/\gamma} & \text{for } 0 \leqslant |\psi| \leqslant c_1\rho^{-1+1/\gamma}, \\ c\rho^{-3/2}|\psi|^{(2-\gamma)/(2\gamma-2)} & \text{for } c_1\rho^{-1+1/\gamma} \leqslant |\psi| \leqslant \varepsilon, \\ c\rho^{-3/2} & \text{for } \varepsilon \leqslant \psi \leqslant \pi. \end{cases} \tag{9.10}$$

This theorem is the basic result in this paper and we are going to write two proofs of it.

In the first proof, we use elementary arguments to estimate the chords introduced in Theorem 9.3.

In the second proof, we apply the divergence theorem to pass from $\widehat{\chi}_{C_\gamma}$ to $\widehat{\mu}_\gamma$, where μ_γ is the measure on $\mathbb{R}^2$, supported on ∂C_γ, where it coincides with the arc length measure. Then we use a partition of unity to split ∂C_γ into dyadic pieces.

First proof of Theorem 9.8. Assume $\psi > 0$ and let $x_0 > 0$ satisfy $\gamma x_0^{\gamma-1} = \tan\psi$, that is, (x_0, x_0^γ) is the point in ∂C_γ with outward unit normal Θ. Let $x_1 < x_2$ be the two solutions of the equation

$$|x|^\gamma = x_0^\gamma + (\rho\cos\psi)^{-1} + \gamma x_0^{\gamma-1}(x - x_0), \tag{9.11}$$

(of course $x_1 < x_0 < x_2$, while the assumption $\psi > 0$ yields $|x_1| < x_2$). We observe that $|\Lambda(\rho^{-1}, \psi)| \leqslant cx_2$ and we now estimate x_2. The inequality $0 \leqslant \psi \leqslant c_1\rho^{-1+1/\gamma}$ implies that the equation (9.11) has no solution when $x > \kappa\rho^{-1/\gamma}$ with a suitably large constant κ. Indeed since $x_0^{\gamma-1} \approx \psi$ we have $x_0 \approx \psi^{1/(\gamma-1)} \leqslant c\rho^{-1/\gamma}$ so that

$$\begin{aligned} &x^\gamma - x_0^\gamma - (\rho\cos\psi)^{-1} - \gamma x_0^{\gamma-1}(x - x_0) \\ &> x^\gamma - c\rho^{-1} - (\rho\cos\psi)^{-1} - c\rho^{-1+1/\gamma}x \\ &> \rho^{-1}((\rho^{1/\gamma}x)^\gamma - c - (\cos\psi)^{-1} - c\rho^{1/\gamma}x) > 0 \end{aligned}$$

provided that $\rho^{1/\gamma}x$ is large enough.

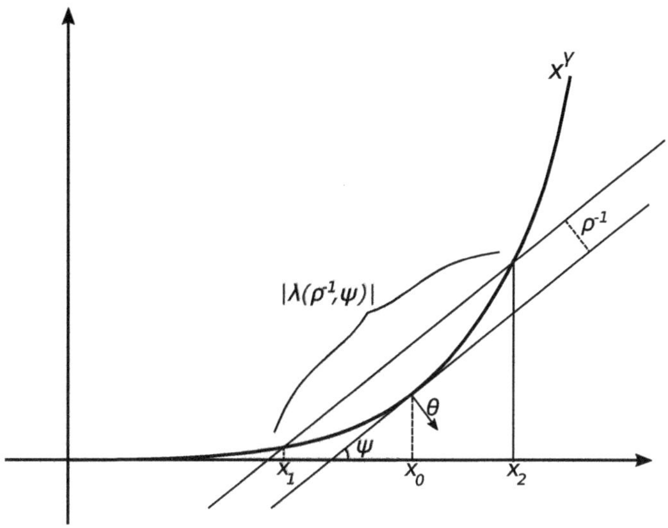

Let us now assume $c\rho^{-1+1/\gamma} \leqslant \psi \leqslant \varepsilon$ with a suitable constant c. Since $x_0^{\gamma-1} \approx \psi$, we have $x_1 > 0$. Indeed, let

$$y(x) = x_0^\gamma + (\rho\cos\psi)^{-1} + \gamma x_0^{\gamma-1}(x - x_0)\,.$$

Let $\psi \geqslant \tilde{c}\,\rho^{-1+1/\gamma}$ (we shall choose $\tilde{c}$ later). Then

$$y(0) = (1-\gamma)x_0^\gamma + (\rho\cos\psi)^{-1} \leqslant (1-\gamma)\,c\,\tilde{c}\,\rho^{-1} + (\rho\cos\psi)^{-1} < 0$$

if $\tilde{c}$ is large enough. Then we observe that, assuming $|x - x_0| \geqslant c'\rho^{-1/2}x_0^{1-\gamma/2}$ with a suitable choice of c', we obtain

$$\begin{aligned}
&x^\gamma - x_0^\gamma - (\rho\cos\psi)^{-1} - \gamma x_0^{\gamma-1}(x - x_0)\\
&= (x_0 + (x - x_0))^\gamma - x_0^\gamma - (\rho\cos\psi)^{-1} - \gamma x_0^{\gamma-1}(x - x_0)\\
&= x_0^\gamma\left(\left(1 + \frac{x - x_0}{x_0}\right)^\gamma - \gamma\frac{x - x_0}{x_0} - 1\right) - (\rho\cos\psi)^{-1}\\
&\geqslant x_0^\gamma\,\frac{\gamma}{2}\left(\frac{x - x_0}{x_0}\right)^2 - (\rho\cos\psi)^{-1} \geqslant \rho^{-1}\left(c\,c'\,\frac{\gamma}{2} - (\cos\psi)^{-1}\right) > 0\,,
\end{aligned}$$

since $\frac{x-x_0}{x_0} > -1$. Observe that we have used the inequality

$$(1+u)^\gamma - \gamma u - 1 \geqslant \gamma u^2/2\,.$$

Then $|x - x_0| \leqslant c\rho^{-1/2}x_0^{1-\gamma/2}$ for every $x_1 \leqslant x \leqslant x_2$. Therefore,

$$|\lambda(\rho,\psi)| \leqslant c\rho^{-1/2}x_0^{1-\gamma/2} \leqslant c\rho^{-1/2}\psi^{(2-\gamma)/(2\gamma-2)}\,.$$

Finally, let $\varepsilon \leqslant \psi \leqslant \pi$. Then Remark 9.7 yields $|\lambda(\rho,\psi)| \leqslant c\rho^{-1/2}$. Collecting the above results and applying Theorem 9.3, we complete the proof. □

For the second proof of Theorem 9.8, we need some well-known lemmas (see, e. g., [29, 31, 38]).

Lemma 9.9. *Let $f \in C^1([a,b])$ be a convex function such that*

$$f'(x) \geqslant \lambda > 0$$

and let φ be a smooth function $[a,b]$. Then

$$\left|\int_a^b e^{2\pi i f(x)}\varphi(x)dx\right| \leqslant \frac{1}{\lambda}\left[|\varphi(b)| + \int_a^b |\varphi'(x)|dx\right].$$

Proof. Integration by parts yields

$$\begin{aligned}\int_a^b e^{2\pi i f(x)}dx &= \int_a^b \frac{1}{2\pi i f'(x)}\frac{d}{dx}(e^{2\pi i f(x)})dx \\ &= \frac{1}{2\pi i f'(b)}e^{2\pi i f(b)} - \frac{1}{2\pi i f'(a)}e^{2\pi i f(a)} \\ &\quad - \int_a^b \frac{d}{dx}\left(\frac{1}{2\pi i f'(x)}\right)e^{2\pi i f(x)}dx.\end{aligned}$$

Hence

$$\begin{aligned}\left|\int_a^b e^{2\pi i f(x)}dx\right| &\leqslant \frac{1}{2\pi f'(b)} + \frac{1}{2\pi f'(a)} - \frac{1}{2\pi}\int_a^b \frac{d}{dx}\left(\frac{1}{f'(x)}\right)dx \\ &= \frac{1}{2\pi f'(b)} + \frac{1}{2\pi f'(a)} + \frac{1}{2\pi}\frac{1}{f'(b)} - \frac{1}{2\pi}\frac{1}{f'(a)} \\ &= \frac{1}{\pi f'(b)} \leqslant \frac{1}{\lambda}.\end{aligned}$$

Let now

$$G(x) = \int_a^x e^{2\pi i f(t)}dt.$$

Then

$$\int_a^b e^{2\pi i f(x)}\varphi(x)dx = [G(x)\varphi(x)]_a^b - \int_a^b G(x)\varphi'(x)dx$$

and, therefore,

$$\left|\int_a^b e^{2\pi i f(x)}\varphi(x)dx\right| \leqslant |G(b)\varphi(b)| + \int_a^b |G(x)||\varphi'(x)|dx$$
$$\leqslant \frac{1}{\lambda}|\varphi(b)| + \frac{1}{\lambda}\int_a^b |\varphi'(x)|dx\,. \qquad \square$$

Lemma 9.10. *Let $f \in C^2([a,b])$ satisfy $f''(x) \geqslant \kappa > 0$ and let φ be a smooth function on $[a,b]$. Then*

$$\left|\int_a^b e^{2\pi i f(x)}\varphi(x)dt\right| \leqslant \frac{4\|\varphi\|_\infty}{\sqrt{\kappa}} + \frac{2\|\varphi'\|_1}{\sqrt{\kappa}}\,.$$

Proof. Let

$$I_1 = \{x \in [a,b] : |f'(x)| \leqslant \sqrt{\kappa}\}$$

and

$$I_2 = \{x \in [a,b] : |f'(x)| > \sqrt{\kappa}\}\,.$$

The convexity of $f(x)$ implies that I_1 is either an interval or the empty set. I_2 is the union of at most two intervals. Let $I_1 = [\alpha, \beta]$. Then the mean value theorem yields

$$(\beta - \alpha)\kappa \leqslant f'(\beta) - f'(\alpha) \leqslant 2\sqrt{\kappa}\,.$$

Hence

$$\int_\alpha^\beta e^{2\pi i f(x)}\varphi(x)dt \leqslant (\beta - \alpha)\|\varphi\|_\infty \leqslant \frac{2\|\varphi\|_\infty}{\sqrt{\kappa}}\,.$$

To end the proof, we observe that the previous lemma yields

$$\left|\int_{I_2} e^{2\pi i f(x)}\varphi(x)dt\right| \leqslant \frac{2}{\sqrt{\kappa}}\|\varphi\|_\infty + \frac{2}{\sqrt{\kappa}}\|\varphi'\|_1\,. \qquad \square$$

Lemma 9.11. *Let $\epsilon \in C^1(\mathbb{R})$ such that $\epsilon(x) \equiv 0$ for $|x| < \frac{1}{2}$ and $|x| \geqslant 1$. Then*

$$\left|\int_{-\infty}^{+\infty} e^{-2\pi i(au+b|u|^\gamma)}\epsilon(u)\,du\right| \leqslant \frac{c}{(1+|(a,b)|)^{1/2}}$$

(where c is independent of a, b, but depends on $\|\epsilon\|_\infty$ and $\|\epsilon'\|_\infty$).

Proof. It is enough to consider the integral on $(0, +\infty)$. Let $f(u) = au + bu^\gamma$ and let

$$J(a,b) = \int_0^{+\infty} e^{-2\pi i f(u)} \epsilon(u)\, du\,.$$

If $|(a,b)| \leqslant 1$, we have the trivial estimate

$$|J(a,b)| \leqslant \int_{1/2}^{1} |\epsilon(u)|\, du \leqslant \frac{1}{2}\|\epsilon\|_\infty\,.$$

Assume $|(a,b)| > 1$ and $\gamma|b| \leqslant \frac{1}{2}|a|$. Then

$$|f'(u)| = |a + b\gamma u^{\gamma-1}| \geqslant |a| - \gamma|b| \geqslant \frac{1}{2}|a|$$

so that, by Lemma 9.9,

$$|J(a,b)| \leqslant 2\frac{\|\epsilon\|_\infty + \|\epsilon'\|_\infty}{|a|} \leqslant \frac{c_2}{|(a,b)|} \leqslant \frac{c_2}{|(a,b)|^{1/2}}\,.$$

Finally, if $\gamma|b| > \frac{1}{2}|a|$ then

$$|f''(u)| = |b\gamma(\gamma-1)u^{\gamma-2}| \geqslant c_3|b|$$

so that by Lemma 9.10

$$|J(a,b)| \leqslant c_4\frac{\|\epsilon\|_\infty + \|\epsilon'\|_\infty}{|b|^{1/2}} \leqslant \frac{c_5}{|(a,b)|^{1/2}}\,. \qquad \square$$

Second proof of Theorem 9.8. For $t \in \mathbb{R}^2$, let $\eta(t)$ be a smooth function supported in a disc U centered at the origin and such that $\eta(t) = 1$ for each $t \in \frac{1}{2}U$. Observe that for U small enough

$$\partial C_\gamma \cap U = \{(t_1, t_2) \in \mathbb{R}^2 : t_2 = |t_1|^\gamma\} \cap U\,.$$

For $t, \xi \in \mathbb{R}^2$, let

$$\omega(t) = \frac{e^{-2\pi i t\cdot\xi}}{-2\pi i|\xi|^2}\xi\,,$$

so that

$$\operatorname{div}\omega(t) = \frac{\partial\omega_1}{\partial t_1} + \frac{\partial\omega_2}{\partial t_2} = e^{-2\pi i t\cdot\xi}\,.$$

Let us write $\xi = \rho\Theta$ in polar coordinates and for every point $t \in \partial C_\gamma$ let $\nu(t)$ be the outward unit normal. Then application of the divergence theorem yields

$$\begin{aligned}
\widehat{\chi}_{C_\gamma}(\xi) &= \int_{C_\gamma} e^{-2\pi i \xi \cdot t}\, dt \qquad (9.12)\\
&= \int_{C_\gamma} \operatorname{div}\omega(t)\, dt\\
&= \frac{-1}{2\pi i \rho} \int_{\partial C_\gamma} e^{-2\pi i \rho \Theta \cdot t}\Theta \cdot \nu(t)\, d\mu_\gamma(t)\\
&= \frac{-1}{2\pi i \rho} \int_{\partial C_\gamma} e^{-2\pi i \rho \Theta \cdot t}\Theta \cdot \nu(t)\eta(t)\, d\mu_\gamma(t)\\
&\quad - \frac{1}{2\pi i \rho} \int_{\partial C_\gamma} e^{-2\pi i \rho \Theta \cdot t}\Theta \cdot \nu(t)(1-\eta(t))\, d\mu_\gamma(t)\\
&=: \frac{-1}{2\pi i \rho} H_1(\xi) - \frac{1}{2\pi i \rho} H_2(\xi)\,,
\end{aligned}$$

where μ_γ is the arc length measure on ∂C_γ.

We first estimate $H_2(\xi)$. Let

$$s \mapsto \Gamma(s)$$

be the parametrization of ∂C_γ by its arc length. Then

$$H_2(\xi) = \int_a^b e^{-2\pi i \rho \Theta \cdot \Gamma(s)}\Theta \cdot \nu(\Gamma(s))(1-\eta(\Gamma(s)))ds\,.$$

Since $\Gamma'(s)$ and $\Gamma''(s)$ are orthogonal vectors with norms $\geqslant c_1 > 0$, then either

$$\left|\frac{d}{ds}(\rho\Theta \cdot \Gamma(s))\right| \geqslant c_2\rho$$

or

$$\left|\frac{d^2}{ds^2}(\rho\Theta \cdot \Gamma(s))\right| \geqslant c_2\rho\,.$$

Therefore, we can split the integral in $H_2(\xi)$ as the sum of a finite number of integrals that satisfy either the assumption of Lemma 9.9 or Lemma 9.10. Hence

$$|H_2(\xi)| \leqslant c_2\rho^{-1/2}.$$

Let us consider the integral $H_1(\xi)$. By our assumption on the support of $\eta(t)$, we can write

$$H_1(\xi) = \int_{\mathbb{R}} e^{-2\pi i(\xi_1 x + \xi_2 |x|^\gamma)}\delta(x)\tau(x)\, dx\,,$$

where $\tau(x)$ is compactly supported and takes value 1 in a neighborhood of $0 \in \mathbb{R}$ (say $\tau(x) = 1$ when $|x| \leqslant 1/2$ and $\tau(x) = 0$ when $|x| > 1$) and $\delta(x)$ is a C^2 function (recall that $\gamma > 2$).

Assume first $|\xi_1| > |\xi_2|$. Since

$$\left|\frac{d}{dx}(\xi_1 x + \xi_2|x|^\gamma)\right| = \left|\xi_1 + \gamma\xi_2|x|^{\gamma-1}\mathrm{sign}(x)\right| \approx |\xi_1| \approx \rho,$$

by Lemma 9.9 we have

$$|H_1(\xi)| \leqslant \frac{c}{\rho}.$$

Let now $|\xi_1| < |\xi_2|$ and let $\epsilon(x) = \tau(x) - \tau(2x)$. Observe that $\epsilon(x)$ is positive and supported in the interval $(-1, -1/4) \cup (1/4, 1)$. The key step in the proof is a dyadic decomposition with the change of variables

$$\int_{\mathbb{R}} e^{-2\pi i(\xi_1 x + \xi_2|x|^\gamma)}\delta(x)\tau(x)\,dx = \sum_{j=1}^{+\infty}\int_{\mathbb{R}} e^{-2\pi i(\xi_1 x + \xi_2|x|^\gamma)}\delta(x)\epsilon(2^j x)\,dx$$
$$= \sum_{j=1}^{+\infty} 2^{-j}\int_{\mathbb{R}} e^{-2\pi i((\xi_1 2^{-j})u + (\xi_2 2^{-\gamma j})|u|^\gamma)}\delta(2^{-j}u)\epsilon(u)\,du\,.$$

By Lemma 9.11, we have

$$\left|\int_{\mathbb{R}} e^{-2\pi i((\xi_1 2^{-j})u + (\xi_2 2^{-\gamma j})u^\gamma)}\delta(2^{-j}u)\epsilon(u)\,du\right| \leqslant c\left[1 + \left|(\xi_1 2^{-j}, \xi_2 2^{-\gamma j})\right|\right]^{-1/2}.$$

Hence

$$\left|\int_{\mathbb{R}} e^{-2\pi i(\xi_1 x + \xi_2 x^\gamma)}\tau(x)\,dx\right| \leqslant c\sum_{j=1}^{+\infty} 2^{-j}\left[1 + \left|(\xi_1 2^{-j}, \xi_2 2^{-\gamma j})\right|\right]^{-1/2}.$$

We recall that we are assuming $|\xi_2| > |\xi_1|$, that is, we are considering only the directions close to be perpendicular to the part of ∂C_γ about the origin. Then

$$\begin{aligned}
&\sum_{j=1}^{+\infty} 2^{-j}\left[1 + \left|(\xi_1 2^{-j}, \xi_2 2^{-\gamma j})\right|\right]^{-1/2} \qquad (9.13)\\
&\leqslant \sum_{2^j \leqslant (|\xi_2|/|\xi_1|)^{1/(\gamma-1)}} 2^{-j}\left|(\xi_1 2^{-j}, \xi_2 2^{-\gamma j})\right|^{-1/2}\\
&+ \sum_{2^j > (|\xi_2|/|\xi_1|)^{1/(\gamma-1)}} 2^{-j}\left|(\xi_1 2^{-j}, \xi_2 2^{-\gamma j})\right|^{-1/2}\\
&\leqslant c \sum_{2^j \leqslant (|\xi_2|/|\xi_1|)^{1/(\gamma-1)}} 2^{j(\gamma/2-1)}|\xi_2|^{-1/2} + c \sum_{2^j > (|\xi_2|/|\xi_1|)^{1/(\gamma-1)}} 2^{-j/2}|\xi_1|^{-1/2}
\end{aligned}$$

$$
\begin{aligned}
&\leqslant c|\xi_2|^{-1/2}\left(\frac{|\xi_2|}{|\xi_1|}\right)^{(\gamma-2)/(2\gamma-2)} + |\xi_1|^{-1/2}\left(\frac{|\xi_1|}{|\xi_2|}\right)^{1/(2\gamma-2)} \\
&\leqslant c|\xi_2|^{-1/2}\left(\frac{|\xi_2|}{|\xi_1|}\right)^{(\gamma-2)/(2\gamma-2)} \\
&\approx \rho^{-1/2}\psi^{-(\gamma-2)/(2\gamma-2)},
\end{aligned}
$$

where $\psi = \pi/2 + \arctan(|\xi_2|/|\xi_1|)$. Hence

$$
|\widehat{\chi}_{C_\gamma}(\xi)| \leqslant c\rho^{-3/2}\psi^{(2-\gamma)/(2\gamma-2)}. \tag{9.14}
$$

Finally, we prove the inequality

$$
|\widehat{\chi}_{C_\gamma}(\rho\Theta)| \leqslant c\rho^{-1-1/\gamma}.
$$

Observe that (9.14) yields the above upper bound when $\psi \geqslant c\rho^{-1+1/\gamma}$. We still have to prove that the same bound is correct when $0 \leqslant \psi \leqslant c\rho^{-1+1/\gamma}$, that is, $|\xi_1|/|\xi_2| \leqslant c\rho^{-1+1/\gamma}$. Finally, we deal with the first inequality. We can assume $|\xi_1| < c|\xi_2|$. By the previous computation, we have to bound

$$
\begin{aligned}
&\sum_{j=1}^{+\infty} 2^{-j}\left[1 + |(\xi_1 2^{-j}, \xi_2 2^{-\gamma j})|\right]^{-1/2} \\
&\leqslant c\sum_{j=1}^{+\infty} 2^{-j}(1 + |\xi_2| 2^{-\gamma j})^{-1/2} \\
&\leqslant c \sum_{2^j \leqslant |\xi_2|^{1/\gamma}} 2^{-j}(|\xi_2| 2^{-\gamma j})^{-1/2} + c \sum_{2^j > |\xi_2|^{1/\gamma}} 2^{-j} \leqslant c|\xi_2|^{-1/\gamma},
\end{aligned}
$$

which yields the first inequality in (9.10). □

9.4 Average decay of $\widehat{\chi}_{C_\gamma}(\xi)$

We shall consider both L^p average discrepancies when C_γ is translated, and L^p average discrepancies when C_γ is translated and rotated. For the latter problem, we shall need estimates for the *L^p (spherical) average decay* of $\widehat{\chi}_{C_\gamma}(\xi)$, that is,

$$
\left\{\int_0^{2\pi} |\widehat{\chi}_{C_\gamma}(\rho\Theta)|^p \, d\theta\right\}^{1/p}
$$

(where $\Theta = (\cos\theta, \sin\theta)$ and $\rho \geqslant 2$). To illustrate the relevance of these averages, we point out that the above estimate (9.7) for the discrepancy of a polygon P is a consequence of the estimate

$$
\int_0^{2\pi} |\widehat{\chi}_P(\rho\Theta)| \, d\theta \leqslant c\frac{\log^2(\rho)}{\rho^2},
$$

which in turn follows from Theorem 9.3. We refer the interested reader to [10, 11], and [39] for more details and applications.

In the next theorem (see [15]), we obtain estimates for the L^p *(spherical) average decay* of $\widehat{\chi}_{C_\gamma}(\xi)$.

Theorem 9.12. *We have*

$$\left\{\int_0^{2\pi} |\widehat{\chi}_{C_\gamma}(\rho\Theta)|^p \, d\theta\right\}^{1/p} \leqslant \begin{cases} c_p \rho^{-3/2} & \textit{for } p < \frac{2\gamma-2}{\gamma-2}, \\ c\rho^{-3/2} \log^{(\gamma-2)(2\gamma-2)}(\rho) & \textit{for } p = \frac{2\gamma-2}{\gamma-2}, \\ c_p \rho^{-1-\frac{1}{p}-\frac{1}{\gamma}+\frac{1}{\gamma p}} & \textit{for } p > \frac{2\gamma-2}{\gamma-2}. \end{cases}$$

Proof. It is enough to integrate between $-\pi/2$ and $\pi/2$. The estimates in Theorem 9.8 yield

$$\begin{aligned} &\left\{\int_{-\pi/2}^{\pi/2} |\widehat{\chi}_{C_\gamma}(\rho\Theta)|^p \, d\theta\right\}^{1/p} \\ &\leqslant \left\{\int_{-\pi/2}^{-\pi/2+c\rho^{-1+1/\gamma}} |\widehat{\chi}_{C_\gamma}(\rho\Theta)|^p \, d\theta\right\}^{1/p} \\ &+ \left\{\int_{-\pi/2+c\rho^{-1+1/\gamma}}^{-\pi/2+\varepsilon} |\widehat{\chi}_{C_\gamma}(\rho\Theta)|^p \, d\theta\right\}^{1/p} \\ &+ \left\{\int_{-\pi/2+\varepsilon}^{\pi/2} |\widehat{\chi}_{C_\gamma}(\rho\Theta)|^p \, d\theta\right\}^{1/p} \\ &\leqslant c\left\{\int_0^{c\rho^{-1+1/\gamma}} |\rho^{-1-1/\gamma}|^p \, d\theta\right\}^{1/p} \\ &+ c\left\{\int_{c\rho^{-1+1/\gamma}}^{\varepsilon} |\rho^{-3/2}\psi^{(2-\gamma)/(2\gamma-2)}|^p \, d\psi\right\}^{1/p} \\ &+ c\left\{\int_{-\pi/2+\varepsilon}^{\pi/2} |\rho^{-3/2}|^p \, d\theta\right\}^{1/p} \\ &= A + B + C, \end{aligned}$$

say. Finally, we have

$$A \leqslant c\rho^{-1-\frac{1}{p}-\frac{1}{\gamma}+\frac{1}{\gamma p}},$$

$$B \leqslant \begin{cases} \rho^{-3/2} & \text{for } p < \frac{2\gamma-2}{\gamma-2}, \\ \rho^{-3/2}\log^{(\gamma-2)/(2\gamma-2)}(\rho) & \text{for } p = \frac{2\gamma-2}{\gamma-2}, \\ \rho^{-1-\frac{1}{p}-\frac{1}{\gamma}+\frac{1}{\gamma p}} & \text{for } p > \frac{2\gamma-2}{\gamma-2}, \end{cases}$$

$$C \leqslant c\rho^{-3/2}. \qquad \square$$

It can be proved that the above estimates are sharp (see [15]).

9.5 Integer points in C_γ

We consider two different averages of the discrepancy function.

9.5.1 Discrepancy over translations

We now prove a few L^p estimates for the discrepancy function

$$\begin{aligned} \mathcal{D}_R(t) = \mathcal{D}(RC_\gamma + t) &= -R^d|C_\gamma| + \operatorname{card}((RC_\gamma + t) \cap \mathbb{Z}^2) \\ &= -R^2|C_\gamma| + \sum_{n\in\mathbb{Z}^2} \chi_{RC_\gamma}(n-t), \end{aligned}$$

which we recall to have Fourier series

$$\sum_{0\neq m\in\mathbb{Z}^2} \widehat{\chi}_{RC_\gamma}(m)e^{2\pi i m\cdot t}.$$

We consider the L^p norms

$$\|\mathcal{D}_R\|_p = \begin{cases} \{\int_{\mathbb{T}^2} |\mathcal{D}_R(RC_\gamma + t)|^p dt\}^{1/p} & \text{for } p < \infty, \\ \sup_{t\in\mathbb{T}^2} |\mathcal{D}_R(RC_\gamma + t)| & \text{for } p = \infty. \end{cases}$$

Our estimates are the following (see [8]).

Theorem 9.13. *For $2 < \gamma \leqslant 3$, we have*

$$\|\mathcal{D}_R\|_p \leqslant \begin{cases} cR^{1-1/\gamma} & \text{for } 1 \leqslant p \leqslant 4/(3-\gamma), \\ cR^{\frac{2}{3}(1-\frac{2}{\gamma p})} & \text{for } p > 4/(3-\gamma). \end{cases} \tag{9.15}$$

Theorem 9.14. *For $\gamma > 3$ and every $p \geqslant 1$, we have*

$$\|\mathcal{D}_R\|_p \leqslant cR^{1-1/\gamma}.$$

Remark 9.15. The proof of Theorem 9.13 follows Hlawka's smoothing argument that is usually used when the curvature of the boundary is strictly positive (i. e., $\gamma = 2$).

Anyway, it takes no extra effort to apply it to the case $\gamma \leqslant 3$. Roughly speaking here, we have to consider two cases. First, the integer points close to the origin, where vertical translations yield discrepancy $\leqslant cR^{1-1/\gamma}$. Second, the integer points away from the origin, where the smoothing argument yields discrepancy $\leqslant cR^{2/3}$. Therefore, $\gamma \leqslant 3$ works as well. The bound $cR^{2/3}$ for $\gamma \leqslant 3$ has been first obtained in [18].

We need the following lemma (see [11]).

Lemma 9.16. *Let $\varphi(t)$ be a smooth nonnegative function supported in a small neighborhood of the origin and such that $\int_{\mathbb{R}^2} \varphi = 1$. Then for every small $\varepsilon > 0$ and $R > 1$, we have*

$$\varepsilon^{-2}\varphi(\varepsilon^{-1}\cdot) * \chi_{(R-\varepsilon)C_\gamma}(t) \leqslant \chi_{RC_\gamma}(t) \leqslant \varepsilon^{-2}\varphi(\varepsilon^{-1}\cdot) * \chi_{(R+\varepsilon)C_\gamma}(t),$$

where $*$ denotes the convolution

$$(f * g)(t) = \int f(t-s)g(s)\,ds.$$

In particular,

$$\begin{aligned} &|C_\gamma|((R-\varepsilon)^2 - R^2) + D_{\varepsilon,R-\varepsilon}(t) \\ &\leqslant \mathcal{D}_R(t) \leqslant |C_\gamma|((R+\varepsilon)^2 - R^2) + D_{\varepsilon,R+\varepsilon}(t), \end{aligned} \tag{9.16}$$

where

$$D_{\varepsilon,R}(t) = R^2 \sum_{0 \neq m \in \mathbb{Z}^2} \widehat{\varphi}(\varepsilon m)\widehat{\chi}_{C_\gamma}(Rm)e^{2\pi i m\cdot t}.$$

Proof. First, we observe that the convexity of C_γ yields

$$\frac{R}{R+\varepsilon}C_\gamma + \frac{\varepsilon}{R+\varepsilon}C_\gamma \subseteq C_\gamma$$

so that

$$(R+\varepsilon)C_\gamma \supseteq RC_\gamma + \varepsilon C_\gamma \supseteq RC_\gamma + B(0,\varepsilon) \tag{9.17}$$

and, therefore, $B(q,\varepsilon) \subseteq (R+\varepsilon)C_\gamma$ for every $q \in \partial(RC_\gamma)$. Applying (9.17) to Interior(C_γ) with R in place of $R+\varepsilon$, we obtain

$$\text{Interior}(RC_\gamma) \supseteq \text{Interior}(R-\varepsilon)C_\gamma + B(0,\varepsilon).$$

Assume there exists $y \in B(q,\varepsilon) \cap \text{Interior}(R-\varepsilon)C_\gamma$. It follows that

$$q \in \text{Interior}(R-\varepsilon)C_\gamma + B(0,\varepsilon) \subseteq \text{Interior}(RC_\gamma)$$

so that $q \notin \partial(RC_\gamma)$. Hence for large R and small ε we have

$$B(q,\varepsilon) \subseteq (R+\varepsilon)C_\gamma \setminus \text{Interior}(R-\varepsilon)C_\gamma$$

for every $q \in \partial(RC_\gamma)$. Then

$$\varepsilon^{-2}\varphi(\varepsilon^{-1}\cdot) * \chi_{(R-\varepsilon)B}(t) \leqslant \chi_{RB}(t) \leqslant \varepsilon^{-2}\varphi(\varepsilon^{-1}\cdot) * \chi_{(R+\varepsilon)B}(t)$$

and, therefore, (9.16). □

Proof of Theorem 9.13. By Lemma 9.16, we have

$$\begin{aligned} \|\mathcal{D}_R\|_p &\leqslant |C_\gamma| \max_{\pm}\left|(R\pm\varepsilon)^2 - R^2\right| + \max_{\pm} \|D_{\varepsilon,R\pm\varepsilon}\|_p \\ &\leqslant cR\varepsilon + \max_{\pm}\left\|D_{\varepsilon,R\pm\varepsilon}(t)\right\|_p . \end{aligned}$$

We write $m = (m_1, m_2)$ and we choose $\varphi(t)$ as in Lemma 9.16, so that, in particular,

$$\left|\widehat{\varphi}(\xi)\right| \leqslant \frac{c_N}{1+|\xi|^N}$$

for every N. Then

$$\begin{aligned} \|D_{\varepsilon,R}\|_p &= \left\{\iint_{\mathbb{T}^2} \left|R^2 \sum_{0\neq m\in\mathbb{Z}^2} \widehat{\varphi}(\varepsilon m)\widehat{\chi}_{C_\gamma}(Rm)e^{2\pi i m\cdot t}\right|^p dt\right\}^{1/p} \\ &\leqslant \left\{\iint_{\mathbb{T}^2} \left|R^2 \sum_{|\arctan(m_1/m_2)|\leqslant c_1|m|^{-1+1/\gamma}} \widehat{\varphi}(\varepsilon m)\widehat{\chi}_{C_\gamma}(Rm)e^{2\pi i m\cdot t}\right|^p dt\right\}^{1/p} \\ &\quad + \left\{\iint_{\mathbb{T}^2} \left|R^2 \sum_{c_1|m|^{-1+1/\gamma}\leqslant|\arctan(m_1/m_2)|<c_2} \widehat{\varphi}(\varepsilon m)\widehat{\chi}_{C_\gamma}(Rm)e^{2\pi i m\cdot t}\right|^p dt\right\}^{1/p} \\ &\quad + \left\{\iint_{\mathbb{T}^2} \left|R^2 \sum_{c_2\leqslant|\arctan(m_1/m_2)|} \widehat{\varphi}(\varepsilon m)\widehat{\chi}_{C_\gamma}(Rm)e^{2\pi i m\cdot t}\right|^p dt\right\}^{1/p} \\ &= I + II + III , \end{aligned}$$

say. In order to prove the first inequality in (9.15), it is enough to consider the case $p = 4/(3-\gamma)$ (observe that for $\gamma = 3$ we have $p = \infty$). We are going to deduce the estimates of I, II, III from Theorem 9.8. We have

$$I \leqslant cR^2 \sum_{|\arctan(m_1/m_2)|\leqslant c_1|m|^{-1+1/\gamma}} \frac{1}{1+|\varepsilon m|}|Rm|^{-1-1/\gamma} .$$

A modification of the above constant c_1 allows us to replace the sum

$$\sum_{|\arctan(m_1/m_2)|\leqslant c_1|m|^{-1+1/\gamma}}$$

with an integral, but for a finite number of unit squares close to the origin and centered on the vertical axis. We write

$$I \leqslant cR^{1-1/\gamma} + cR^{1-1/\gamma} \int_1^{+\infty} \int_0^{c(R\rho)^{-1+1/\gamma}} d\psi \frac{1}{1+\varepsilon\rho} \rho^{-1-1/\gamma} \rho \, d\rho$$

$$\leqslant cR^{1-1/\gamma} + c \int_1^{+\infty} \rho^{-1} \frac{1}{1+\varepsilon\rho} \, d\rho = cR^{1-1/\gamma} + c\log(1/\varepsilon) .$$

By the Hausdorff–Young inequality, we have for $\frac{1}{p} + \frac{1}{q} = 1$,

$$II \leqslant \left\{ \int_{\mathbb{T}^2} \left| R^2 \sum_{c_1|m|^{-1+1/\gamma} \leqslant |\arctan(m_1/m_2)| < c_2} \widehat{\varphi}(\varepsilon m) \widehat{\chi}_{C_\gamma}(Rm) e^{2\pi i m \cdot t} \right|^p dt \right\}^{1/p}$$

$$\leqslant cR^2 \left\{ \sum_{c_1|m|^{-1+1/\gamma} \leqslant |\arctan(m_1/m_2)| < c_2} \left| \widehat{\varphi}(\varepsilon m) \widehat{\chi}_{C_\gamma}(Rm) \right|^q \right\}^{1/q}$$

$$\leqslant cR^2 \left\{ \sum_{c_1|m|^{-1+1/\gamma} \leqslant |\arctan(m_1/m_2)| < c_2} \left| \frac{1}{1+|\varepsilon m|} |Rm|^{-3/2} \left| \frac{m_1}{m_2} \right|^{(2-\gamma)/(2\gamma-2)} \right|^q \right\}^{1/q}$$

$$\leqslant cR^{1/2} \left\{ \int_1^{+\infty} \int_{c\rho^{-1+1/\gamma}}^{c_2} \frac{1}{(1+\varepsilon\rho)^q} \rho^{-3q/2} \psi^{q(2-\gamma)/(2\gamma-2)} \, d\psi \rho d\rho \right\}^{1/q}$$

$$= cR^{1/2} \left\{ \int_1^{+\infty} \frac{1}{(1+\varepsilon\rho)^q} \rho^{1-3q/2} \int_{c\rho^{-1+1/\gamma}}^{c_2} \psi^{q(2-\gamma)/(2\gamma-2)} \, d\psi d\rho \right\}^{1/q}$$

$$= cR^{1/2} \left\{ \int_1^{+\infty} \frac{1}{(1+\varepsilon\rho)^q} \rho^{1-3q/2} \, d\rho \right\}^{1/q}$$

$$= cR^{1/2} \left\{ \int_\varepsilon^{+\infty} \frac{1}{(1+s)^q} \left(\frac{s}{\varepsilon} \right)^{1-3q/2} \frac{1}{\varepsilon} \, ds \right\}^{1/q}$$

$$= cR^{1/2} \varepsilon^{3/2-2/q}$$

$$= cR^{1/2} \varepsilon^{(2-\gamma)/2} ,$$

because $q = 4/(\gamma+1) < 4/3$.

$$III \leqslant cR^2 \left\{ \sum_{c_2 \leqslant |\arctan(m_1/m_2)|} \left(\frac{1}{1+\varepsilon|m|} |Rm|^{-3/2} \right)^q \right\}^{1/q}$$

$$\leqslant cR^{1/2} \int_1^{+\infty} \frac{1}{(1+\varepsilon\rho)^q} \rho^{1-3q/2} \, d\rho = cR^{1/2} \varepsilon^{3/2-2/q} = cR^{1/2} \varepsilon^{(2-\gamma)/2} .$$

Then

$$\|\mathcal{D}_R\|_p \leqslant cR\varepsilon + cR^{1-1/\gamma} + c\log(1/\varepsilon) + cR^{1/2}\varepsilon^{(2-\gamma)/2}.$$

By choosing $\varepsilon = R^{-1/\gamma}$, we obtain

$$\|\mathcal{D}_R\|_p \leqslant cR^{1-1/\gamma}.$$

A similar computation shows that

$$\|\mathcal{D}_R\|_\infty = \sup_t |\mathcal{D}(RC_\gamma + t)| \leqslant cR^{2/3}.$$

To end the proof we need to show that $\|\mathcal{D}_R\|_p \leqslant cR^{(2\gamma p-4)/(3\gamma p)}$ for $4/(3-\gamma) < p < \infty$. Interpolation between the previous two cases yields

$$\begin{aligned}\|\mathcal{D}_R\|_p &= \left\{\int_{\mathbb{T}^2} |\mathcal{D}_R(t)|^p \, dt\right\}^{1/p} \\ &\leqslant \left\{\int_{\mathbb{T}^2} \|\mathcal{D}_R\|_\infty^{p-4/(3-\gamma)} |\mathcal{D}_R(t)|^{4/(3-\gamma)} \, dt\right\}^{1/p} \\ &= \|\mathcal{D}_R\|_\infty^{1-4/(3p-\gamma p)} \|\mathcal{D}_R\|_{4/(3-\gamma)}^{4/(3p-\gamma p)} \leqslant cR^{(2\gamma p-4)/(3\gamma p)}.\end{aligned}$$

□

Proof of Theorem 9.14. It is enough to consider the case $p = +\infty$. Arguing as in the previous proof, we write $\|\mathcal{D}_R\|_\infty \leqslant I + II + III$ and we obtain

$$I \leqslant cR^{1-1/\gamma}, \quad II \leqslant R^{1/2}\varepsilon^{-1/2}, \quad III \leqslant R^{1/2}\varepsilon^{-1/2}.$$

Since now $1 - 1/\gamma > 2/3$, choosing $\varepsilon = R^{-1+2/\gamma}$ we obtain

$$\|\mathcal{D}_R\|_\infty \leqslant cR^{1-1/\gamma}.$$

□

9.5.2 Discrepancy over translations and rotations

We obtain better estimates by averaging the discrepancy over translations and rotations. Here is a result from [21].

Theorem 9.17. *Let $2 < \gamma \leqslant 3$ and $p < 4$ (hence $p \leqslant (2\gamma - 2)/(\gamma - 2)$). Then*

$$\left\{\int_{SO(2)} \int_{\mathbb{T}^2} |\mathcal{D}(R\sigma(C_\gamma) + t)|^p \, dt \, d\sigma\right\}^{1/p} \leqslant c\, R^{1/2}, \tag{9.18}$$

where the constant c depends on γ and on p.

Proof. Let q be the conjugate index of p (that is $1/p + 1/q = 1$). By the inequalities of Hausdorff–Young and Minkowski, and by Theorem 9.12, we have

$$\left\{\int_{SO(2)}\int_{\mathbb{T}^2}|\mathcal{D}(R\sigma(C_\gamma)+t)|^p\,dt\,d\sigma\right\}^{1/p}$$
$$=\left\{\int_{SO(2)}\left[\left(R^2\int_{\mathbb{T}^2}\left|\sum_{m\neq 0}\widehat{\chi}_{C_\gamma}(R\sigma(m))e^{2\pi i m\cdot t}\right|^p dt\right)^{1/p}\right]^p d\sigma\right\}^{1/p}$$
$$\leqslant R^2\left\{\int_{SO(2)}\left\{\sum_{m\neq 0}|\widehat{\chi}_{C_\gamma}(R\sigma(m))|^q\right\}^{p/q}d\sigma\right\}^{1/p}$$
$$=R^2\left(\left\|\sum_{m\neq 0}|\widehat{\chi}_{C_\gamma}(R\sigma(m))|^q\right\|_{L^{p/q}(SO(2))}\right)^{1/q}$$
$$\leqslant R^2\left(\sum_{m\neq 0}\left\||\widehat{\chi}_{C_\gamma}(R\sigma(m))|^q\right\|_{L^{p/q}(SO(2))}\right)^{1/q}$$
$$\leqslant R^2\left\{\sum_{m\neq 0}\left\{\int_{SO(2)}|\widehat{\chi}_{C_\gamma}(R\sigma(m))|^p\,d\sigma\right\}^{q/p}\right\}^{1/q}$$
$$\leqslant cR^2\left\{\sum_{m\neq 0}|Rm|^{-3q/2}\right\}^{1/q}=cR^{1/2},$$

because $q > 4/3$. □

It is known that (9.18) can be reversed (see [15] for a proof). Here, we propose a different proof which depends on a general argument. We need a few preliminary results which are essentially known (see [38] and [24]).

Proposition 9.18. *Let $\phi \in C^\infty(-\infty,+\infty)$ be a convex function such that $\phi(0) = \phi'(0) = 0$, $\phi''(0) > 0$. Let $\delta = \frac{1}{5}\frac{\phi''(0)}{\|\phi'''\|_\infty}$, let $\psi \in C_0^\infty(-\delta,\delta)$ and let*

$$I(\lambda)=\int_{\mathbb{R}} e^{i\lambda\phi(x)}\psi(x)\,dx. \tag{9.19}$$

Then there exists $c > 0$ such that

$$\left|I(\lambda)-\psi(0)\sqrt{\frac{2\pi}{\lambda\phi''(0)}}e^{i\pi/4}\right|\leqslant c\,\frac{1}{\lambda}.$$

The constant c depends on κ_1 and κ_2, where $\phi''(0) \geqslant \kappa_1$, $\|\phi\|_{C^5} \leqslant \kappa_2$ and $\|\psi\|_{C^2} \leqslant \kappa_2$.

The proof is not short just because we want a constant c that depends on the norms of the functions and not on the functions themselves.

The proof of Proposition 9.18 needs a few lemmas.

Lemma 9.19. *Let $\phi \in C^\infty(-\delta, \delta)$ be a smooth function, and let for $|x| < \delta$,*

$$\omega(x) = x^{-k} \int_0^x (x-t)^n \phi(t)dt$$

with $n, k \geqslant 0$. Then for $0 \leqslant r \leqslant n+1-k$ there exists c, independent of ϕ, such that

$$\left|\omega^{(r)}(x)\right| \leqslant c\,\delta^{n+1-k-r} \|\phi\|_\infty .$$

Proof. Clearly,

$$\left|\omega(x)\right| \leqslant |x|^{-k} |x|^{n+1} \|\phi\|_\infty \leqslant \delta^{n+1-k} \|\phi\|_\infty .$$

We claim that, for $1 \leqslant r \leqslant n+1-k$, the derivative $\omega^{(r)}(x)$ is a linear combination of terms of the form

$$x^{-\alpha} \int_0^x \frac{(x-t)^\beta}{\beta!} \phi(t)dt$$

with $\beta - \alpha = n - k - r$ and $\beta \geqslant 0$. The proof is by induction and it is enough to observe that

$$\frac{d}{dx}\left(x^{-\alpha} \int_0^x \frac{(x-t)^\beta}{\beta!} \phi(t)dt \right)$$
$$= -\alpha x^{-\alpha-1} \int_0^x \frac{(x-t)^\beta}{\beta!} \phi(t)dt + \beta x^{-\alpha} \int_0^x \frac{(x-t)^{\beta-1}}{\beta!} \phi(t)\, dt .$$

Hence

$$\left|\omega^{(r)}(x)\right| \leqslant c \sum_{\alpha+\beta=n-k-r,\ \beta\geqslant 0,} |x|^\alpha \int_0^{|x|} |x|^\beta |\phi(t)| dt$$
$$\leqslant c \sum_{\alpha+\beta=n-k-r,\ \beta\geqslant 0,} \delta^{\alpha+\beta+1} \|\phi\|_\infty \leqslant c\delta^{n+1-k-r} \|\phi\|_\infty .$$

□

Lemma 9.20. *Let $\phi \in C^\infty(-\delta, +\delta)$ such that*

$$\phi(0) = \phi'(0) = \cdots = \phi^{(k-1)}(0) = 0 .$$

Then the function,

$$\psi(x) = \frac{\phi(x)}{x^k},$$

is smooth and for every integer $n \geqslant 0$, we have

$$\|\psi\|_{C^n} \leqslant c \|\phi\|_{C^{n+k}} .$$

Proof. By the integral form of the remainder in Taylor's theorem, for every n we can write

$$\phi(x) = \frac{\phi^{(k)}(0)}{k!}x^k + \cdots + \frac{\phi^{(n+k-1)}(0)}{(n+k-1)!}x^{n+k-1} + \int_0^x \frac{(x-t-1)^{n+k-1}}{(n+k-1)!}\phi^{(n+k)}(t)dt$$

(if $n = 0$ then only the integral appears). Let

$$\omega(x) = x^{-k}\int_0^x \frac{(x-t-1)^{n+k-1}}{(n+k-1)!}\phi^{(n+k)}(t)dt .$$

Then, by Lemma 9.19 we have

$$\begin{aligned}\|\psi\|_{C^n} &\leqslant c\|\phi\|_{C^{n+k-1}} + \|\omega\|_{C^n} \\ &\leqslant c\|\phi\|_{C^{n+k-1}} + c\|\phi^{(n+k)}\|_\infty \leqslant c\|\phi\|_{C^{n+k}} .\end{aligned}$$

□

Lemma 9.21. *There exist absolute constants $c_1, c_2 > 0$ such that if $\phi \in C^\infty(-\infty, +\infty)$ is a convex function satisfying $\phi(0) = \phi'(0) = 0$, $\phi''(0) > 0$, and $\delta = \frac{\phi''(0)}{\|\phi'''\|_\infty}$. Then*

$$g(x) = x\sqrt{\frac{\phi(x)}{x^2}}$$

is smooth and invertible in $(-\delta, \delta)$. Moreover,

$$g'(0) = \sqrt{\frac{\phi''(0)}{2}} \tag{9.20}$$

and, for $|x| < \delta$,

$$c_1\sqrt{\phi''(0)} \leqslant g'(x) \leqslant c_2\sqrt{\phi''(0)} .$$

Finally, $\|g\|_{C^n}$ can be bounded from above by a constant that depends only on $\|\phi\|_{C^{2+n}}$, and from below by a constant that depends only on $\phi''(0)$.

Proof. The integral form of the remainder in Taylor's theorem and Lemma 9.19 yield

$$\frac{\phi(x)}{x^2} = \frac{\phi''(0)}{2} + x^{-2}\int_0^x \frac{(x-t)^2}{2}\phi'''(t)\, dt ,$$

so that, for $|x| < \delta$, we have

$$\left|\frac{\phi(x)}{x^2} - \frac{\phi''(0)}{2}\right| \leqslant \frac{1}{6}\delta\|\phi'''\|_\infty = \frac{1}{6}\phi''(0) .$$

Hence, for $|x| < \delta$,

$$\frac{1}{3}\,\phi''(0) \leqslant \frac{\phi(x)}{x^2} \leqslant \frac{2}{3}\,\phi''(0)\,.$$

Observe that this and Lemma 9.20 imply that $g(x)$ is smooth. Similarly,

$$\left|\frac{\phi'(x)}{x} - \phi''(0)\right| = \left|x^{-1}\int_0^x (x-t)\phi'''(t)dt\right| \leqslant \frac{1}{2}\delta\|\phi'''\|_\infty = \frac{1}{2}\phi''(0)\,,$$

so that

$$\frac{1}{2}\phi''(0) \leqslant \frac{\phi'(x)}{x} \leqslant \frac{3}{2}\phi''(0)\,.$$

Finally, since

$$g'(x) = \frac{1}{2}\frac{\phi'(x)}{x}\left(\frac{\phi(x)}{x^2}\right)^{-1/2},$$

there are absolute constants $c_1, c_2 > 0$ such that

$$c_1\sqrt{\phi''(0)} \leqslant |g'(x)| \leqslant c_2\sqrt{\phi''(0)}\,.$$

Observe that

$$\frac{d^n}{dx^n}\left(x\sqrt{\frac{\phi(x)}{x^2}}\right) \le c^*\,,$$

where the constant c^* depends on a lower bound for $\frac{\phi(x)}{x^2}$ and a lower bound for $\frac{d^k}{dx^k}(\frac{\phi(x)}{x^2})$, when $k \le n$. Then, by Lemma 9.20, c^* depends on a lower bound of $\phi''(0)$ and on $\|\phi\|_{C^{n+2}}$. □

Proof of Proposition 9.18. Let $I(\lambda)$ be as in (9.19). Again let $g = x\sqrt{\frac{\phi(x)}{x^2}}$. Then $[g(x)]^2 = \phi(x)$, so that the change of variables $u = g(x)$ and Lemma 9.21 yield

$$I(\lambda) = \int_{\mathbb{R}} e^{i\lambda u^2}\frac{\psi(g^{-1}(u))}{g'(g^{-1}(u))}du = \int_{\mathbb{R}} e^{i\lambda u^2}h(u)du\,,$$

with $h(u)$ smooth and compactly supported. Let $\eta \in C_0^\infty(-\infty, +\infty)$ such that $\eta(u) \equiv 1$ on the support of $h(u)$ and let

$$R(u) = \frac{h(u)e^{u^2} - h(0)}{u}\,.$$

Then

$$I(\lambda) = \int_{\mathbb{R}} e^{i\lambda u^2}h(u)\eta(u)du$$

$$
\begin{aligned}
&= \int_{\mathbb{R}} e^{i\lambda u^2} e^{-u^2} [h(u)e^{u^2}] \eta(u)du \\
&= \int_{\mathbb{R}} e^{i\lambda u^2} e^{-u^2} [h(0) + uR(u)] \eta(u)du \\
&= h(0) \int_{\mathbb{R}} e^{i\lambda u^2} e^{-u^2} \eta(u)du + \int_{\mathbb{R}} e^{i\lambda u^2} e^{-u^2} uR(u)\eta(u)du \\
&= h(0) \int_{\mathbb{R}} e^{i\lambda u^2} e^{-u^2} du + h(0) \int_{\mathbb{R}} e^{i\lambda u^2} e^{-u^2} [1 - \eta(u)]du \\
&+ \int_{\mathbb{R}} e^{i\lambda u^2} e^{-u^2} uR(u)\eta(u)du \\
&= I_1(\lambda) + I_2(\lambda) + I_3(\lambda).
\end{aligned}
$$

The integral in $I_1(\lambda)$ can be computed through a familiar trick:

$$
\begin{aligned}
\left(\int_{-\infty}^{+\infty} e^{i\lambda u^2} e^{-u^2} du \right)^2 &= \int_{-\infty}^{+\infty} \int_{-\infty}^{+\infty} e^{i\lambda(u^2+v^2)} e^{-(u^2+v^2)} dudv \\
&= \int_0^{2\pi} \int_0^{+\infty} e^{(i\lambda-1)\rho^2} \rho d\rho d\theta = \frac{\pi}{1 - i\lambda}.
\end{aligned}
$$

Hence (9.20) yields

$$
I_1(\lambda) = h(0) \frac{\sqrt{\pi}}{(1 - i\lambda)^{1/2}} = \frac{\psi(0)}{\sqrt{\phi''(0)}} \frac{\sqrt{2\pi}}{(1 - i\lambda)^{1/2}}
$$

(here we consider the branch of $z^{1/2}$ that for $z > 0$ agrees with $\sqrt{z}$). Then, for $\lambda > 1$,

$$
\begin{aligned}
I_1(\lambda) &= \frac{\psi(0)}{\sqrt{\phi''(0)}} \sqrt{2\pi} \left(-i\lambda \left(1 + \frac{1}{-i\lambda} \right) \right)^{-1/2} \\
&= \frac{\psi(0)}{\sqrt{\phi''(0)}} \frac{\sqrt{2\pi}}{\sqrt{\lambda}} e^{i\pi/4} + \frac{\psi(0)}{\sqrt{\phi''(0)}} O\left(\frac{1}{\lambda} \right).
\end{aligned}
$$

Integration by parts in $I_2(\lambda)$ yields

$$
\begin{aligned}
I_2(\lambda) &= \frac{\sqrt{2}\psi(0)}{\sqrt{\phi''(0)}} \int_{-\infty}^{+\infty} e^{i\lambda u^2} e^{-u^2} [1 - \eta(u)]du \\
&= \frac{\psi(0)}{i\lambda\sqrt{2\phi''(0)}} \int_{-\infty}^{+\infty} 2i\lambda u e^{i\lambda u^2} \frac{e^{-u^2}[1 - \eta(u)]}{u} du
\end{aligned}
$$

$$= \frac{\psi(0)}{i\lambda\sqrt{2\phi''(0)}} \int_{-\infty}^{+\infty} e^{i\lambda u^2} \frac{d}{du}\left[\frac{e^{-u^2}[1-\eta(u)]}{u}\right] du,$$

so that

$$|I_2(\lambda)| \leqslant c\,\frac{1}{\lambda}\,\frac{|\psi(0)|}{\sqrt{\phi''(0)}}$$

(note that we can always assume that $\eta(u) \equiv 1$ in a given neighborhood of the origin). Finally,

$$\begin{aligned} I_3(\lambda) &= \frac{1}{2i\lambda}\int 2i\lambda u e^{i\lambda u^2} e^{-u^2} R(u)\eta(u)\,du \\ &= \frac{1}{2i\lambda}\int e^{i\lambda u^2} \frac{d}{du}[e^{-u^2} R(u)\eta(u)]\,du \end{aligned}$$

so that

$$|I_3(\lambda)| \leqslant \frac{1}{2\lambda}\int \left|\frac{d}{du}[e^{-u^2}R(u)\eta(u)]\right| du.$$

Since

$$h(u)e^{u^2} = h(0) + h'(0)u + \int_0^u (u-t)\frac{d^2}{dt^2}[e^{t^2}h(t)]\,dt,$$

we have

$$R(u) = h'(0) + \frac{1}{u}\int_0^u (u-t)\frac{d^2}{dt^2}[e^{t^2}h(t)]\,dt,$$

$$|R(u)| \leqslant |h'(0)| + \sup\left|\frac{d^2}{dt^2}[e^{t^2}h(t)]\right|,$$

where the supremum is on the support of $h(t)$. We also have

$$R'(u) = \frac{1}{u^2}\int_0^u t\frac{d^2}{dt^2}[e^{t^2}h(t)]\,dt,$$

so that

$$|R'(u)| \leqslant \sup\left|\frac{d^2}{dt^2}[e^{u^2}h(t)]\right| \leqslant c\|h\|_{C^2}.$$

Since

$$h(t) = \frac{\psi(g^{-1}(t))}{g'(g^{-1}(t))}$$

and, by Lemma 9.21,

$$g'(u) \approx c_1 \sqrt{\phi''(0)},$$

we can control $\|h\|_{C^2}$ through an upper bound on $\|\psi\|_{C^2}$ and $\|g\|_{C^3}$, and a lower bound on $\phi''(0)$. In turns, by Lemma 9.21, $\|g\|_{C^3}$ can be bounded by $\|\phi\|_{C^5}$. □

Asymptotic estimates for the Fourier transform of the characteristic function of a convex body with smooth boundary having everywhere strictly positive curvature are well known (see [23] and [22]). In the next lemma, we replace the above global assumption on the curvature with a local one.

Lemma 9.22. *Let C be a strictly convex planar body with smooth boundary but for a single point that we assume to be the origin where we only assume C^2 regularity. Let I be a small closed interval contained in $(0,\pi)$. For every direction $\theta \in I$, let $\sigma_1(\theta)$ and $\sigma_2(\theta)$ be the two points in ∂C where the tangents are perpendicular to Θ. We assume that the curvatures $K(\sigma_1(\theta))$ and $K(\sigma_2(\theta))$ are positive. Then*

$$\begin{aligned}\widehat{\chi}_C(\rho\Theta) = & -\frac{1}{2\pi i}\rho^{-3/2}\left[e^{-2\pi i\rho\Theta\cdot\sigma_1(\theta)+\pi i/4}K^{-1/2}(\sigma_1(\theta))\right. \\ & \left.-e^{-2\pi i\rho\Theta\cdot\sigma_2(\theta)-\pi i/4}K^{-1/2}(\sigma_2(\theta))\right] \\ & +\mathcal{O}(\rho^{-2}),\end{aligned}$$

with the implicit constant in $\mathcal{O}(\rho^{-2})$ depending only on $\inf_{\theta\in I} K(\sigma_j(\theta))$.

Proof. By the divergence theorem, we have

$$\widehat{\chi}_C(\rho\Theta) = \frac{-1}{2\pi i\rho}\int_{\partial C} e^{-2\pi i\rho\Theta\cdot t}\,\Theta\cdot\nu(t)\,d\mu(t),$$

where $d\mu$ is the arc length measure on ∂C. Let

$$s \mapsto \Gamma(s)$$

be the arc length parametrization of ∂C. Then

$$\widehat{\chi}_C(\rho\Theta) = \frac{-1}{2\pi i\rho}\int_0^1 e^{-2\pi i\rho\Theta\cdot\Gamma(s)}\,\Theta\cdot\nu(\Gamma(s))ds$$

(without loss of generality we can assume that the arc length of ∂C is 1). Observe that in the above integral the phase $\Theta\cdot\Gamma(s)$ is stationary when $\Gamma(s) = \sigma_j(\theta)$. Let

$$J_j = \{s \in [0,1] : \Gamma(s) = \sigma_j(\theta) \text{ for some } \theta \in I\}$$

and let $\varphi_1(s)$ and $\varphi_2(s)$ be cut-off functions that take value 1 in J_1 and J_2, respectively. Then

$$\begin{aligned}\widehat{\chi}_C(\rho\Theta) &= \frac{-1}{2\pi i\rho}\int_0^1 e^{-2\pi i\rho\Theta\cdot\Gamma(s)}\Theta\cdot\nu(\Gamma(s))\varphi_1(s)ds \\ &+ \frac{-1}{2\pi i\rho}\int_0^1 e^{-2\pi i\rho\Theta\cdot\Gamma(s)}\Theta\cdot\nu(\Gamma(s))\varphi_2(s)ds \\ &+ \frac{-1}{2\pi i\rho}\int_0^1 e^{-2\pi i\rho\Theta\cdot\Gamma(s)}\Theta\cdot\nu(\Gamma(s))[1-\varphi_1(s)-\varphi_2(s)]ds \\ &= A_1 + A_2 + A_3\,,\end{aligned}$$

say. The integral in A_3 can be easily estimated since in the support of $[1-\varphi_1(s)-\varphi_2(s)]$ the phase is not stationary and we can integrate by parts. Therefore, we obtain

$$|A_3| \leqslant c\rho^{-2}\,.$$

In the integral in A_1, the phase is stationary at one point, say $\overline{s}$ where

$$\Theta\cdot\Gamma'(\overline{s}) = 0\,.$$

Observe that at the point $\overline{s}$ we have

$$\Theta\cdot\Gamma''(\overline{s}) = \left|\Gamma''(\overline{s})\right| = K(\sigma_1(\theta))\,,$$

where $K(\sigma_1(\theta))$ denotes the curvature of ∂C at $\sigma_1(\theta) = \Gamma(\overline{s})$. By Proposition 9.18, we have

$$\begin{aligned}A_1 &= -\frac{e^{-2\pi i\rho\Theta\cdot\Gamma(\overline{s})}}{2\pi i\rho}\int_0^1 e^{2\pi i\rho[\Theta\cdot\Gamma(\overline{s})-\Theta\cdot\Gamma(s)]}\Theta\cdot\nu(\Gamma(s))\varphi_1(s)ds \\ &= -\frac{e^{-2\pi i\rho\Theta\cdot\sigma_1(\theta)}}{2\pi i\rho}\sqrt{\frac{2\pi}{2\pi\rho K(\sigma_1(\theta))}}\,e^{i\pi/4} + O(\rho^{-2}) \\ &= -\frac{1}{2\pi i}\rho^{-3/2}e^{-2\pi i\rho\Theta\cdot\sigma_1(\theta)+i\pi/4}K^{-1/2}(\sigma_1(\theta)) + O(\rho^{-2})\,.\end{aligned} \tag{9.21}$$

Similarly,

$$\begin{aligned}A_2 &= -\frac{e^{-2\pi i\rho\Theta\cdot\Gamma(\overline{s})}}{2\pi i\rho}\int_0^1 e^{-2\pi i\rho[\Theta\cdot\Gamma(s)-\Theta\cdot\Gamma(\overline{s})]}\Theta\cdot\nu(\Gamma(s))\varphi_2(s)ds \\ &= \frac{e^{-2\pi i\rho\Theta\cdot\sigma_2(\theta)}}{2\pi i\rho}\sqrt{\frac{2\pi}{2\pi\rho K(\sigma_2(\theta))}}\,e^{-i\pi/4} + O(\rho^{-2}) \\ &= \frac{e^{-2\pi i\rho\Theta\cdot\sigma_2(\theta)}}{2\pi i}\rho^{-3/2}K^{-1/2}(\sigma_2(\theta))e^{-i\pi/4} + O(\rho^{-2})\,.\end{aligned}$$

□

We can now prove the following result (see [15] for a different proof).

Theorem 9.23. *For every $\gamma > 2$ and $p \geq 1$, we have*

$$\left\{ \int_{SO(2)} \int_{\mathbb{T}^2} |\mathcal{D}(R\sigma(C_\gamma) + t)|^p \, dt d\sigma \right\}^{1/p} \geqslant c\, R^{1/2}.$$

Proof. By our assumptions on C_γ, there is a positive constant κ and an interval $I \subset [-\pi/2 - \varepsilon, -\pi/2 + \varepsilon]$ such $K(\sigma_2(\theta)) > \kappa$ whenever $\theta \notin I$. Since (on the side close to the origin) $K(\sigma_1(\theta)) \to 0$ as $\theta \to 0$ there is an interval $J \subset I$ such that $K(\sigma({}_1\theta)) < \kappa/2$ for all $\theta \in J$. Then the asymptotic expansion in Lemma 9.22 yields

$$\begin{aligned} \int_0^{2\pi} |\widehat{\chi}_C(\rho\Theta)| \, d\theta &> \int_{J \cup (J+\pi)} |\widehat{\chi}_C(\rho\Theta)| \, d\theta \\ &\geqslant c\rho^{-3/2} \int_{J \cup (J+\pi)} |K^{-1/2}(\sigma_1(\theta)) - K^{-1/2}(\sigma_2(\theta))| - c_1\rho^{-2} \geqslant c\rho^{-3/2}. \end{aligned} \tag{9.22}$$

Then, for every $0 \neq k \in \mathbb{Z}^2$, (9.22) and an orthogonality argument yield

$$\begin{aligned} &\left\{ \int_{SO(2)} \int_{\mathbb{T}^2} |\mathcal{D}(R\sigma(C_\gamma) + t)|^p \, dt d\sigma \right\}^{1/p} \\ &= \left\{ \int_{SO(2)} \left(\left\{ \int_{\mathbb{T}^2} |\mathcal{D}(R\sigma(C_\gamma) + t)|^p \, dt \right\}^{1/p} \right)^p d\sigma \right\}^{1/p} \\ &\geqslant R^2 \left\{ \int_{SO(2)} \left| \int_{\mathbb{T}^2} \left(\sum_{m \neq 0} \widehat{\chi}_{C_\gamma}(R\sigma(m)) e^{2\pi i m \cdot t} \right) e^{-2\pi i k \cdot t} dt \right|^p d\sigma \right\}^{1/p} \\ &\geqslant cR^2 \left\{ \int_{SO(2)} |\widehat{\chi}_{C_\gamma}(R\sigma(k))|^p \, d\sigma \right\}^{1/p} \geqslant cR^{1/2}. \end{aligned}$$

□

The upper bound $R^{1/2}$ still holds true for suitable rotations of C_γ; see [8].

Theorem 9.24. *Let $\widetilde{C}_\gamma$ be a rotated copy of C_γ and we assume that the outward unit normal (α, β) at the flat point satisfies the following Diophantine condition: for every given $\delta < 2/(\gamma - 2)$, there exists $c > 0$ such that for every positive integer n we have*

$$\left\| n \frac{\alpha}{\beta} \right\| \geqslant \frac{c}{n^{1+\delta}},$$

where $\|x\|$ is the distance of the real number x from the integers. Then

$$\left\{ \int_{\mathbb{T}^2} |\mathcal{D}(R\widetilde{C}_\gamma + t)|^2 \, dt \right\}^{1/2} \leqslant cR^{1/2}.$$

Proof. Of course, we may assume $|\alpha| < |\beta|$. We write

$$\begin{aligned}
\int_{\mathbb{T}^2} \left|\mathcal{D}(R\widetilde{C}_\gamma + t)\right|^2 dt &= R^4 \sum_{(m_1,m_2)\neq(0,0)} \left|\widehat{\chi}_{\widetilde{C}_\gamma}(Rm_1, Rm_2)\right|^2 \\
&\leqslant R^4 \sum_{0<|-\beta m_1+\alpha m_2|<1/2} \left|\widehat{\chi}_{\widetilde{C}_\gamma}(Rm_1, Rm_2)\right|^2 \\
&+ R^4 \sum_{1/2\leqslant|-\beta m_1+\alpha m_2|<|\alpha m_1+\beta m_2|} \left|\widehat{\chi}_{\widetilde{C}_\gamma}(Rm_1, Rm_2)\right|^2 \\
&+ R^4 \sum_{0<|\alpha m_1+\beta m_2|\leqslant|-\beta m_1+\alpha m_2|} \left|\widehat{\chi}_{\widetilde{C}_\gamma}(Rm_1, Rm_2)\right|^2 \\
&= A + B + C\,,
\end{aligned}$$

say. We are going to apply the estimates in Theorem 9.8, with

$$\psi \approx \frac{|-\beta m_1 + \alpha m_2|}{\sqrt{m_1^2 + m_2^2}}\,.$$

In order to estimate A, we observe that $0 < |-\beta m_1 + \alpha m_2| < 1/2$ implies $m_1^2 + m_2^2 \approx m_2^2$ and, therefore,

$$\begin{aligned}
A &\leqslant cR \sum_{0<|-\beta m_1+\alpha m_2|<1/2} \psi^{-(\gamma-2)/(\gamma-1)} \left(m_1^2 + m_2^2\right)^{-3} \\
&\leqslant cR \sum_{0<|-\beta m_1+\alpha m_2|<1/2} |-\beta m_1 + \alpha m_2|^{-(\gamma-2)/(\gamma-1)} |m_2|^{-2-1/(\gamma-1)} \\
&\leqslant cR \sum_{0<|-\beta m_1+\alpha m_2|<1/2} \left\| m_2 \frac{\alpha}{\beta} \right\|^{-(\gamma-2)/(\gamma-1)} |m_2|^{-2-1/(\gamma-1)} \\
&\leqslant cR \sum_{0<|-\beta m_1+\alpha m_2|<1/2} |m_2|^{(1+\delta)(\gamma-2)/(\gamma-1)} |m_2|^{-2-1/(\gamma-1)} = cR\,,
\end{aligned}$$

because $\delta < 2/(\gamma - 2)$. As for B we can replace the sum with an integral and have

$$\begin{aligned}
B &\leqslant cR \sum_{1/2\leqslant|-\beta m_1+\alpha m_2|<|\alpha m_1+\beta m_2|} |-\beta m_1 + \alpha m_2|^{-(\gamma-2)/(\gamma-1)} \\
&\times |\alpha m_1 + \beta m_2|^{-2-1/(\gamma-1)} \\
&\leqslant cR \int_{1/2\leqslant|\xi|\leqslant|s|} |\xi|^{-(\gamma-2)/(\gamma-1)} |s|^{-2-1/(\gamma-1)}\, d\xi ds \\
&\leqslant cR\,.
\end{aligned}$$

Finally,

$$\begin{aligned}
C &\leqslant R \sum_{0<|\alpha m_1+\beta m_2|\leqslant|-\beta m_1+\alpha m_2|} \left|(m_1, m_2)\right|^{-3} \\
&\leqslant cR \sum_{(m_1,m_2)\neq(0,0)} \left|(m_1, m_2)\right|^{-3} = cR\,.
\end{aligned}$$

□

Remark 9.25. We recall that if ω is an irrational algebraic number, then Roth's theorem [35] says that for every $\varepsilon > 0$ there exists $c > 0$ such that

$$\|n\omega\| \geq \frac{1}{n^{1+\varepsilon}} .$$

9.6 Irregularities of distribution for C_γ

The above upper bound $R^{1/2}$ for the discrepancy is best possible in the following sense. Let the integer N be a square,[1] say $N = M^2$. Then the set

$$\frac{1}{M}\mathbb{Z}^2 \cap [-\frac{1}{2}, \frac{1}{2})^2$$

contains N points and, for a convex planar body $C \subset [-\frac{1}{2}, \frac{1}{2})^2$, we have

$$\operatorname{card}(\mathbb{Z}^2 \cap MC) = \operatorname{card}\left(\frac{1}{M}\mathbb{Z}^2 \cap C\right).$$

Then the study of integer points in large convex bodies is a counterpart to a classical "irregularities of distribution" problem (see [4, 30]). In other words, it is a particular answer to the problem of choosing N points in $[-1/2, 1/2)^2$ to approximate the area of a given family of sets.

We have the following result.

Theorem 9.26. *Let C_γ be as in the Introduction. Let N be a positive large integer. Then there exists a constant $c > 0$ such that for every finite set*

$$\{u(j)\}_{j=1}^{N} \subset [-1/2, 1/2)^2$$

we have

$$\left\{\int_{1/2}^{1}\int_{\mathbb{T}^2}\left|-N|C_\gamma| + \sum_{j=1}^{N}\chi_{\tau C_\gamma}(u(j)+t)\right|^2 dt\,d\tau\right\}^{1/2} \geqslant c\,N^{1/4} . \tag{9.23}$$

Corollary 9.27. *Let C_γ and N be as in the previous theorem. Then there exists a dilated and translated copy $\widetilde{C}_\gamma$ of C_γ such that*

$$\left|-N|C_\gamma| + \sum_{j=1}^{N}\chi_{\widetilde{C}_\gamma}(u(j))\right| \geqslant c\,N^{1/4} .$$

1 Actually, it is not necessary to choose N to be a square, see [14, p. 3533].

Note that in order to compare (9.23) with the results in the previous section, we should take $R = N^{1/2}$.

To prove Theorem 9.26, we first need a mild variant of a classical result due to J. W. S. Cassels (see, e. g., [31]). For every positive real number K, let we consider the square

$$Q_K = \{m = (m_1, m_2) \in \mathbb{Z}^2 : |m_1| \leqslant K, |m_2| \leqslant K\}.$$

Lemma 9.28. *For every choice of positive integers H, N, and L, such that $H < \sqrt{L}$, let*

$$\widetilde{Q}_N = Q_{\sqrt{LN}} \setminus Q_H . \tag{9.24}$$

Then for every finite set $\{u(j)\}_{j=1}^N \subset \mathbb{T}^2$, we have

$$\sum_{0 \neq m \in \widetilde{Q}_N} \left| \sum_{j=1}^N e^{2\pi i m \cdot u(j)} \right|^2 \geqslant (L - H^2) N^2 . \tag{9.25}$$

Proof. Since

$$\sum_{|m_1| \leqslant H} \sum_{|m_2| \leqslant H} \left| \sum_{j=1}^N e^{2\pi i m \cdot u(j)} \right|^2 \leqslant N^2 H^2$$

it is enough to show that

$$\sum_{|m_1| \leqslant \sqrt{LN}} \sum_{|m_2| \leqslant \sqrt{LN}} \left| \sum_{j=1}^N e^{2\pi i m \cdot u(j)} \right|^2 \geqslant LN^2 ,$$

and this will follow from the inequality

$$\sum_{|m_1| \leqslant [\sqrt{LN}]} \sum_{|m_2| \leqslant [\sqrt{LN}]} \left| \sum_{j=1}^N e^{2\pi i m \cdot u(j)} \right|^2 \geqslant N([\sqrt{LN}] + 1)^2 . \tag{9.26}$$

Indeed let $u(\ell) = (u_1(\ell), u_2(\ell))$. Then the LHS of (9.26) is larger than

$$\begin{aligned} &\sum_{|m_1| \leqslant [\sqrt{LN}]} \sum_{|m_2| \leqslant [\sqrt{LN}]} \left(1 - \frac{|m_1|}{[\sqrt{LN}] + 1}\right) \\ &\times \left(1 - \frac{|m_2|}{[\sqrt{LN}] + 1}\right) \left| \sum_{j=1}^N e^{2\pi i m \cdot u(j)} \right|^L \\ &= \sum_{|m_1| \leqslant [\sqrt{LN}]} \sum_{|m_2| \leqslant [\sqrt{LN}]} \left(1 - \frac{|m_1|}{[\sqrt{LN}] + 1}\right) \left(1 - \frac{|m_2|}{[\sqrt{LN}] + 1}\right) \end{aligned} \tag{9.27}$$

$$
\begin{aligned}
&\times \sum_{j=1}^{N}\sum_{k=1}^{N} e^{2\pi i m\cdot(u(j)-u(k))} \\
&= \sum_{j=1}^{N}\sum_{k=1}^{N} \sum_{|m_1|\leqslant[\sqrt{LN}]} \left(1-\frac{|m_1|}{[\sqrt{LN}]+1}\right) e^{2\pi i m_1(u_1(j)-u_1(k))} \\
&\times \sum_{|m_2|\leqslant[\sqrt{LN}]} \left(1-\frac{|m_2|}{[\sqrt{LN}]+1}\right) e^{2\pi i m_2(u_2(j)-u_2(k))} \\
&= \sum_{j=1}^{N}\sum_{k=1}^{N} K_{[\sqrt{LN}]}(u_1(j)-u_1(k)) K_{[\sqrt{LN}]}(u_2(j)-u_2(k)) , \qquad (9.28)
\end{aligned}
$$

where

$$
K_M(x) = \sum_{j=-M}^{M} \left(1-\frac{|j|}{M+1}\right) e^{2\pi i j x} = \frac{1}{M+1}\left(\frac{\sin(\pi(M+1)x)}{\sin(\pi x)}\right)^2
$$

is the Fejér kernel on $\mathbb{T}$. Since $K_M(x) \geqslant 0$ for every x, the last term in (9.28) is not smaller than the "diagonal"

$$
\begin{aligned}
&\sum_{j=1}^{N} K_{[\sqrt{LN}]}(u_1(j)-u_1(j)) K_{[\sqrt{LN}]}(u_2(j)-u_2(j)) \\
&\quad = N\, K_{[\sqrt{LN}]}(0)\, K_{[\sqrt{LN}]}(0) = N([\sqrt{LN}]+1)^2 . \qquad \square
\end{aligned}
$$

Now we need an estimate from below of $\int_{1/2}^{1} |\widehat{\chi}_{sC_\gamma}(k)|^2\, ds$, for $0 \neq k \in \mathbb{Z}^2$.

Lemma 9.29. *Let C_γ be as in the Introduction. Then there exist constants $c_1, c_2 > 0$ such that for $|\xi| \geqslant c_1$ we have*

$$
\left\{ \int_{1/2}^{1} |\widehat{\chi}_{C_\gamma}(\tau\xi)|^2\, d\tau \right\}^{1/2} \geqslant c_2 |\xi|^{-3/2} .
$$

Proof. Let $\xi = \rho\Theta$, arguing as in the proof of Lemma 9.22 we write

$$
\begin{aligned}
\widehat{\chi}_{C_\gamma}(\tau\rho\Theta) &= \frac{-1}{2\pi i\tau\rho} \int_0^1 e^{-2\pi i\tau\rho\Theta\cdot\Gamma(s)}\Theta\cdot\nu(\Gamma(s))\varphi_1(s)ds \\
&+ \frac{-1}{2\pi i\tau\rho} \int_0^1 e^{-2\pi i\tau\rho\Theta\cdot\Gamma(s)}\Theta\cdot\nu(\Gamma(s))\varphi_2(s)ds \\
&+ \frac{-1}{2\pi i\tau\rho} \int_0^1 e^{-2\pi i\tau\rho\Theta\cdot\Gamma(s)}\Theta\cdot\nu(\Gamma(s))[1-\varphi_1(s)-\varphi_2(s)]ds \\
&= A_1(\tau\rho) + A_2(\tau\rho) + A_3(\tau\rho) .
\end{aligned}
$$

We have

$$\left\{\int_{1/2}^{1}\left|\widehat{\chi}_{C_y}(\tau\xi)\right|^2 d\tau\right\}^{1/2}$$
$$\geqslant \left\{\int_{1/2}^{1}\left|A_1(\tau\rho)+A_2(\tau\rho)\right|^2 d\tau\right\}^{1/2} - \left\{\int_{1/2}^{1}\left|A_3(\tau\rho)\right|^2 d\tau\right\}^{1/2}.$$

Since (in A_3) in the support of $[1-\varphi_1(s)-\varphi_2(s)]$ the phase is not stationary, integration by parts yields

$$\left|A_3(\tau\rho)\right| \leqslant c\tau^{-2}\rho^{-2},$$

and, therefore,

$$\left\{\int_{1/2}^{1}\left|\widehat{\chi}_{C_y}(\tau\xi)\right|^2 d\tau\right\}^{1/2} \geqslant \left\{\int_{1/2}^{1}\left|A_1(\tau\rho)+A_2(\tau\rho)\right|^2 d\tau\right\}^{1/2} - c\rho^{-2}.$$

By our assumptions on C_y, we know that at least one (say the first one) of the two integrals in A_1 and A_2 corresponds to a part of ∂C_y where the curvature is bounded away from zero. Let $\eta \in C_0^{\infty}(1/2,1)$ be a cut-off function such that $0 \leqslant \eta(\tau) \leqslant 1$ and $\eta(\tau) \equiv 1$ for $5/8 \leqslant \tau \leqslant 7/8$. Then

$$\int_{1/2}^{1}\left|A_1(\tau\rho)+A_2(\tau\rho)\right|^2 d\tau \geqslant \int_{1/2}^{1}\left|A_1(\tau\rho)+A_2(\tau\rho)\right|^2 \eta(\tau)d\tau$$
$$= \int_{1/2}^{1}\left(\left|A_1(\tau\rho)\right|^2 + \left|A_2(\tau\rho)\right|^2 + 2\operatorname{Re}(A_1(\tau\rho)\overline{A_2(\tau\rho)})\right)\eta(\tau)d\tau$$
$$\geqslant \int_{1/2}^{1}\left|A_1(\tau\rho)\right|^2 \eta(\tau)d\tau + 2\operatorname{Re}\int_{1/2}^{1}(A_1(\tau\rho)\overline{A_2(\tau\rho)})\eta(\tau)d\tau$$

For the second integral, we have

$$\int_{1/2}^{1} A_1(\tau\rho)\overline{A_2(\tau\rho)}\eta(\tau)d\tau$$
$$= \frac{-1}{4\pi^2\rho^2}\int_{1/2}^{1}\tau^{-2}\int_0^1\int_0^1 e^{2\pi i\tau\rho\Theta\cdot[\Gamma(w)-\Gamma(s)]}\left[\Theta\cdot\nu(\Gamma(s))\Theta\cdot\nu(\Gamma(w))\right]$$
$$\times\, \varphi_1(s)\varphi_2(w)ds\,dw\;\eta(\tau)d\tau$$

$$\frac{-1}{4\pi^2\rho^2}\int_0^1\int_0^1\int_{1/2}^1 e^{2\pi i\tau\rho\Theta\cdot[\Gamma(w)-\Gamma(s)]}\frac{\eta(\tau)}{\tau^2}d\tau$$
$$\times\left[\Theta\cdot\nu(\Gamma(s))\Theta\cdot\nu(\Gamma(w))\right]\varphi_1(s)\varphi_2(w)dsdw$$

Observe that if $\ell(\tau)=\eta(\tau)/\tau^2$, then

$$\int_{1/2}^1 e^{2\pi i\tau\rho\Theta[\Gamma(w)-\Gamma(s)]}\frac{\eta(\tau)}{\tau^2}\,d\tau=\widehat{\ell}(\rho\Theta[\Gamma(w)-\Gamma(s)])\,.$$

Since $|\Theta\cdot[\Gamma(w)-\Gamma(s)]|\geqslant c>0$ for every w,s in the supports of φ_1 and φ_2, respectively, integration by parts gives

$$\int_{1/2}^1 e^{2\pi i\tau\rho\Theta\cdot[\Gamma(w)-\Gamma(s)]}\frac{\eta(\tau)}{\tau^2}d\tau=O(\rho^{-L})$$

for every L. It follows that

$$\left\{\int_{1/2}^1\left|A_1(\tau\rho)+A_2(\tau\rho)\right|^2d\tau\right\}^{1/2}\geqslant c\left\{\int_{1/2}^1\left|A_1(\tau\rho)\right|^2\eta(\tau)d\tau\right\}^{1/2}+O(\rho^{-L}).$$

Also, by our choice of A_1, we have

$$A_1(\tau\rho)=-\frac{1}{2\pi i}(\tau\rho)^{-3/2}e^{-2\pi i\tau\rho\Theta\cdot\sigma_1(\theta)+i\frac{\pi}{4}}K^{-1/2}(\sigma_1(\theta))+O(\tau^{-2}\rho^{-2})$$

so that

$$\left\{\int_{1/2}^1\left|A_1(\tau\rho)\right|^2\eta(\tau)d\tau\right\}^{1/2}\geqslant c_1\rho^{-3/2}K^{-1/2}(\sigma_1(\theta))-c_2\rho^{-2}\,.$$

Finally,

$$\left\{\int_{1/2}^1\left|\widehat{\chi}_{C_y}(\tau\xi)\right|^2\,d\tau\right\}^{1/2}\geqslant c_1\rho^{-3/2}-c_2\rho^{-2}\geqslant c_3\rho^{-3/2}$$

for ρ large enough. □

Proof of Theorem 9.26. We apply the Parseval theorem, (9.25), and Lemma 9.29, where we choose $H=c_1$. Then, for $\widetilde{Q}_N$ as in (9.24), we have

$$\int_{1/2}^1\int_{\mathbb{T}^2}\left|-N|C_y|+\sum_{j=1}^N\chi_{\tau C_y}(u(j)+t)\right|^2\,dtd\tau$$

$$= \int_{1/2}^{1} \sum_{m \neq 0} \left| \sum_{j=1}^{N} e^{2\pi i m \cdot u(j)} \right|^2 |\widehat{\chi}_{\tau C_\gamma}(m)|^2 \, d\tau$$

$$\geqslant \sum_{m \in Q_N} \left| \sum_{j=1}^{N} e^{2\pi i m \cdot u(j)} \right|^2 \int_{1/2}^{1} \tau^2 |\widehat{\chi}_{C_\gamma}(\tau m)|^2 \, d\tau$$

$$\geqslant c |\sqrt{N}|^{-3} \sum_{m \in Q_N} \left| \sum_{j=1}^{N} e^{2\pi i m \cdot u(j)} \right|^2 \geqslant c\, N^{1/2} . \qquad \square$$

Remark 9.30. We have already pointed out that the discrepancy results for C_γ are "intermediate" between the case of a convex body with smooth boundary having everywhere positive curvature, and the case of a polygon (just send $\gamma \to 2$ or $\gamma \to +\infty$). This is not the case for the main result in this section. Indeed we know that for a polygon we have a logarithmic lower bound (see [31]) which has a counterpart in Davenport's paper [19]. The "explanation" is that a polygon does not have points on the boundary with positive curvature, while for every $\gamma < +\infty$ the convex body C_γ has such points.

9.7 Remarks on higher dimensional cases

Kendall's upper bound works in higher dimensions as well. Indeed, let $B = \{t \in \mathbb{R}^d : |t| \leqslant 1\}$ and let $t \in \mathbb{T}^d = \mathbb{R}^d/\mathbb{Z}^d$. Let

$$D_R(\sigma, t) = -R^d |B| + \operatorname{card}\left((\sigma(RB) + t) \cap \mathbb{Z}^d\right) .$$

Then (see, e. g., [11])

$$\left\{ \int_{\mathbb{T}^d} |D_R(\sigma, t)|^2 \, dt \right\}^{1/2} \leqslant c\, R^{(d-1)/2} .$$

Interestingly (see [32]), its converse

$$\left\{ \int_{\mathbb{T}^d} |D_R(\sigma, t)|^2 \, dt \right\}^{1/2} \geqslant c_1\, R^{(d-1)/2}$$

holds if and only if $d \not\equiv 1 (\mathrm{mod}\ 4)$.

Theorem 9.3 does not extend to the case $d \geqslant 3$. Indeed, consider the cube Q in the following figure and the Fourier transform $\widehat{\chi}_Q(\xi)$ in the direction of ξ. Then $|\widehat{\chi}_Q(\xi)|$ cannot be controlled by the area of the triangle (i. e., the section) perpendicular to ξ

(at distance $1/|\xi|$).

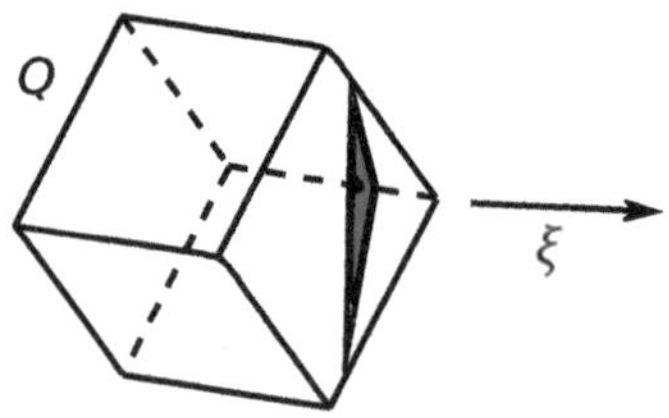

Indeed the area of the triangle decays of order 2, so that the "parallel section function" $\mathbb{R} \ni x \mapsto h(x)$, which measures the areas of the sections of C perpendicular to ξ, has a shape similar to the following one:

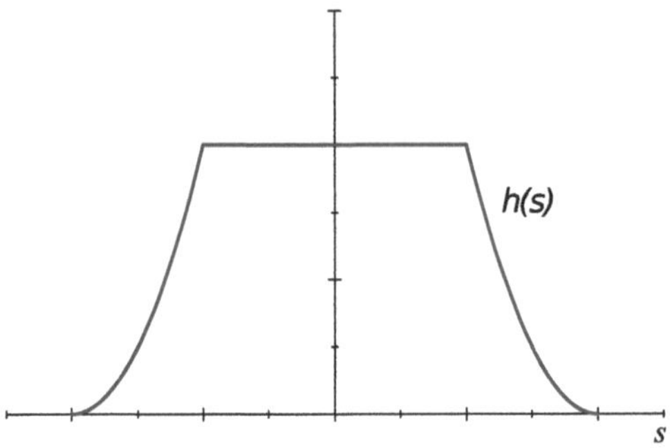

The above figure shows that the parallel section function $h(x)$ is more regular at the boundary of its support than inside it. Since the Fourier transform is mostly affected by the "irregular" points, the decay of $\widehat{\chi}_Q(\xi)$ cannot be controlled by a geometric estimate around the boundary of Q. Anyhow this may not be an obstacle. Indeed, in the case of a ball B or in the case of a convex body C with smooth boundary having positive curvature, we can still use the asymptotics of Bessel functions (or more refined estimates introduced by E. Hlawka and C. Herz) to estimate $\widehat{\chi}_C(\xi)$. In the case of a polyhedron, we may obtain fairly precise estimates working by induction on its faces. See also [16, 1] for general results concerning convexity and geometric estimates of Fourier transforms.

The dyadic argument in the second proof of Theorem 9.8 holds true in several variables as well (see [12]).

Theorems 9.13 and 9.14 can be extended to several variables with the following more general assumption on ∂C_γ.

Definition 9.31. Let U be a bounded open neighborhood of the origin in $\mathbb{R}^{d-1}$, let $\Phi \in C^\infty(U \setminus \{0\})$, and let $\gamma > 1$. For every $x \in U \setminus \{0\}$, let $\mu_1(x), \ldots, \mu_{d-1}(x)$ be the eigenvalues of the Hessian matrix of Φ. We say that $\Phi \in S_\gamma(U)$ if for $j = 1, \ldots, d-1$,

$$0 < \inf_{x \in U \setminus \{0\}} |x|^{2-\gamma} \mu_j(x)$$

and, for every multiindex α,

$$\sup_{x\in U\setminus\{0\}} |x|^{|\alpha|-\gamma}\left|\frac{\partial^{|\alpha|}\Phi}{\partial x^{\alpha}}(x)\right| < +\infty.$$

Let B be a convex body in $\mathbb{R}^d$, let $t \in \partial B$, and let $\gamma > 2$. We say that t is an isolated flat point of order γ if, in a neighborhood of t and in a suitable Cartesian coordinate system with the origin in t, ∂B is the graph of a function $\Phi \in S_\gamma(U)$.

Also Theorem 9.17 can be extended to several variables; see [21].

Bibliography

[1] J. Bak, D. McMichael, J. Vance, S. Wainger. Fourier transforms of surface area measure on convex surfaces in R^3. Am. J. Math. **111**, 633–668 (1989).

[2] A. Barvinok. Integer Points in Polyhedra. European Mathematical Society (2008).

[3] J. Beck. Irregularities of point distribution I. Acta Math. **159**, 1–49 (2008).

[4] J. Beck, W. W. L. Chen. Irregularities of Distribution. Cambridge Univ. Press (1987).

[5] M. Beck, S. Robins. Computing the Continuous Discretely. Integer-Point Enumeration in Polyhedra. Springer (2015).

[6] J. Bourgain, N. Watt. Mean square of zeta function, circle problem and divisor problem revisited. arXiv.org/pdf/1709.04340.pdf, preprint.

[7] L. Brandolini, L. Colzani. Localization and convergence of eigenfunction expansions. J. Fourier Anal. Appl. **5**, 431–447 (1999).

[8] L. Brandolini, L. Colzani, B. Gariboldi, G. Gigante, G. Travaglini. Discrepancy for convex bodies with isolated flat points. arXiv.org/pdf/1807.07059.pdf, preprint.

[9] L. Brandolini, L. Colzani, G. Gigante, G. Travaglini. L^p and weak-L^p estimates for the number of integer points in translated domains. Math. Proc. Camb. Philos. Soc. **159**, 471–480 (2015).

[10] L. Brandolini, L. Colzani, G. Travaglini. Average decay of Fourier transforms and integer points in polyhedra. Ark. Mat. **35**, 253–275 (1997).

[11] L. Brandolini, G. Gigante, G. Travaglini. Irregularities of distribution and average decay of Fourier transforms. In: W. W. L. Chen, A. Srivastav, G. Travaglini (eds.) A Panorama of Discrepancy Theory, pp. 159–220. Springer (2014).

[12] L. Brandolini, A. Greenleaf, G. Travaglini. $L^p - L^{p'}$ estimates for overdetermined Radon transforms. Trans. Am. Math. Soc. **359**, 2559–2575 (2007).

[13] L. Brandolini, S. Hofmann, A. Iosevich. Sharp rate of average decay of the Fourier transform of a bounded set. Geom. Funct. Anal. **13**, 671–680 (2003).

[14] L. Brandolini, A. Iosevich, G. Travaglini. Planar convex bodies, Fourier transform, lattice points, and irregularities of distribution. Trans. Am. Math. Soc. **355**, 3513–3535 (2003).

[15] L. Brandolini, M. Rigoli, G. Travaglini. Average decay of Fourier transforms and geometry of convex sets. Rev. Mat. Iberoam. **14**, 519–560 (1998).

[16] J. Bruna, A. Nagel, S. Wainger. Convex hypersurfaces and Fourier transforms. Ann. Math. **127**, 333–365 (1988).

[17] B. Chazelle. The Discrepancy Method. Cambridge Univ. Press (2000).

[18] Y. Colin de Verdiere. Nombre de points entiers dans une famille homothétique de domaines de $\mathbb{R}^n$. Invent. Math. **43**, 15–52 (1977).

[19] H. Davenport. Note on irregularities of distribution. Mathematika **3**, 131–135 (1956).
[20] M. Drmota, R. Tichy. Sequences, Discrepancies and Applications. Springer (1997).
[21] B. Gariboldi. Discrepancy of a convex set with zero curvature at one point. To appear in Mathematika
[22] C. S. Herz. Fourier transforms related to convex sets. Ann. Math. **75**, 81–92 (1962).
[23] E. Hlawka. Über Integrale auf konvexen Körpern. I and II. Monatshefte Math. **54**, 1–36, and 81–99 (1950).
[24] L. Hörmander. The Analysis of Linear Partial Differential Operators. I. Distribution Theory and Fourier Analysis. Springer-Verlag (2003).
[25] M. Huxley. Area, Lattice Points and Exponential Sums. Oxford Science Publ. (1996).
[26] M. Huxley. A fourth power discrepancy mean. Monatshefte Math. **73**, 231–238 (2014).
[27] H. Iwaniec, E. Kowalski. Analytic Number Theory. Amer. Math. Soc. (2004).
[28] D. G. Kendall. On the number of lattice points inside a random oval. Quart. J. Math. Oxford **19**, 1–26 (1948).
[29] E. Kratzel. Lattice Points. Kluwer Acad. Publ. (1988).
[30] J. Matoušek. Geometric Discrepancy. Algorithms and Combinatorics, vol. 18. Springer-Verlag (1999).
[31] H. L. Montgomery. Ten Lectures on the Interface between Analytic Number Theory and Harmonic Analysis. Amer. Math. Soc. (1994).
[32] L. Parnovski, A. Sobolev. On the Bethe-Sommerfeld conjecture for the polyharmonic operator. Duke Math. J. **107**, 209–238 (2001).
[33] A. N. Podkorytov. The asymptotic of a Fourier transform on a convex curve. Vestn. Leningr. Univ., Math. **24**, 57–65 (1991).
[34] K. Roth. On irregularities of distribution. Mathematika **1**, 73–79 (1954).
[35] K. Roth. Rational approximations to algebraic numbers. Mathematika **2**, 1–20 (1955).
[36] W. M. Schmidt. Irregularities of distribution IV. Invent. Math. **7**, 55–82 (1968).
[37] A. Schrijver. Theory of Linear and Integer Programming. John Wiley & Sons (1986).
[38] E. Stein. Harmonic Analysis: Real Variable Methods, Orthogonality and Oscillatory Integrals. Princeton Univ. Press (1993).
[39] G. Travaglini. Average decay of the Fourier transform. In: L. Brandolini, L. Colzani, A. Iosevich, G. Travaglini (eds.) Fourier Analysis and Convexity, pp. 245–268. Birkhäuser (2004).
[40] G. Travaglini. Number Theory, Fourier Analysis and Geometric Discrepancy. Cambridge Univ. Press (2014).
[41] G. Travaglini, M. R. Tupputi. A characterization theorem for the L^2-discrepancy of integer points in dilated polygons. J. Fourier Anal. Appl. **22**, 675–693 (2016).

Radon Series on Computational and Applied Mathematics

Volume 25
Space-Time Methods. Applications to Partial Differential Equations
Ulrich Langer, Olaf Steinbach (Eds.), 2019
ISBN: 978-3-11-054787-0, e-ISBN: 978-3-11-054848-8

Volume 24
Maxwell's Equations. Analysis and Numerics
Ulrich Langer, Dirk Pauly, Sergey I. Repin (Eds.), 2019
ISBN: 978-3-11-054264-6, e-ISBN: 978-3-11-054361-2

Volume 23
Combinatorics and Finite Fields. Difference Sets, Polynomials, Pseudorandomness and Applications
Kai-Uwe Schmidt, Arne Winterhof (Eds.), 2019
ISBN: 978-3-11-064179-0, e-ISBN: 978-3-11-064209-4

Volume 22
The Radon Transform. The First 100 Years and Beyond
Ronny Ramlau, Otmar Scherzer (Eds.), 2019
ISBN: 978-3-11-055941-5, e-ISBN: 978-3-11-056085-5

Volume 21
Hamilton-Jacobi-Bellman Equations. Numerical Methods and Applications in Optimal Control
Dante Kalise, Karl Kunisch, Zhiping Rao (Eds.), 2018
ISBN: 978-3-11-054263-9, e-ISBN: 978-3-11-054359-9

Volume 20
Fluid-Structure Interaction. Modeling, Adaptive Discretisations and Solvers
Stefan Frei, Bärbel Holm, Thomas Richter, Thomas Wick, Huidong Yang (Eds.), 2017
ISBN: 978-3-11-049527-0, e-ISBN: 978-3-11-049425-9

Volume 19
Tensor Numerical Methods in Scientific Computing
Boris N. Khoromskij, 2018
ISBN: 978-3-11-037013-3, e-ISBN: 978-3-11-036591-7

www.degruyter.com

Bei Fragen zur Produktsicherheit wenden Sie sich bitte an:
If you have any questions regarding product safety,
please contact:

Walter de Gruyter GmbH
Genthiner Straße 13
10785 Berlin
productsafety@degruyterbrill.com